Knowledge Accumulation and Industry Evolution

Written by internationally acclaimed experts in the economics of innovation, this volume examines how the biotechnology and pharmaceutical sector is affected by the dynamics of innovation, institutions, and public policy. It contributes both theoretically and empirically to the increasingly influential Schumpetarian framework in industrial economics, which places innovation at the centre of the analysis of competition. Both quantitative and qualitative studies are included, and this varied perspective adds to the richness of the volume's insights. The contributors explore different ideas regarding the historical evolution of technology in the sector, and how firms and industry structure have coevolved with innovation dynamics. Important policy questions are considered regarding the future of innovation in this sector and its impact on the economy.

MARIANA MAZZUCATO is Professor of Economics at the Open University, where she is also Director of the inter-faculty research centre, Innovation, Knowledge and Development (IKD).

GIOVANNI DOSI is Professor of Economics at the Sant'Anna School of Advanced Studies in Pisa.

Knowledge Accumulation and Industry Evolution

The Case of Pharma-Biotech

Edited by

MARIANA MAZZUCATO AND GIOVANNI DOSI

 CAMBRIDGE
UNIVERSITY PRESS

CAMBRIDGE UNIVERSITY PRESS
Cambridge, New York, Melbourne, Madrid, Cape Town, Singapore, São Paulo

CAMBRIDGE UNIVERSITY PRESS
The Edinburgh Building, Cambridge CB2 2RU, UK

Published in the United States of America by Cambridge University Press, New York

www.cambridge.org
Information on this title: www.cambridge.org/9780521858229

First published 2006

Printed in the United Kingdom at the University Press, Cambridge

A catalogue record for this book is available from the British Library

ISBN-13 978-0-521-85822-9 hardback
ISBN-10 0-521-85822-4 hardback

We dedicate this book to the memory of
our dear colleagues and mentors
Keith Pavitt and Paul Geroski

Contents

Figures

Tables

Contributors

ASHISH ARORA	Carnegie Mellon University, Pittsburgh, Pennsylvania, United States
GIULIO BOTTAZZI	Laboratory of Economics and Management, Sant'Anna School of Advanced Studies, Pisa, Italy
ELENA CEFIS	Utrecht University, Netherlands, and Bergamo University, Italy
JOANNA CHATAWAY	Economic and Social Research Council Innogen Centre, Technology, Open University, Milton Keynes, United Kingdom
MATTEO CICCARELLI	European Central Bank, Frankfurt, Germany
WESLEY M. COHEN	Duke University, Durham, North Carolina, United States
BENJAMIN CORIAT	Institutions, Innovation et Dynamiques Economiques, Centre d'Economie de Paris Nord, Centre National de la Recherche Scientifique, Paris 13 University, France
GIOVANNI DOSI	Laboratory of Economics and Management, Sant'Anna School for Advanced Studies, Pisa, Italy
LOUIS GALAMBOS	Institute of Applied Economics and the Study of Business Enterprise, Johns Hopkins University, Baltimore, Maryland, United States
CHRISTIAN GARAVAGLIA	Centro di Ricerca sui Processi di Innovazione e Internazionalizzazione, Bocconi University, Milan, Italy, and Cattaneo University-Liuc, Castellanza, Italy

BOYAN JOVANOVIC | New York University and National Bureau of Economic Research, Cambridge, Massachusetts, United States

FRANK R. LICHTENBERG | Graduate School of Business, Columbia University, New York, and National Bureau of Economic Research, Cambridge, Massachusetts, United States

CATHERINE LYALL | Economic and Social Research Council Innogen Centre, Edinburgh University, United Kingdom

SURYA MAHDI | Science Policy Research Unit, University of Sussex, Brighton, United Kingdom

FRANCO MALERBA | Centro di Ricerca sui Processi di Innovazione e Internazionalizzazione, Bocconi University, Milan, Italy

MYRIAM MARIANI | Istituto di Economia Politica and Centro di Ricerca sui Processi di Innovazione e Internazionalizzazione, Bocconi University, Milan, Italy

MARIANA MAZZUCATO | Open University, Milton Keynes, and Economic and Social Research Council Innogen Centre, Open University, Milton Keynes, United Kingdom

PAUL NIGHTINGALE | Science Policy Research Unit, University of Sussex, Brighton, United Kingdom

LUIGI ORSENIGO | Department of Engineering, University of Brescia, and Centro di Ricerca sui Processi di Innovazione e Internazionalizzazione, Bocconi University, Milan, Italy

FABIENNE ORSI | Institutions, Innovation et Dynamiques Economiques, Centre d'Economie de Paris Nord, Centre National de la Recherche Scientifique, Paris 13 University, France

FABIO PAMMOLLI | Institutions, Markets, Technologies, Lucca Institute for Advanced Studies, and University of Florence, Italy

PIERRE REGIBEAU   University of Essex and Centre for Economic Policy Research, London, United Kingdom

KATHARINE ROCKETT   University of Essex and Centre for Economic Policy Research, London, United Kingdom

ANGELO SECCHI   Laboratory of Economics and Management, Sant'Anna School of Advanced Studies, Pisa, Italy

CHRISTINE SEVILLA   Institut National de la Santé et de la Recherche Médicale U379, Marseille 2 University, France

JOYCE TAIT   Economic and Social Research Council Innogen Centre, Edinburgh University, United Kingdom

JOHN P. WALSH   University of Illinois at Chicago, United States

DAVID WIELD   Economic and Social Research Council Innogen Centre, Technology, Open University, Milton Keynes, United Kingdom

Acknowledgments

We thank both the Economics and Social Research Council's (ESRC's) centre for social and economic research on innovation in genomics (Innogen) and the Open University's centre for Innovation, Knowledge and Development (IKD) for funding the workshop on "Innovation, growth and market structure in pharma-biotech" (March 2003, London), which stimulated the production of this book. In addition, Mariana Mazzucato gratefully acknowledges (partial) support from the European Commission Key Action "Improving the socio-economic knowledge base" through contract no. HPSE-CT-2002-00146. Giovanni Dosi gratefully acknowledges (partial) support from the Italian Ministry for Education, Higher Education and Research (MIUR) through projects "Structures and dynamics of industries and markets" (FIRB 2003, RBAU01BM9F) and "Learning, competitive interactions and the ensuing patterns of industrial evolution" (PRIN 2004).

1 | Introduction

GIOVANNI DOSI AND
MARIANA MAZZUCATO

T HIS book addresses, from a variety of perspectives, the patterns of innovation through the history of the pharmaceutical industry (which we take to include nowadays both pharmaceutical and biotechnology firms) and the ways that knowledge has coevolved with the dynamics of firm growth, industry structure and the broader socio-economic environment.

Ever since it began the pharmaceutical industry has indeed been an example of a "science-based" industry, whereby innovation is driven, to a large extent, by joint advances in pure and applied sciences, together with complementary progress in research technologies – undertaken both within public research institutions and business firms. As such, it is a fascinating industry to study. And it is even more so in the light of the profound changes that have occurred recently in the underlying knowledge bases – associated with the so-called "biotech" revolution – as well as in the broad institutional regimes governing the generation and appropriation of the economic benefit of innovations. The welfare and policy implications are equally paramount, given the socio-economic importance of the industry for health, agriculture and food production.

Let us start with a brief overview of the history of the industry, in order to give some background for the analyses that follow, and then proceed by flagging some of the central issues addressed by the chapters below.

1.1 The evolution of the industry: an overview

The history of the international pharmaceutical industry has already been extensively analyzed by several scholars.[1] Here let us first mention

---

[1] See Aftalion (1959); Ackerknecht (1973); Arora, Landau, and Rosenberg (1998); Bovet (1988); Chandler (1990); Freeman (1982); Gambardella (1995);

a few major characteristics of innovation processes, competition and industrial structures. The origin of the pharmaceutical industry dates back to the late nineteenth century, and is one of the earliest examples of commercial exploitation in an organized manner of scientific research and discovery, beginning with the emergence of synthetic dye production and the discovery of the therapeutic effects of dyestuff components and other organic chemicals.

Before the 1930s pharmaceutical innovation was almost completely dependent upon a few large, diversified and vertically integrated German and Swiss firms, such as Hoechst, Bayer, Ciba, and Sandoz. These firms entered the industry in the late nineteenth century and manufactured drugs based on synthetic dyes, being able to leverage their scientific and technical competencies in organic chemistry, chemical synthesis, and medicinal chemistry.

The early emergence of a restricted group of firms with large-scale in-house research and development (R&D) capabilities was a consequence of the nature of pharmaceutical R&D in the chemical synthesis paradigm, and, in particular, of the appearance of a dominant "routinized regime of search" – paraphrasing Nelson and Winter (1982) – based on extensive exploration of chemical compounds and on incremental structural modifications of *drug prototypes*, organized around highly structured processes for mass screening programs (see Schwartzman, 1976). Throughout the evolution of the industry, "the organizational capabilities developed to manage the process of drug development and delivery – competencies in the management of large-scale clinical trials, the process of gaining regulatory approval, and marketing and distribution – have also acted as powerful barriers to entry into the industry" (Henderson, Orsenigo, and Pisano, 1999).

World War II induced a massive jump in research and production efforts, sponsored by the US government (especially in the field of antibiotics), which fostered the accumulation of vast search capabilities in US firms and their entry into the international oligopolistic core.

Following World War II and the commercialization of penicillin, pharmaceutical companies embarked on a period of massive investment in R&D and built large-scale internal R&D capabilities (Henderson, Orsenigo, and Pisano, 1999). Also benefiting from the dramatic

Henderson, Orsenigo, and Pisano (1999); Orsenigo (1989); and Pammolli (1996). This overview draws upon Bottazzi et al. (2001).

increase of public support for biomedical research and health care expenditure in the post-war period, the international pharmaceutical industry experienced a significant wave of discovery, with the introduction of a few hundred new chemical entities in the 1950s and 1960s, from hydrocortisone and several other corticoids, to thiazide diuretic drugs, to major and minor tranquilizers, to the initial birth control products (Grabowski and Vernon, 2000).

Throughout its evolution the industry has been characterized by a significant heterogeneity in terms of firms' strategic orientations and innovative capabilities. Competition, in the top segment of the industry, has always centered around new product introductions, often undertaken by the oligopolistic core of the industry, subject both to incremental advances over time and to imitation and generic competition after patent expiration (allowing a large "fringe" of firms to thrive). In a good approximation, the "oligopolistic core" of the industry has been composed of the early innovative entrants, joined after World War II by a few American and British firms. At the same time, until the mid-1970s a relatively small number of new firms entered the industry, and even fewer its "core."

However, things have begun to change since then, with a major transition in the technological paradigm underlying search activities, from one based on pragmatic knowledge and quasi-random screening to one of "guided discovery" (or "discovery by design"). This has been linked with major advances in computational techniques and the biological sciences, including molecular and cell biology, biochemistry, protein and peptide chemistry and physiology.[2]

All in all, the "molecular biology" revolution has had (and is having) major consequences on the patterns of division of "innovative labor" (Arora and Gambardella, 1994; Gambardella, 1995), fostering the emergence of specialized suppliers and pure *search firms*; moreover, the dramatic increase of plausibly explorable disease targets offered novel opportunities of entry for a few new small firms into new product markets.

More precisely, Gambardella (1995) identifies two subsequent technological paradigms in the pharmaceutical industry. In the first one, dominant before about 1980, the search for innovation was carried out

---

[2] See Sneader (1996) and Orsenigo, Pammolli, and Riccaboni (2001).

principally through "random screening," driven by relatively tacit search heuristics and involving a great deal of serendipity (e.g. the search for one therapy leading to the unexpected discovery of a different one). Conversely, the paradigm increasingly dominant after 1980, resting on major theoretical advances in molecular biology and, jointly, in biotechnologies and computational chemistry, has made the search process more "guided," also entailing a higher degree of path dependency in the search process (Gambardella, 1995). In a similar vein, Orsenigo, Pammolli, and Riccaboni (2001) argue that the pharmaceutical industry went through a sort of *transitional regime* from the late 1970s until the early 1990s, and has been going through a new one ever since. The former began to define new "biological hypotheses" and new molecular targets, while search relied on heuristic/search methods and co-specialized technologies that tended to be specific to given fields of application. Conversely, the contemporary regime – they argue – is characterized by the emergence of new generic tools (transversal technologies), such as combinatorial chemistry.[3]

Finally, at the institutional level, the "guided search" paradigm has come together (causality is a trickier matter) with major changes in the legal conditions for the appropriation of new knowledge, including prominently in the United States the Bayh–Dole Act (1980), which allowed universities and small businesses to patent discoveries emanating from research sponsored by the National Institutes of Health (NIH), and then to grant exclusive licenses to drug companies. This institutional change led to two phenomena that are fundamental to understanding the recent changes in the structure of the pharmaceutical industry: first, a boom in biotech startups; and, second, a new division of labor between small and large firms. In the new division of labor, dedicated biotechnology firms (DBFs) and publicly funded labs (NIH and universities) typically concentrate on upstream research, while

---

[3] Moreover, Padua and Orsenigo (2003) claim that, more recently, the division of labor in the pharmaceutical industry has been evolving towards a hierarchical structure, whereby large firms "decompose the search space" by defining general hypotheses of work whereas small firms work on the sub-hypotheses. They also claim that in recent years biotech has been developing an intense net of collaborations between large and small firms whereby, more so than in previous regimes, the pattern of innovative activity follows a process of "creative accumulation," in which the overall coordination takes place at the level of the large firms.

"big pharma" buy from them initial drug compounds and concentrate on bringing the drugs to market through more costly clinical trials and marketing campaigns.

There are several themes running through the interpretation of the history of the pharmaceutical industry, sketched above.

1.2 Mechanisms of knowledge generation

We have made only brief mention of the profound changes that the industry has undergone recently, amounting to what we have called a paradigm change. By that we mean changes in the knowledge bases, know-how, search procedures, and characteristics of physical equipment (Dosi, 1982, 1988; Nelson and Winter, 1982), in turn implying significant discontinuities in the ways that knowledge is generated and economically exploited.

The chapters by Nightingale and Mahdi, and by Mariani, address different properties of such paradigmatic discontinuities. The former (chapter 3) analyzes the extent to which new biotechnologies have indeed affected the search process for new drugs and, in particular, how far the industry has been able to "industrialize" search and experimentation. The answer they offer is that the industry did undergo a series of incremental changes, which also increased economies of scale and scope in R&D, but this fell short of the "revolution" in drug discovery and development heralded by many industrialists and policy-makers. In line with the similar argument in Nightingale and Martin (2004), the interpretation is rather "incrementalist," whereby multiple technological bottlenecks make changes in search patterns slow and uneven. The authors also interpret in this light the division of labor between "big pharma" and biotech firms, involving the painstaking absorption by the former of any array of process (and product) technologies.

Mariani's chapter 4 tackles the issue of knowledge accumulation from a complementary perspective, finding that there might indeed be solid evidence for a discontinuity between a "big pharma/chemical mode" and a "biotech mode" of knowledge accumulation. In the former, her evidence shows that innovative capabilities are firm-specific and cumulative within firms, while relatively shielded from the characteristics of the context in which they operate. This appears to apply notwithstanding the fact, noted by Galambos in his comments (chapter 5), that traditional "big pharma" also continues to be heavily concentrated in a few locations, at least in the case of the United States – a phenomenon that dates back to the beginning of the industry. Conversely, under the "biotech mode," the statistical evidence suggests that the innovative capabilities of individual firms are fueled by the knowledge spillovers from other firms located in the same areas.

1.3 Organizational capabilities, "creative destruction," and "creative accumulation"

How do the processes of search and knowledge accumulation relate to the behaviors and life profiles of individual firms? This represents a crucial link between those investigations addressing the general *characteristics of search and problem-solving procedures*, on the one hand, and those focusing on the patterns of *corporate learning and growth*, on the other (Geroski and Mazzucato, 2002).

This is an absolutely central concern of *knowledge-based* (i.e. *capability-based* or, largely overlapping, *resource-based*) theories of the firm (see Teece, Pisano, and Shuen, 1997; Teece et al., 1994; Montgomery, 1995; Dosi, Nelson, and Winter, 2000; and Dosi and Marengo, 2000). The bottom line of this perspective is that the proximate boundaries of the firms are shaped by the knowledge that they embody and roughly map into the "decompositions" of the problem-solving space over which they operate. What one firm does along the whole chain from raw material to final products is "cut" along the boundaries of the *sub-problems* that a particular firm happens to address. So, for example, some firms might embody specific capabilities concerning the exploration of either gene properties or gene manipulation (in our case, *in primis*, biotech firms), while others might have accumulated abilities in the search for therapeutical benefits (and side effects), together with complementary assets in production and marketing (i.e. typically "big pharma").

The evidence mentioned above does indeed suggest patterns of *complementary knowledge accumulation*. Given that, what is the role, if any, for the "creative destruction" emphasized early in his career by Joseph Schumpeter as the driving force of capitalist innovation?

Lichtenberg (chapter 2) suggests that such a process takes place primarily at the level of *individual products*, and such evidence is corroborated by Bottazzi, Pammolli, and Secchi (chapter 7). More generally, the history of the pharmaceutical industry can be read – as emphasized in Galambos' comments (chapter 5) – by a multi-level evolutionary process driven by scientific advances and discoveries of new chemical entities (undertaken to a large extent by public, non-profit institutions) coupled with the transformation of the former into new, hopefully effective and safe drugs (most often undertaken by the pharmaceutical companies).

The ensuing market evolution is, in fact, made up of two dynamics, operating on different timescales (see Lichtenberg, chapter 2 of this volume, and Bottazzi et al., 2001). The first one concerns the opening of new markets addressing a new pathology or an already "cured" pathology through different biochemical routes. The second evolutionary process regards competition *stricto sensu* amongst firms within each market on the grounds of similar chemical compounds.

Clearly, the growth of firms depends on both dynamics, with the timing of entry and the types of product introduced being important factors in shaping corporate competitive profiles. Let us turn to these issues.

1.4 Innovation, competition, and firm growth

The pharmaceutical industry offers a privileged point of observation in order to tackle a long-lasting but crucial question in industrial economics: *what determines the observed rates of innovation across firms?*

The evidence from the industry does indeed add to the evidence against the old conjecture, according to which innovation ought to be driven by firm size and by the intensity of market competition. In fact, most empirical studies show that the intensity of R&D spending is not influenced in any statistically significant way by the size of the firm, while at the same time both *ex ante* and *ex post* proxies for "market power" explain very little of the inter-industry differences in the propensity to innovate (see Cohen, Levin, and Mowery, 1987; Cohen and Levin, 1989; and Geroski, 1994).

Rather, there is mounting evidence suggesting that a good deal of the inter-industry differences in the propensity to search and to innovate relate to the characteristics of the underlying technological knowledge – including the opportunities it entails and the mechanisms of appropriability it offers (Dosi, 1988; Levin, Cohen, and Mowery, 1985; and

Klevorick et al., 1995) – which in turn often vary along the life cycle of each industry (Klepper, 1997). In fact, in this emerging tradition, well in tune with the seminal analysis by Pavitt (1984), one has begun to "map" intersectoral differences in the propensity to innovate and the modes through which innovation is pursued into underlying differences in the sources of innovative opportunities themselves (e.g. direct scientific advances, versus equipment-embodied technical advances, versus learning by interacting with suppliers and customers, etc.).

In this respect, the pharmaceutical industry conforms well with such a knowledge-driven view of the preparedness to undertake innovative activities. The steady emergence of new opportunities fueled both by science and by the development of new instruments and techniques supports a quite high *average* R&D propensity and a persistent flow of innovations (with some qualifications, as discussed below).

What about *inter-firm* differences in the revealed rates of innovation? Indeed, as some of our contributors show in another work (Bottazzi et al., 2002), in the international pharmaceutical industry one observes the *persistent* coexistence of two basic types of firms, mapping into distinct technological competencies and competitive strategies. The first group, closely corresponding to the *core*, undertakes what is sometimes called "pioneering R&D" (Grabowski and Vernon, 1987); generates the overwhelming majority of new chemical entities (NCEs); when successful enjoys big, albeit not very long-lasting, first-mover advantages; and charges premium prices. The second group undertakes primarily imitative R&D; generates incremental innovations and more competitively priced "me too" drugs; takes up licenses from the core; and is present to different degrees in the generic markets, after patent expirations.

Ultimately, what one sees is a *long-term ecology* of the industry relying on competition, but also the complementarity, between two organizational populations, the relative sizes of which are shaped by diverse competencies in accessing innovative opportunities (and, to some extent, also by intellectual property right (IPR) regimes, influencing the span and length of legal protection for temporary monopolies on innovation).

In fact, *heterogeneity* in the abilities to innovate across firms, *even within the same lines of business*, is an increasingly accepted "stylized fact." And the pharmaceutical industry stands as a sound corroboration of a more general phenomenon. In turn, such persistent heterogeneity is circumstantial but robust evidence for *idiosyncratic differences* in technological capabilities across firms that persistently hold over time (see Teece, Pisano, and Shuen, 1997, and Nelson, 1991, and – on the empirical side – Geroski, van Reenen, and Walters, 1997, and Cefis, 2003).

1.5 Firm growth patterns

Granted all that, what are the effects of different corporate innovative capabilities upon the growth profiles of individual firms? In order to start answering the question, one should look at the statistical properties of firm growth profiles themselves. Here, a useful, even if biased, yardstick is the so-called "Gibrat law" (see Ijiri and Simon, 1977) of growth. In shorthand, the "law" in its *weak* version simply states that growth rates are uncorrelated with initial sizes (no matter where the "initial" time is set). Or, putting it another way, there are no systematic scale advantages or disadvantages in the growth process. A *strong* version of the same conjecture fully endorses the former proposition, and adds further that corporate growth is generally driven by small, uncorrelated, independently distributed random events.

For the purpose of the investigation of the growth properties of the pharmaceutical industry, one ought primarily to investigate whether the *strong* version of the "law" holds. After all, concerning the *weak* version, the overwhelming evidence does not bend either towards an unlimited tendency to monopoly or, conversely, towards some mythical "optimal scale." So, even when one finds – as one often does – a unit root growth process, a serious issue remains concerning the properties of growth rate distributions and their temporal profiles.

Suggestive evidence shows that more volatile, largely intertemporally uncorrelated growth rates characterize in particular the early phases of the life cycles of (micro-) industries and markets (Geroski and Mazzucato, 2002; Mazzucato, 2002, 2003). However, in the drugs industry each firm holds portfolios of products that happen to be at different stages of their "life cycle." Hence one should ideally distinguish between the properties of growth at the level of single products/markets, on the one hand, and growth patterns of the firm as a whole. In fact, Bottazzi et al. (2001), as well as Bottazzi, Pammolli and Secchi (chapter 7 in this volume), show that growth dynamics in the world pharmaceutical industry display a significant autocorrelation structure both at the level of single therapeutical markets and at the level of the firm. Moreover, one observes the rather ubiquitous property of growth rates to be exponentially distributed. As discussed at greater length in Bottazzi et al. (2001), an implication of such evidence is that one can hardly attribute the growth dynamics to the addition of small, uncorrelated events. Rather, the frequent appearance of "big" events implied by an exponential distribution is well in tune with the notions that, first, discrete (possibly innovation-driven) events are at the root of the growth process, and, second, the very nature of the competition process is likely to reinforce the correlation structure underlying "big" positive and negative changes in the market share of the various firms. (The simple intuition here is that the very competitive process implies correlation in firm growth: a big rise in market advantage for one player necessarily implies a corresponding fall for the competitors.)

In this book, the contribution by Cefis, Ciccarelli, and Orsenigo (chapter 6) adds other important pieces of evidence regarding the structure of corporate growth, highlighting, again, *idiosyncratic* firm-specific (but *not* size-specific) growth profiles whereby (i) differences in growth rates do not seem to disappear over time, (ii) the notion of mean reversion finds only weak corroboration (i.e. firms with a higher initial size do not necessarily grow more slowly than smaller firms), and (iii) estimated steady states of firm sizes and growth rates do not converge to a common limit distribution (i.e. heterogeneity is a long-run phenomenon).

Chapter 7, by Bottazzi, Pammolli, and Secchi, also addresses the relationship between the variance in growth rates and firm sizes, identifying a negative correlation quite similar to that often found by other scholars in connection with different industries. What accounts for such a relation? The authors put forward an interpretation in terms of diversification patterns. In fact, the evidence suggests that there is a log-linear relation between firm size and the number of "micro-markets" in which a firm operates. Such a relation, the authors show, is fully able to account for the observed scaling of growth variance versus size. In turn, as discussed in Bottazzi et al. (2001) and Bottazzi (2001), the observed properties of diversification profiles can be well explained by a *branching process*. Such a dynamic finds intuitive roots in the underlying processes of *capability accumulation*, whereby knowledge is augmented incrementally and put to use in interrelated areas of application, cutting across different products and therapeutical classes.

1.6 Alternative theoretical accounts of industrial evolution

The chapters by Cefis, Ciccarelli, and Orsenigo (chapter 6) and by Bottazzi, Pammolli, and Secchi (chapter 7) are both written with an "inductive" interpretation of the properties of the panel data on the international pharmaceutical industry. But how does one theoretically account for the general evidence on industrial evolution to which the above chapters offer important additions?

In this book we have two nearly opposite archetypes of theorizing styles. Garavaglia, Malerba, and Orsenigo (chapter 8) try to account for a few evolutionary regularities, including the historical stability of the "oligopolistic core" of the industry combining with a relatively low level of concentration – on the grounds of a "history-friendly model" (HFM) (on such a modeling approach, see also Malerba et al., 1999, and Malerba and Orsenigo, 2002). In brief, an HFM is a particular style of evolutionary modeling built, so to speak, "bottom-up," on the grounds of observed regularities in the mechanism of technical and institutional change and in the patterns of interaction. Formalizing such mechanisms, HFMs typically explore via computer simulations their ability to reproduce and "explain" a set of "stylized facts," including some of those recalled above. And, indeed, the chapter by Garavaglia, Malerba, and Orsenigo does so quite convincingly, albeit in ways that pay very little tribute to the "rationality" and forecasting ability of individual firms and with no commitment to any notion of industry-level equilibrium of any sort. Conversely, these latter notions are the very building blocks of the comments by Jovanovic (chapter 9), which amount to a sophisticated attempt to interpret some of the regularities highlighted by Cefis et al., Bottazzi et al., and Garavaglia et al. on the grounds of the equilibrium behavior by a family of representative agents that differ, in one case, in terms of response rates to technological shocks and, in another case, in terms of the technological vintages, which they master.

As editors of the volume we do not want to take a position here on the relative merits and weaknesses of the different approaches. The reader is invited to appreciate the overlappings and also the unresolved tensions and challenges entailed in the different ways of understanding what "industrial evolution" is all about.

1.7 Institutions and policies shaping the evolution and the socio-economic impact of the industry

Lichtenberg's chapter (chapter 2) allows us to appreciate the socio-economic importance of the pharmaceutical industry in terms of health, the quality of life, food availability, the quality of the environment, etc. His empirical analysis, which focuses on the elasticity of life expectancy to pharmaceutical innovation, distinguishes between truly innovative drugs (new molecular entities (NMEs) with priority status) and less innovative drugs that have therapeutic qualities similar to already marketed drugs, or "me too" drugs (this distinction is central to many of the critical evaluations, briefly reviewed in our concluding chapter, of the industry's innovativeness; see Angell, 2004). Indeed, Lichtenberg finds a highly significant positive relationship between the introduction of priority drugs and a rise in the mean age at death.

Such an impact is profoundly affected by the institutions that govern the generation of new knowledge, the terms on which profit-seeking organizations can access it and the terms under which the users are able to access the final output of the industry. The whole area entails highly controversial issues (some of which will be reconsidered in chapter 14 of this book).

A *first* major controversy concerns the impact of different patent systems upon (i) the *rates* of innovation, (ii) the *directions* of innovation, and (iii) the ultimate *distributive* and *welfare effects*.

Chapter 10, by Walsh, Arora, and Cohen, asks whether recent changes in patenting laws and practices, and licensing behaviors (e.g. those prompted by the Bayh–Dole Act, mentioned above), pose a threat to future innovation. In particular, does the increase in patenting of "research tools" in the drug discovery process hinder drug discovery itself? (For a more popular argument along these lines, see Angell, 2004). The authors are especially concerned with the possibility of a "tragedy of the anticommons" (Heller and Eisenberg, 1998), which can emerge when "there are numerous property right claims to separate building blocks for some product or line of research. When these property rights are held by numerous claimants, the negotiations necessary to their combination may fail, quashing the pursuit of otherwise promising lines of research or product development" (Walsh, Arora, and Cohen, below). Yet, if the authors of the chapter are right, there is little to worry about. They suggest that: (a) drug discovery has not been impeded by the increase in patenting of research tools; (b) likewise, university research has not been hindered by concerns about patents (except in the case of genetic diagnostics); and (c) this is, notwithstanding some delays and restrictions, due to access negotiation to patented research tools.

Chapter 11, by Orsi, Sevilla, and Coriat, offers a less optimistic view about the effects of patenting, at least as far as "upstream technologies" are concerned. The authors argue that, in fact, patenting in upstream technologies (i.e. basic research, or tools for research) is threatening the future of basic research itself, and has dangerous consequences all the way down to public health care. To make the argument they first provide a brief but stimulating history of the economic foundations of patent law, highlighting the dual role of patents in stimulating innovation – namely, on the one hand, to offer an incentive to innovate by rewarding the inventor with a monopoly on the invention (otherwise why bother?) and, on the other hand, to provide detailed public information on the new invention for future inventors once the patent expires. The authors argue that the first role – the ability of patents to stimulate innovation – is threatened when patents provide a monopoly on upstream research. This is because, following Arrow (1962) and Nelson (1959), basic research is meant to provide common knowledge bases with "public good" features, which, if patented, would hurt future innovative developments "downstream" in so far as their use is restricted. This, the authors argue, is exactly what has happened. Since 1995 the patentability domain has widened in the United States so that it no longer restricts patents to "inventions" but also includes "discoveries." This means that patents are not restricted to inventions with "practical or commercial utility," but also encompass discoveries that allow the *exploration* of future innovative possibilities – i.e. the knowledge base itself. As an example, the authors study the effect of granting patents on a breast cancer gene (BRCA1), where the patent not only covered the DNA sequence of the gene but also all the diagnostic and therapeutic applications. Giving the patent holder a monopoly on the "discovery" of the gene and the related techniques of diagnosis and treatment inevitably narrowed both the possibilities of breast cancer research and the ability of patients to be objectively diagnosed and treated – both being at this point at the discretion of the monopolist itself. This, they argue, is just one example of a broader trend entailing serious threats to the future of research and to the future of the public health system.

The last two chapters address another highly controversial issue, namely genetically modified (GM) food crops. Chapter 12, by Regibeau and Rockett, focuses on the structure and behavior of the GM food industry, while chapter 13, by Tait, Chataway, Lyall, and Wield, assesses different regulatory instruments and their likely effects.

The GM food industry – which turns out to be a nearly separate industry halfway between the traditional food industry and biopharma – is, obviously, heavily influenced by the regulatory environment. First, regulation – Regibeau and Rockett argue – happens to foster industrial concentration, due to fixed costs, economies of scope and specific elements of expertise in the approval process. Second, licensing is used to control regulatory costs. Third, patents and their associated returns do not appear to be the primary driver of research into GM food: rather, patents are viewed as "defensive" tools making imitation more difficult. Fourth, licensing is used extensively in this industry, and one of its roles is to substitute for litigation costs. Fifth, somewhat counter-intuitively, an increase in litigation costs might actually *increase* R&D expenditure in this area, rather than deter entry into the field. The reason appears to be that, by making more research efforts, firms may obtain stronger patent portfolios, to be used as "bargaining chips" in order to reduce litigation costs.

Clearly, as the chapter shows, the impact of the regulatory environment upon the structure and dynamics of the industry is pervasive. But, in turn, what forms and "philosophies" of regulation is one talking about? The question is addressed by the chapter by Tait et al., which offers a sort of taxonomy of regulatory measures and policy instruments, classified as "enabling" versus "constraining" and "discriminating" versus "indiscriminate." Examples (including comparisons of pesticide regulation in the United States and the European Union) illustrate the differential impact of diverse categories of policies upon corporate strategies and potential buyers.

Overall, the book leads the reader from an examination of the patterns of knowledge generation in the pharmaceutical, biopharma and, to some extent, GM food industries to an analysis of the ensuing dynamics of markets and firms, all the way to an appreciation of the importance of the non-market institutions, laws, and regulations. In that, we trust, the various contributions of this book illuminate important aspects of the evolutionary processes guiding one of the core industries in modern economies. At the same time, we hope that the book will help to stir up healthy debates concerning both the interpretation of the history and performance of the industry and the institutional arrangements by which it should be governed.

References

Ackerknecht, E. (1973), *Therapeutics from the Primitives to the Twentieth Century*, Hafner, New York.

Aftalion, J. (1959), *History of the International Chemical Industry*, University of Pennsylvania Press, Philadelphia.

Angell, M. (2004), *The Truth about the Drug Companies: How They Deceive Us and What to Do about It*, Random House, New York.

Arora, A., and A. Gambardella (1994), "The changing technology of technical change: general and abstract knowledge and the division of innovative labour," *Research Policy*, 23, 523–32.

Arora, A., R. Landau, and N. Rosenberg (1998), *Chemicals and Long-Term Economic Growth: Insights from the Chemical Industry*, Wiley, New York.

Arrow, K. J. (1962), "Economic welfare and the allocation of resources for innovation," in R. R. Nelson (ed.), *The Rate and Direction of Inventive Activity*, Princeton University Press, Princeton, NJ, 609–25.

Bottazzi, G. (2001), *Firm Diversification and the Law of Proportionate Effects*, Working Paper no. 2001/01, Laboratory of Economics and Management, Sant' Anna School for Advanced Studies, Pisa.

Bottazzi, G., E. Cefis, and G. Dosi (2002), "Corporate growth and industrial structures: some evidence from the Italian manufacturing industry," *Industrial and Corporate Change*, 11 (4), 705–23.

Bottazzi, G., G. Dosi, M. Lippi, F. Pammolli, and M. Riccaboni (2001), "Innovation and corporate growth in the evolution of the drug industry," *International Journal of Industrial Organization*, 19, 1161–87.

Bovet, D. (1988), *Une Chemie qui Guérit: Histoire de la Découverte des Sulfamides*, Editions Payot, Paris.

Cefis, E. (2003), "Is there persistence in innovative activities?," *International Journal of Industrial Organization*, 21, 489–515.

Chandler, A. (1990), *Scale and Scope*, Harvard University Press, Cambridge, MA.

Cohen, W. M., and R. Levin (1989), "Empirical studies of innovation and market structure," in R. Schmalensee and R. Willig (eds.), *The Handbook of Industrial Organization*, North-Holland, Amsterdam, 1060–107.

Cohen, W. M., R. Levin, and D. C. Mowery (1987), "Firm size and R&D intensity: a re-examination," *Journal of Industrial Economics*, 35 (4), 543–65.

Dosi, G. (1982), "Technological paradigms and technological trajectories," *Research Policy*, 11, 147–62.

  (1988), "Sources, procedures and microeconomic effects of innovation," *Journal of Economic Literature*, 26 (3), 1120–71.

Dosi, G., and L. Marengo (2000), "On the tangled intercourses between transaction cost economics and competence-based views of the firm: a comment," in N. Foss and V. Mahnke (eds.), *Competence, Governance and Entrepreneurship*, Oxford University Press, Oxford, 80–92.

Dosi, G., R. R. Nelson, and S. G. Winter (eds.) (2000), *The Nature and Dynamics of Organizational Capabilities*, Oxford University Press, Oxford and New York.

Freeman, C. (1982), *The Economics of Industrial Innovation*, Francis Pinter, London.

Gambardella, A. (1995), *Science and Innovation: The US Pharmaceutical Industry in the 1980s*, Cambridge University Press, Cambridge.

Geroski, P. (1994), *Market Structure, Corporate Performance and Innovative Activity*, Oxford University Press, New York.

Geroski, P., and M. Mazzucato (2002), "Learning and the sources of corporate growth," *Industrial and Corporate Change*, 11 (4), 623–44.

Geroski, P., J. van Reenen, and C. F. Walters (1997), "How persistently do firms innovate?," *Research Policy*, 26 (1), 33–48.

Grabowski, H., and J. Vernon (1987), "Pioneers, imitators, and generics: a simulation model of Schumpeterian competition," *Quarterly Journal of Economics*, 102 (3), 491–525.

  (2000), "The determinants of pharmaceutical research and development expenditures," *Journal of Evolutionary Economics*, 10, 201–15.

Heller, M. A., and R. S. Eisenberg (1998), "Can patents deter innovation? The anti-commons in biomedical research," *Science*, 280, 698–701.

Henderson, R. M., L. Orsenigo, and G. Pisano (1999), "The pharmaceutical industry and the revolution in molecular biology: exploring the interactions between scientific, institutional, and organizational change," in D. C. Mowery and R. R. Nelson (eds.), *Sources of Industrial Leadership: Studies of Seven Industries*, Cambridge University Press, Cambridge, 267–311.

Ijiri, Y., and H. Simon (1977), *Skew Distributions and Sizes of Business Firms*, North-Holland, Amsterdam.

Klepper, S. (1997), "Industry life cycles," *Industrial and Corporate Change*, 6 (1), 145–81.

Klevorick, A., R. Levin, R. R. Nelson, and S. G. Winter (1995), "On the sources and interindustry differences in technological opportunities," *Research Policy*, 24, 185–205.

Levin, R. C., W. M. Cohen, and D. C. Mowery (1985), "R&D appropriability, opportunity, and market structure: new evidence on some Schumpeterian hypotheses," *American Economic Review*, 75 (May), 20–4.

Malerba, L., R. R. Nelson, L. Orsenigo, and S. G. Winter (1999), "History-friendly models of industry evolution: the case of the computer industry," *Industrial and Corporate Change*, 8 (1), 3–40.

Malerba, F., and L. Orsenigo (2002), "Innovation and market structure in the dynamics of the pharmaceutical industry and biotechnology: toward a history-friendly model," *Industrial and Corporate Change*, 11 (4), 667–703.

Mazzucato, M. (2002), "The PC industry: new economy or early life-cycle," *Review of Economic Dynamics*, 5, 318–45.

(2003), "Risk, variety and volatility: innovation, growth and stock prices in old and new industries," *Journal of Evolutionary Economics*, 13 (5), 491–512.

Montgomery, C. A. (1995), *Resource-Based and Evolutionary Theories of the Firm: Toward a Synthesis*, Kluwer Academic, Boston.

Nelson, R. R. (1959), "The simple economics of basic economic research," *Journal of Political Economy*, 67 (3), 297–306.

(1991), "Why do firms differ, and why does it matter?," *Strategic Management Journal*, 12 (8), 61–75.

Nelson, R. R., and S. G. Winter (1982), *An Evolutionary Theory of Economic Change*, Harvard University Press, Cambridge, MA.

Nightingale, P., and P. Martin (2004), "The myth of the biotech revolution," *Trends in Biotechnology*, 22 (11), 564–9.

Orsenigo, L. (1989), *The Emergence of Biotechnology*, St Martin's Press, New York.

Orsenigo, L., F. Pammolli, and M. Riccaboni (2001), "Technological change and network dynamics: lessons from the pharmaceutical industry," *Research Policy*, 30, 485–508.

Pádua, M., and L. Orsenigo (2003), *Is Biotechnology an Ill-Structured or a Complex Problem? Towards a Methodology for the Analysis of Properties underlying Knowledge Production within Ill-Structured Problems and Implications for Industrial Dynamics*, paper presented at the third European Meeting on Applied Evolutionary Economics, 9–12 April, University of Augsburg.

Pammolli, F. (1996), *Innovazione, Concorrenza e Strategie di Sviluppo nell'Industria Farmaceutica*, Guercini, Milan.

Pavitt, K. (1984), "Sectoral patterns of technological change: towards a taxonomy and a theory," *Research Policy*, 3, 343–73.

Schwartzman, D. (1976), *Innovation in the Pharmaceutical Industry*, Johns Hopkins University Press, Baltimore.

Sneader, W. (1996), *Drug Prototypes and their Exploitation*, Wiley, New York.

Teece, D. J., G. Pisano, and A. Shuen (1997), "Dynamic capabilities and strategic management," *Strategic Management Journal*, 18 (7), 509–33.

Teece, D. J., R. Rumelt, G. Dosi, and S. G. Winter (1994), "Understanding corporate coherence: theory and evidence," *Journal of Economic Behavior and Organization*, 23 (1), 1–30.

PART I

Innovation and industry evolution

2 | Pharmaceutical innovation as a process of creative destruction

FRANK R. LICHTENBERG

2.1 Introduction

In *Capitalism, Socialism and Democracy*, Joseph Schumpeter (1942) argued that "the fundamental impulse that sets and keeps the capitalist engine in motion comes from the new consumers' goods, the new methods of production or transportation, the new markets, the new forms of industrial organization that capitalist enterprise creates." Entrepreneurs discover and implement innovations that "incessantly revolutionize ... the economic structure from within, incessantly destroying the old one, incessantly creating a new one" – a process that he labeled *creative destruction*. The innovation process is creative in the sense that it creates value (increases consumer welfare), and destructive in the sense that it reduces or eliminates returns to capital and labor producing obsolete products.

The characterization of innovation as a process of creative destruction was an important insight, but there has been little direct, systematic empirical evidence about this process. This chapter provides an econometric analysis of pharmaceutical innovation as a process of creative destruction.

The pharmaceutical industry is an attractive sector for studying this process, for several reasons. As table 2.1 indicates, the industry is one of the most research-intensive industries in the US economy: its ratio of R&D expenditure to sales is over 10 percent, compared to less than 3 percent for the manufacturing sector as a whole.[1] Econometric investigations of innovation are usually hampered by a lack of reliable data, but because the pharmaceutical industry has been strictly regulated by the Food and Drug Administration (FDA) since 1939 we can reconstruct the complete, detailed history of pharmaceutical innovation over the last sixty years or

---

[1] The Pharmaceutical Research and Manufacturers Association (PhRMA) estimates that the average pre-tax R&D cost of developing a new molecular entity was $359 million in 1990.

Table 2.1 *Manufacturing industries ranked by company-funded R&D expenditure as a percentage of net sales, 1994*

Communication equipment	10.3
Drugs and medicines	10.2
Office, computing and accounting machines	7.9
Electronic components	7.3
Optical, surgical, photographic and other instruments	7.2
Scientific and mechanical measuring instruments	5.8
Aircraft and missiles	5.3
Motor vehicles and motor vehicle equipment	3.4
Industrial chemicals	3.3
Other chemicals	2.5
Other machinery, except electrical	2.5
Rubber products	2.3
Other electrical equipment	2.1
Stone, clay and glass products	1.5
Other transportation equipment	1.2
Other manufacturing industries	1.1
Paper and allied products	1.0
Fabricated metal products	1.0
Radio and TV receiving equipment	1.0
Non-ferrous metals and products	0.9
Petroleum refining and extraction	0.8
Textiles and apparel	0.6
Lumber, wood products and furniture	0.6
Food, kindred and tobacco products	0.5
Ferrous metals and products	0.3

Source: National Science Foundation.

so.[2] Moreover, large surveys conducted by the National Center for Health Statistics enable us to estimate the number of times doctors prescribed each of about 800 drugs in both 1980 and 1994.

[2] I obtained from the FDA (by submitting a Freedom of Information Act request) a computerized list of all new drug approvals (NDAs) and abbreviated new drug approvals (ANDAs) since 1939. The NDA and ANDA files included 4370 and 6024 records, respectively, and each record indicates the NDA or ANDA number, the approval date, the generic and trade names of the drug, the dosage form, route of administration, strength, applicant name, "therapeutic potential" (priority or standard) and "chemical type" (new molecular entity, new formulation, new manufacturer, etc.).

The first part of the chapter (section 2.2) is an attempt to assess *how* destructive creative destruction is. In particular, we examine the impact of the introduction of new drugs in a therapeutic class on the demand for old drugs within the same class and on the total size of the market. We also briefly consider the effect of new drug introduction on market structure.

Most of the remainder of this chapter attempts to determine how *creative* creative destruction is: it is an examination of the consumer benefits of, or social returns to, pharmaceutical innovation. During the period 1970–91 the life expectancy of Americans increased by about 5.4 years. Our goal is to perform detailed econometric tests of the hypothesis that this decline in mortality is due, to an important extent, to the introduction and use of new drugs – i.e. to "changes in pharmaceutical technology." Trends in mortality and their role in economic growth are discussed in section 2.3. Section 2.4 briefly reviews some existing (primarily anecdotal) evidence about the impact of pharmaceutical innovation on mortality. In section 2.5 we present a simple econometric model for estimating this impact, describe our procedures for constructing the variables included in this model and discuss issues pertaining to the estimation and interpretation of parameter estimates. Estimates of the model and their economic implications are discussed in section 2.6.

There have been many important government policy events affecting the pharmaceutical industry in the last fifteen years. A failure to account for the effects of policies on innovation may lead to seriously distorted perceptions of the long-term consequences of these policies for social and consumer welfare. Section 2.7 outlines an econometric framework for examining the impact of government policy changes on R&D investment in the pharmaceutical industry. The final section contains a summary and conclusions.

2.2 The impact of new drug introduction on market size and structure

Since 1980 the FDA has approved over forty new molecular entities, on average, per year (see table 2.2). More than half of the 1352 NMEs that the FDA has approved since its inception were approved after 1980. It is therefore not surprising that, as table 2.3 shows, the distribution of drugs prescribed by physicians changed considerably between 1980 and 1994. Ten of the drugs that were among the top

Table 2.2 Numbers of new molecular entities approved by the FDA

Period	Frequency	Percentage	Cumulative frequency	Cumulative percentage
1940–44	4	0.3	4	0.3
1945–49	21	1.6	25	1.8
1950–54	68	5.0	93	6.9
1955–59	86	6.4	179	13.2
1960–64	101	7.5	280	20.7
1965–69	70	5.2	350	25.9
1970–74	107	7.9	457	33.8
1975–79	157	11.6	614	45.4
1980–84	174	12.9	788	58.3
1985–89	189	14.0	977	72.3
1990–94	228	16.9	1205	89.1
1995–97	147	10.9	1352	100.0

Note: Year of approval unknown for fifteen NMEs.

Source: Author's calculations, based on unpublished FDA data.

twenty drugs prescribed in 1980 were no longer top twenty drugs by 1994. Similarly, ten of the top twenty drugs in 1994 were not in the top twenty in 1980. Some of them, such as albuterol (for which Schering received FDA approval in 1981), had not yet been approved by the FDA in 1980.

Table 2.3 contains data for drugs from a number of different drug classes. However, there is likely to be a separate market for each class of drugs, so we should assess the impact of new drug introductions on market size and structure within drug classes. The top twenty drug classes, ranked by number of 1994 prescriptions, are shown in table 2.4. Table 2.5 presents, for two important drug classes (anti-hypertensives and antidepressants), the number of 1980 and 1994 prescriptions for each drug, arranged in reverse chronological order (by FDA approval date).[3] We divide the drugs in each list into two groups: drugs approved in or after 1980, and drugs approved before 1980.

[3] These lists include only drugs for which the FDA approval date is known. Unfortunately, we were unable to determine the approval date for some of the drugs coded in the NAMCS Drug Mentions files.

Table 2.3 Top twenty drugs prescribed in doctor-office visits in 1980 and 1994

Rank in 1980	Percentage share of 1980 prescriptions	Percentage share of 1994 prescriptions	Drug	Class
1	2.9	1.4	Hydrochlorothiazide	Diuretics
2	2.3	1.5	Aspirin	General analgesics
3	1.9	X	Penicillin	Penicillins
4	1.8	1.0	Phenylpropanolamine	Nasal decongestants
5	1.8	X	Alcohol	Antitussives, expectorants, mucolytics
6	1.7	1.2	Erythromycin	Erythromycins and lincosamides
7	1.7	1.0	Phenylephrine	Nasal decongestants
8	1.7	3.0	Acetaminophen	General analgesics
9	1.5	0.9	Codeine	Antitussives, expectorants, mucolytics
10	1.4	X	Tetracycline	Tetracyclines
11	1.4	X	Pseudoephedrine	Nasal decongestants
12	1.3	X	Riboflavin	Vitamins, minerals
13	1.3	1.1	Digoxin	Cardiac glycosides
14	1.3	X	Chlorpheniramine	Nasal decongestants
15	1.3	X	Ampicillin	Penicillins
16	1.2	3.8	Amoxicillin	Penicillins
17	1.2	X	Propranolol	Antihypertensive agents

Table 2.3 Top twenty drugs prescribed in doctor–office visits in 1980 and 1994 (continued)

Rank in 1980	Percentage share of 1980 prescriptions	Percentage share of 1994 prescriptions	Drug	Class
18	1.1	1.2	Furosemide	Diuretics
19	1.1	X	Ergocalciferol	Vitamins, minerals
20	1.1	X	Neomycin	Ocular anti-infective and anti-inflammatory agents
	31.0		**Sum of top twenty drugs**	
1	3.8	1.2	Amoxicillin	Penicillins
2	3.0	1.7	Acetaminophen	General analgesics
3	1.6	X	Albuterol	Bronchodilators, antiasthmatics
4	1.5	2.3	Aspirin	General analgesics
5	1.4	X	Ibuprofen	Antiarthritics
6	1.4	2.9	Hydrochlorothiazide	Diuretics
7	1.3	X	Multivitamins general	Vitamins, minerals
8	1.2	1.1	Furosemide	Diuretics
9	1.2	1.7	Erythromycin	Erythromycins and lincosamides
10	1.2	X	Guaifenesin	Antitussives, expectorants, mucolytics
11	1.1	X	Estrogens	Estrogens and progestins

Rank in 1980	Percentage share of 1980 prescriptions	Percentage share of 1994 prescriptions	Drug	Class
12	1.1	1.3	Digoxin	Cardiac glycosides
13	1.1	X	Prednisone	Adrenal corticosteroids
14	1.0	X	Diltiazem	Antianginal agents
15	1.0	X	Beclomethasone	Unclassified
16	1.0	1.7	Phenylephrine	Nasal decongestants
17	1.0	1.8	Phenylpropanolamine	Nasal decongestants
18	0.9	X	Triamcinolone	Adrenal corticosteroids
19	0.9	1.5	Codeine	Antitussives, expectorants, mucolytics
20	0.9	X	Levothyroxine	Agents used to treat thyroid disease
	27.6		**Sum of top twenty drugs**	

X: Not in top twenty in that year.

Note: Total estimated number of prescriptions was 899 million in 1980 and 921 million in 1994.

Table 2.4 Top twenty drug classes, ranked by 1994 National
Ambulatory Medical Care Survey (NAMCS) mentions

Class	Cumulative percentage of prescriptions
Antihypertensive agents	9.3
Penicillins	17.2
Diuretics	25.0
Antiarthritics	31.7
Adrenal corticosteroids	36.7
Antidepressants	41.6
Antianginal agents	45.9
Bronchodilators, antiasthmatics	50.2
Dermatologics	54.0
Blood glucose regulators	57.4
Antihistamines	60.5
Antianxiety agents	63.6
Cephalosporins	66.2
General analgesics	68.5
Unclassified	70.7
Cardiac glycosides	72.8
Antiarrhythmic agents	74.6
Agents used to treat hyperlipidemia	76.3
Agents used in disorders of upper gastrointestinal tract	77.9
Sulfonamides and trimethoprim	79.4

Note that table 2.5 contains a column labeled "therapeutic potential."
In the course of the approval process, the FDA classifies drugs into two
categories: "priority review drugs" and "standard review drugs." The
former are "drugs that appear to represent an advance over available
therapy," and the latter are "drugs that appear to have therapeutic qualities similar to those of already marketed drugs." The results we will report
below indicate that the introduction of priority and standard drugs have
quite different effects both on market size and structure and on mortality.

Table 2.5a indicates that about three-fourths of the antihypertensive
drugs prescribed in 1994 were approved by the FDA in or after 1980.
Between 1980 and 1994 the number of new drugs increased by 32.6

Table 2.5a Antihypertensive drug mentions in 1980 and 1994

FDA approval date	1980 prescriptions (thousands)	1994 prescriptions (thousands)	Drug	Applicant	Therapeutic potential
31JUL92	0	2270.89	Amlodipine	Pfizer Cen Res	Standard
31JUL92	0	218.09	Bisoprolol fumarate	Lederle Labs	Standard
19NOV91	0	1315.81	Quinapril	Parke Davis	Standard
25JUL91	0	691.19	Felodipine	Astra Merck	Standard
25JUN91	0	868.01	Benazepril hydrochloride	Ciba	Standard
16MAY91	0	768.20	Fosinopril sodium	Bristol Myers Squibb	Standard
28JAN91	0	951.41	Ramipril	Hoechst Marion Rssl	Standard
20DEC90	0	642.19	Isradipine	Sandoz Pharm	Standard
02NOV90	0	1318.19	Doxazosin mesylate	Pfizer Labs	Standard
21DEC88	0	315.21	Nicardipine	Syntex Labs	Standard
29DEC87	0	5161.17	Lisinopril	Merck	Standard
27OCT86	0	622.75	Guanfacine Hcl	Robins	Standard
30DEC85	0	47.43	Mexiletine Hcl	Boehringer Ingelheim	Priority
24DEC85	0	7135.88	Enalapril	Merck	Priority
28DEC84	0	308.19	Acebutolol hydrochloride	Wyeth Ayerst Labs	Standard
01AUG84	0	488.01	Labetalol Hcl	Schering	Priority
03SEP82	0	95.23	Pindolol	Sandoz Pharm	Standard
19AUG81	0	5784.23	Atenolol	Zeneca Pharms Grp	Standard

Table 2.5a Antihypertensive drug mentions in 1980 and 1994 (continued)

FDA approval date	1980 prescriptions (thousands)	1994 prescriptions (thousands)	Drug	Applicant	Therapeutic potential
06APR81	0	3558.28	Captopril	Bristol Myers Squibb	Priority
	0.0	32,560.36	All drugs approved after 1979		
10DEC79	885.37	1101.37	Nadolol	E R Squibb	Standard
18OCT79	0.00	366.21	Minoxidil	Pharmacia and Upjohn	Priority
07AUG78	2688.15	3721.65	Metoprolol	Geigy Pharma	Priority
23JUN76	1139.84	557.12	Prazosin	Pfizer Labs	Priority
07DEC67	24.85	49.07	Hydroxyurea	E R Squibb	Priority
13NOV67	10,345.20	3655.61	Propranolol	Wyeth Ayerst Labs	Priority
14JUL67	5197.46	703.54	Chlorthalidone	Rhone Poulenc Rorer	Priority
20DEC62	7695.82	527.59	Methyldopa	Merck Pharm Res	Priority
26SEP61	129.00	13.08	Polythiazide	Pfizer Labs	Standard
05JUL60	182.05	25.44	Guanethidine	Ciba	Priority
19APR57	587.68	16.38	Deserpidine	Abbott Labs	Standard
03MAR53	0.00	83.77	Phentolamine	Ciba	Priority
13MAR52	3625.13	789.46	Hydralazine	Ciba	Priority
	32,500.55	11,610.29	All drugs approved before 1980		

Table 2.5b Antidepressant drug mentions in 1980 and 1994

FDA approval date	1980 prescriptions (thousands)	1994 prescriptions (thousands)	Drug	Applicant	Therapeutic potential
28DEC93	0	660.62	Venlafaxine	Wyeth Ayerst Labs	Standard
29DEC92	0	179.78	Paroxetine	Skb Pharms	Standard
30DEC91	0	5242.50	Sertraline	Pfizer Pharms	Standard
29DEC89	0	462.94	Clomipramine	Ciba Geigy	Priority
26SEP89	0	227.10	Clozapine	Sandoz Pharm	Priority
29DEC87	0	6430.04	Fluoxetine hydrochloride	Lilly Res Labs	Priority
30DEC85	0	927.30	Bupropion	Glaxo Wellcome	Priority
16AUG85	25.88	83.12	Tranylcypromine	Smithkline Beecham	Priority
24DEC81	0	2669.75	Trazodone	Apothecon	Standard
01DEC80	0	10.64	Maprotiline	Ciba	Standard
22SEP80	0	11.10	Amoxapine	Lederle Labs	Standard
	25.88	16,904.89	All drugs approved after 1979		
12JUN79	187.60	7.32	Trimipramine	Wyeth Ayerst Labs	Standard
25FEB75	51.78	52.37	Loxapine	Lederle Labs	Standard
23SEP69	2153.45	1276.30	Doxepin	Pfizer Labs	Standard
27SEP67	134.27	33.40	Protriptyline	Msd	Standard
06NOV64	308.44	2142.91	Nortriptyline	Eli Lilly	Standard
07APR61	1594.97	1983.80	Amitriptyline	Zeneca Pharms Grp	Priority
27FEB57	1701.28	553.74	Perphenazine	Schering	Standard
	6131.79	6049.84	All drugs approved before 1980		

million, and the number of old (pre-1980) drugs declined by 20.9 million. Thus, about two-thirds of the increase in sales of new drugs was offset by a decline in sales of old drugs. But table 2.5b reveals that this kind of displacement did not occur in the market for antidepressants. The consumption of new drugs in this class increased by 16.9 million between 1980 and 1994, but the consumption of old drugs remained approximately constant at about 6 million. Another (possibly related) difference between these two drug classes is that almost a half of the prescriptions for new antidepressant drugs were for priority drugs, whereas only about a third of the prescriptions for new antihypertensive drugs were for priority drugs.

The extent to which sales of new drugs are offset by declines in sales of old drugs is much greater in the case of antihypertensives than it is in the case of antidepressants. To determine the extent to which the creation of new drugs "destroys" the demand for old drugs *in general*, we estimated the following models using data on the full cross-section of seventy-three drug classes:

$$\Delta\text{OLD} = \delta + \gamma\,\Delta\text{NEW} + u \tag{2.1}$$

$$\Delta\text{OLD} = \delta' + \gamma_P\,\Delta\text{NEW\_PRI} + \gamma_S\,\Delta\text{NEW\_STA} + u' \tag{2.2}$$

where

$\Delta\text{OLD} =$	1994 prescriptions for drugs approved before 1980–1980 prescriptions for drugs approved before 1980;
$\Delta\text{NEW} =$	1994 prescriptions for all drugs approved in or after 1980–1980 prescriptions for all drugs approved in or after 1980 (≈ 0);
$\Delta\text{NEW\_PRI} =$	1994 prescriptions for priority drugs approved in or after 1980–1980 prescriptions for priority drugs approved in or after 1980 (≈ 0);
$\Delta\text{NEW\_STA} =$	1994 prescriptions for standard drugs approved in or after 1980–1980 prescriptions for standard drugs approved in or after 1980 (≈ 0).

Estimates of equations (2.1) and (2.2) are presented in the first two columns of table 2.6. The first column indicates that, on average, 40 percent of sales of all new drugs come at the expense of sales of old drugs. But the estimates in the second column imply that the effect of sales of new priority drugs is very different from the effect of sales of

Table 2.6 Estimates of the relationship across drug classes between changes in new drug consumption and changes in old (or total) drug consumption

Dependent variable	ΔOLD	ΔOLD	$\Delta TOTAL$	$\Delta TOTAL$
ΔNEW	− 0.394		0.606	
	(4.25)		(6.55)	
ΔNEW_PRI		0.020		1.020
		(0.10)		(5.32)
ΔNEW_STA		− 0.712		0.288
		(4.50)		(1.82)
Intercept	809.3	773.9	809.3	773.9
	(1.39)	(1.37)	(1.39)	(1.37)
R^2	0.2031	0.2656	0.3767	0.4255
F value (Prob.-value)		5.9531		5.9531
to test H_0: $\gamma_P = \gamma_S$		(0.017)		(0.017)

$N = 73$.
ΔOLD = 1994 prescriptions for drugs approved before 1980–1980 prescriptions for drugs approved before 1980
ΔNEW = 1994 prescriptions for drugs approved after 1979–1980 prescriptions for drugs approved after 1979 (≈ 0)
ΔNEW_PRI = 1994 prescriptions for priority drugs approved after 1979–1980 prescriptions for priority drugs approved after 1979 (≈ 0)
ΔNEW_STA = 1994 prescriptions for standard drugs approved after 1979–1980 prescriptions for standard drugs approved after 1979 (≈ 0)
$\Delta TOTAL = \Delta OLD + \Delta NEW\_PRI + \Delta NEW\_STA$ = 1994 prescriptions – 1980 prescriptions.
t-statistics in parentheses.

new standard drugs.[4] Sales of new priority drugs have *no effect* on sales of old drugs. In contrast, the sale of one hundred new standard drugs reduces the sales of old drugs by seventy; we can hardly reject the hypothesis that it reduces it by a hundred.

From one perspective, this finding may seem surprising and even anomalous. One might expect the demand for existing drugs to be reduced more by the introduction of drugs that "appear to represent an advance over available therapy" than by the introduction of ones that "appear to have therapeutic qualities similar to those of already

---

[4] The hypothesis that $\gamma_P = \gamma_S$ is decisively rejected.

marketed drugs."[5] But suppose that, instead of considering the effect of new drug introductions on the sale of old drugs, we consider their effect on the total consumption of drugs. Let us define

$$\Delta TOTAL = \Delta OLD + \Delta NEW\_PRI + \Delta NEW\_STA$$
$$= 1994 \text{ prescriptions} - 1980 \text{ prescriptions} \qquad (2.3)$$

Substituting (2.2) into (2.3),

$$\Delta TOTAL = (\delta' + \gamma_P \Delta NEW\_PRI + \gamma_S \Delta NEW\_STA + u')$$
$$+ \Delta NEW\_PRI + \Delta NEW\_STA$$
$$= \delta' + (1 + \gamma_P) \Delta NEW\_PRI + (1 + \gamma_S) \Delta NEW\_STA + u'$$

To obtain the coefficients on ΔNEW_PRI and ΔNEW_STA in the $\Delta TOTAL$ regression, we simply add 1 to their respective coefficients in the ΔOLD regression. For convenience, this regression and its restricted counterpart are shown in columns 3 and 4 of table 2.6. The coefficients on ΔNEW_PRI and ΔNEW_STA in the $\Delta TOTAL$ regression are 1.02 and 0.29, respectively. Sales of new priority drugs increase total drug sales virtually one for one (since they do not come at the expense of old drugs), whereas only 30 percent of new standard drug sales represent net additions to total sales. Drugs that represent an advance over available therapy expand the total market for drugs in that class, whereas drugs that have therapeutic qualities similar to those of already marketed drugs mainly reduce sales of already marketed drugs.

We can also explore (for a subset of drug classes) how the degree of displacement of old drugs depends on the extent of overlap between the inventors of the new drugs and the old drugs. Let SHR_OLD80_{ij} = inventor i's share of all old (pre-1980 approval) drugs in class j consumed in 1980, and SHR_NEW94_{ij} = inventor i's share of all new drugs in class j consumed in 1994.[6] We calculated the "inventor similarity" index

---

[5] On the other hand, standard drugs are, by definition, closer in "product space" to existing drugs than priority drugs are; they may therefore be more easily substitutable for old products.

[6] These are shares of drugs by inventor (applicant), not manufacturer. If Lilly manufactures or markets a drug invented by Merck, sales of that drug are counted as Merck sales.

$$\text{SIMIL}_j = \frac{\Sigma_i \, \text{SHR\_OLD80}_{ij} \text{SHR\_NEW94}_{ij}}{[(\Sigma_i \, \text{SHR\_OLD80}_{ij}^2)(\Sigma_i \, \text{SHR\_NEW94}_{ij}^2)]^{1/2}}$$

This index is bounded between 0 and 1. If the distribution of class j's 1994 new drug prescriptions, by inventor, were identical to its distribution of 1980 old drug prescriptions, by inventor, $\text{SIMIL}_j = 1$. If there were no overlap between these distributions, $\text{SIMIL}_j = 0$. To distinguish between the effects on old drug sales of sales of new drugs by previous and new innovators within the drug class, we estimated the model

$$\Delta\text{OLD} = \delta" + \gamma_{\text{SIM}} \, (\text{SIMIL} * \Delta\text{NEW})$$
$$+ \gamma_{\text{DIS}} \, ((1 - \text{SIMIL}) * \Delta\text{NEW}) + u" \qquad (2.4)$$

The estimate of equation (2.4) is (t-statistics in parentheses)

$$\Delta\text{OLD} = 1310 - 0.883 \, (\text{SIMIL} * \Delta\text{NEW}) - 0.165 \, ((1 - \text{SIMIL})$$
$$* \Delta\text{NEW}) + u"$$

$$(0.98) \quad (2.76) \qquad\qquad\qquad (0.65)$$
$$N = 36 \qquad R^2 = 0.2771$$

These estimates imply that sales of old drugs are reduced much more by the introduction of new drugs by firms that have previously invented drugs in the same class than they are by the introduction of new drugs by firms that have not previously invented drugs in the same class. Indeed, the estimated effect is insignificantly different from zero when there is no overlap between old and new inventors ($\text{SIMIL} = 0$).

This finding is consistent with our findings concerning priority versus standard drugs, because first-time innovators are more likely to introduce priority drugs than experienced innovators: 54.2 percent of inventors' first drugs approved within a class are priority drugs, whereas only 34.7 percent of inventors' subsequent drugs approved within a class are priority drugs. Classes with low inventor similarity tend to be those with a high share of priority drugs in all new drugs.

We also analyzed the relationship across drug classes between the rate of new drug introduction and the log change between 1980 and 1994 in the inventor concentration of sales, by estimating the following model:

$$\log (\text{HHI94}/\text{HHI80}) = \pi_0 + \pi_P \, (\text{NEW\_PRI94}/\text{TOT94})$$
$$+ \pi_S \, (\text{NEW\_STA94}/\text{TOT94}) + e$$

where HHI94 and HHI80 are Herfindahl–Hirschman indexes of inventor concentration of prescriptions in 1994 and 1980, respectively, and (NEW_PRI94/TOT94) and (NEW_STA94/TOT94) are the shares of new priority and standard drugs, respectively, in total 1994 prescriptions. The estimated equation is

$$\log(HHI94/HHI80) =$$
$$-.067 + .054\,(NEW\_PRI94/TOT94) - .675\,(NEW\_STA94/$$
$$TOT94) + e$$

$$(0.88)\quad(0.28)\qquad\qquad\qquad\qquad(3.33)$$

$$N = 58 \qquad\qquad R^2 = 0.1683$$

High rates of introduction of standard drugs, but not of priority drugs, significantly reduce concentration within a class. One possible interpretation of this finding is that introduction of a standard drug is a form of *imitation* that reduces the market shares of previous innovators.

Our findings concerning the effect of new drug introduction on market size and structure may be summarized as follows.

• Sales of new priority drugs increase total drug sales virtually one for one (since they do not come at the expense of old drugs), whereas only 30 percent of new standard drug sales represent net additions to total sales. Drugs that represent an advance over available therapy expand the total market for drugs in that class, whereas drugs that have therapeutic qualities similar to those of already marketed drugs mainly reduce sales of already marketed drugs.

• Sales of old drugs are reduced much more by the introduction of new drugs by firms that have previously invented drugs in the same class than they are by the introduction of new drugs by firms that have not previously invented drugs in the same class. First-time innovators are more likely to introduce priority drugs than experienced innovators.

• High rates of introduction of standard drugs, but not of priority drugs, significantly reduce concentration within a class, perhaps because the introduction of a standard drug is a form of *imitation* that reduces the market shares of previous innovators.

We now proceed to an analysis of the effects of pharmaceutical innovation on mortality reduction and economic growth.

2.3 The long-term decline in mortality and life-years lost

The United States experienced substantial improvement in the average health of its population during the twentieth century. The most widely used measure of medical progress is the growth in life expectancy at birth. Life expectancy (E) in selected years since 1920 is shown below:

1920 54
1965 70
1980 74
1995 76
2002 77

The average person born in 2002 expects to live twenty-three years (43 percent) longer than the average person born in 1920.

Surely almost everyone would agree that this represents a significant improvement in the (economic) well-being of the average person. Yet the most widely used measure of long-run economic growth – growth in *annual* per capita GDP (Y_A) – does not reflect this increase in life expectancy. A better measure of economic well-being might be (expected) *lifetime* per capita GDP (Y_L), where $Y_L = Y_A * E$.[7] The growth in lifetime income is the sum of the growth in annual income and the growth in life expectancy: $y_L = y_A + e$ (where lower-case letters denote growth rates).

This identity has two important implications. First, the growth in annual income understates "true" economic growth (i.e. lifetime income growth). During the last three decades the average annual values of y_A and e have been 2.00 percent and 0.27 percent, respectively. Hence Y_L has grown about 14 percent more rapidly than Y_A. Second, a 10 percent increase in life expectancy has the same effect on Y_L as a 10 percent increase in Y_A. Factors that contribute to increased life expectancy also contribute to long-run economic growth.

While mean life expectancy is an extremely important indicator, other mortality statistics are also worthy of consideration. Table 2.7 presents

---

[7] Adopting Y_L as our measure of economic well-being implies assigning a marginal value of an additional year of life of Y_A, since $dY_L/dE = Y_A$. This valuation is roughly consistent with (although somewhat lower than) that suggested by other investigators. In 1985, US per capita income (in current dollars) was $16,217. Cutler et al. (1998), citing Viscusi (1993), use a "benchmark estimate" of the value of a life-year of $25,000, and upper and lower bounds of $50,000 and $10,000.

*Table 2.7 Mean life expectancy and statistics of age
distribution of deaths, 1970, 1980, and 1991*

	1970	1980	1991
Life expectancy at birth (years)	70.8	73.7	75.5
Age at death (years)			
Mean	64.6	67.7	70.0
Standard deviation	21.8	20.4	19.8
Coefficient of variation	.337	.302	.283
Median	70	72	74
25 percent	57	60	63
10 percent	35	41	43
5 percent	10	22	28

Sources: Life expectancy at birth – National Center for Health Statistics
(2004); age at death – author's calculations, based on 1970, 1980, and 1991
Vital Statistics – Mortality Detail files.

moments and quantiles of the age distribution of deaths in the years 1970,
1980, and 1991. Mean age at death increased by about the same amount
(5.4 years) during 1970–91 as life expectancy. This is not surprising, since
life expectancy figures are actuarial extrapolations of the average age of
decedents. A less well-known fact is that the standard deviation of age
at death has been declining. People tend to live longer than they used to,
and there is also less uncertainty about the age of death. This is because
the area in the bottom tail of the age distribution of deaths has sharply
declined. In 1970 5 percent of the population died by age 10 and 10 percent died by age 35; by 1991 the 5 percent and 10 percent values were 28
and 43, respectively. If people are risk-averse, they are made better off by
the reduction in the variance, as well as by the increase in the mean, of the
age at death.

Another mortality measure widely used in health statistics is life-
years lost before age sixty-five per population under sixty-five (LYL).
This measure is defined as follows:

$$\text{LYL} = \{\Sigma_i \max(0, 65 - \text{age\_death}_i)\} / \text{POP\_LT65},$$

where age_death_i is the age of death of the ith decedent, and POP_LT65
is the population under sixty-five. For example, in a population of
1000, if in a given year two people die at age fifty, three people
die at age sixty, and five people die at age sixty-five or later,

Table 2.8a Life-years lost before age sixty-five per 100,000 population under sixty-five years of age, 1970–1991

	1970	1980	1991	1991/1980
Both sexes and races	8596	6416	5556	.866
White male	9757	7612	6406	.842
Black male	20,284	14,382	14,432	1.003
White female	5527	3983	3288	.826
Black female	12,188	7927	7276	.918

Source: National Center for Health Statistics (2004).

Table 2.8b Infant versus total life-years lost, 1970–1991

Year	Infant mortality rate	Infant deaths (thousands)	Infant LYL	Total LYL
1970	20.0	74,620	4,850,300	15,747,872
1980	12.6	45,511	2,958,215	12,896,160
1991	8.9	36,588	2,378,220	12,250,980

Source: National Center for Health Statistics (2004).

$LYL = [2(65 − 50) + 3(65 − 60) + 5(0)]/1000 = 45/1000 = .045$. This measure gives a great deal of weight to deaths that occur at early ages (especially infant mortality), and no weight at all to deaths beyond age sixty-five. The shrinking of the lower tail of the age distribution documented in table 2.7 leads us to expect the rate of decline in LYL to be faster than the rate of increase in life expectancy. The data in table 2.8a confirm this: LYL declined 35 percent between 1970 and 1991. A substantial part of the decline in LYL was due to a reduction in infant mortality (death before the first birthday). The infant mortality rate declined by 55 percent over this twenty-one-year period (see table 2.8b). Infant deaths accounted for over 30 percent of total life-years lost in 1970; by 1991 they accounted for under 20 percent.[8]

---

[8] The rate of medical progress has been declining. Life expectancy increased by 3.6 years per decade between 1920 and 1965, and by only 2.0 years per decade between 1965 and 1995. Similarly, LYL decreased 25.4 percent from 1970 to 1980, and by only 13.4 percent from 1980 to 1991.

Table 2.9 Percentage reduction in life-years lost of selected diseases, 1980–1991

Rickettsioses and other arthropod-borne diseases	136
Diseases of the esophagus, stomach, and duodenum	31
Cerebrovascular disease	18
Chronic obstructive pulmonary disease and allied conditions	−16

Our goal is to perform detailed econometric tests of the hypothesis that the decline in mortality documented above is due, to an important extent, to the introduction and use of new drugs – i.e. to "changes in pharmaceutical technology." The creation and Food and Drug Administration approval of new drugs requires substantial investment in research and development. Numerous econometric studies have shown that R&D investment has a significant positive impact on the growth in *annual* per capita income (y_A), or on total factor productivity growth.[9] And, while there is considerable anecdotal and case study evidence suggesting that pharmaceutical innovation has made important contributions to the other source of lifetime income growth (increases in life expectancy), there is little systematic econometric evidence on this issue. This chapter is an attempt to fill this gap in the literature.

The overall econometric approach is to estimate the relationship, across diseases, between the extent of pharmaceutical innovation and changes in mortality (e.g. the increase in mean age at death or the reduction in life-years lost). For this approach to be successful, there must be significant cross-disease variation in these variables. Diseases are indeed quite heterogeneous with respect to the rate of progress. The percentage reductions in LYL from 1980 to 1991 of four (ICD9 – International Classification of Diseases, Ninth Revision – two-digit) diseases are shown in table 2.9.

There is also substantial variation across drug classes (which to some extent correspond to diseases) with respect to both the amount and timing of drug approvals. Table 2.10 shows the history of new molecular entities approved within several selected drug classes. The first anticoagulant/thrombolytic drug (heparin) approved by the FDA was in 1942. In contrast, the first cephalosporin (cephalexin) was approved

---

[9] See, for example, Griliches and Lichtenberg (1984) and Lichtenberg and Siegel (1991).

Table 2.10 The history of new molecular entities approved within selected drug classes

Date	Molecule	Therapeutic potential	Chemical type	Applicant	Trade name
Drug class: **Antianginals**					
10DEC79	Nadolol	Standard	NME	E R Squibb	Corgard
31DEC81	Nifedipine	Priority	NME	Pfizer	Procardia
05NOV82	Diltiazem	Standard	NME	Hoechst Marion Rssl	Cardizem
22FEB85	Tolazoline			Ciba	Priscoline
13MAY85	Dipyridamole		NME	Danbury Pharma	Dipyridamole
28DEC90	Bepridil	Priority	NME	Johnson Rw	Vascor
31JUL92	Amlodipine	Standard	NME	Pfizer Cen Res	Norvasc
Drug class: **Anticoagulants/Thrombolytics**					
05FEB42	Heparin			Pharmacia and Upjohn	Heparin sodium
21JAN48	Dicumarol			Abbott Labs	Dicumarol
15DEC48	Protamine	Standard	New manufacturer	Eli Lilly	Protamine sulfate
04NOV55	Warfarin	Standard	NME	Dupont Merck	Coumadin
19JUN62	Anisindione	Standard	NME	Schering	Miradon
29MAR93	Enoxaparin	Priority	NME	Rhone Poulenc Rorer	Lovenox
Drug class: **Antigout**					
26APR51	Probenecid	Priority	NME	MSD	Benemid
13MAY59	Sulfinpyrazone	Standard	NME	Ciba	Anturane
08JUN60	Colchicine	Standard	New combination	MSD	Colbenemi
19AUG66	Allopurinol	Priority	NME	Glaxo Wellcome	Zyloprim

in 1971, and the first antianginal drug (nodolol) was approved in 1979. No antigout drugs have been approved since allopurinol was in 1966.

2.4 Previous evidence on the contribution of new drugs to mortality reduction

PhRMA provides an informal, anecdotal account of the contribution of drug innovation to medical progress in this century. We simply quote their account here:

Antibiotics and vaccines played a major role in the near-eradication of major diseases of the 1920s, including syphilis, diphtheria, whooping cough, measles, and polio. Since 1920, the combined death rate from influenza and pneumonia has been reduced by 85 percent. Despite a recent resurgence of tuberculosis (TB) among the homeless and immuno-suppressed populations, antibiotics have reduced the number of TB deaths to one-tenth the levels experienced in the 1960s. Before antibiotics, the typical TB patient was forced to spend three to four years in a sanitarium and faced a 30 to 50 percent chance of death. Today most patients can recover in 6 to 12 months given the full and proper course of antibiotics.

Pharmaceutical discoveries since the 1950s have revolutionized therapy for chronic as well as acute conditions. From 1965 to 1995 cardiovascular drugs such as antihypertensives, diuretics, beta blockers, and ACE (angiotensin-converting enzyme) inhibitors drastically reduced deaths from hypertension, hypertensive heart disease, and ischemic heart disease.[10]

Similarly, H2 blockers, proton pump inhibitors and combination therapies cut deaths from ulcers by more than 60 percent. Anti-inflammatory therapies and bronchodilators reduced deaths from emphysema by 31 percent and provided relief for those with asthma. Had no progress been made against disease between 1960 and 1990, roughly 335,000 more people would have died in 1990 alone.

Since 1960, vaccines have greatly reduced the incidence of childhood diseases – many of which once killed or disabled thousands of American children. Likewise, vaccines for hepatitis B introduced during the 1980s now protect a new generation of American children from a leading cause of liver disease.

Recent evidence indicates that new drug therapies have sharply reduced fatalities from AIDS (*New York Times*, 1998):

---

[10] Dustan, Roccella, and Garrison (1996) arrive at a similar conclusion: "In the past 2 decades, deaths from stroke have decreased by 59% and deaths from heart attack by 53%. An important component of this dramatic change has been the increased use of antihypertensive drugs."

AIDS deaths in New York City plummeted by 48 percent last year, accelerating earlier gains attributed to improved drug therapies ... the declines crossed sex and racial lines, suggesting that the new therapies were reaching all segments of the AIDS population. National figures for the first six months of 1997 also showed a similar sharp decline, 44 percent, from the corresponding period of 1996 ... Theoretically, the decline in AIDS deaths could have resulted from prevention efforts or some unknown factor ... But the likeliest explanation is expanded use of combinations of newer and older drugs that began to be introduced in recent years, New York City and Federal health officials said.

The anecdotal evidence about the impact of new drugs on mortality is in stark contrast to econometric evidence presented by Skinner and Wennberg (2000) about the relationship between medical expenditure *in general* and outcomes.[11] They analyzed this relationship using both a 20 percent sample of all Medicare enrollees and a 5 percent sample of very ill Medicare patients hospitalized with heart attack (AMI – acute myocardial infarction), stroke, gastrointestinal bleeding, and lung cancer. Per capita medical expenditures vary considerably across regions. For example, average Medicare expenditures on elderly patients in the last six months of life are twice as high in Miami as they are in Minneapolis, and the average number of visits to specialists is five times as high. However, intensive econometric analysis provided "no evidence that higher levels of spending translates into extended survival."[12]

Perhaps the best evidence of the impact on mortality of new drugs comes from clinical trials, of which there have undoubtedly been thousands. One such study was the West of Scotland Coronary Prevention Study of 6595 ostensibly healthy men aged forty-five through sixty-four. The results of the study (*New York Times*, 1995) indicated that the cholesterol-lowering drug

pravastatin reduces the risk of heart attack and death in a broad range of people, not just those with established heart disease, but also among those who are at risk for their first heart attack ... Over five years, those [healthy

---

[11] Pharmaceutical expenditure accounts for only about 10 percent of total US health expenditure.

[12] Lichtenberg's (2000) analysis of longitudinal quinquennial country-level data for a sample of seventeen OECD countries during the period 1960–90 also found no significant effect of either in-patient or ambulatory expenditure on mortality. However, I did find significant effects of pharmaceutical consumption on both life expectancy at age forty and life-years lost.

individuals] treated with ... pravastatin suffered 31 percent fewer nonfatal heart attacks and at least 28 percent fewer deaths from heart disease than a comparable group of men who received a placebo ... In previous studies, pravastatin had been shown to reduce the risk of heart attack by 62 percent in patients with high cholesterol who already had heart disease.

Evidence from clinical trials is extremely useful and of great scientific value, but there does not appear to be any way to summarize or combine all this evidence to shed light on the average or aggregate contribution of pharmaceutical innovation to mortality reduction and economic growth, which is our goal.

2.5 Model specification and estimation

To assess the contribution of pharmaceutical innovation to the reduction in mortality of Americans, I will estimate models of the form

$$\ln (\mathrm{MORT}_{t-k}/\mathrm{MORT}_t) = \alpha + \beta \, (\mathrm{DRUGS}_{t-k,t}/\mathrm{DRUGS}_{.t}) + \varepsilon \quad (2.5)$$

using data on a cross-section of diseases, where MORT_t = a measure of mortality (e.g. mean age at death or total life-years lost) in year t; $\mathrm{DRUGS}_{t-k,t}$ = the number of drugs prescribed in year t that received FDA approval in year t−k or later; and $\mathrm{DRUGS}_{.t} = \sum_k \mathrm{DRUGS}_{t-k,t}$ = the total number of drugs prescribed in year t. For example, if t = 1980 and k = 10, the equation becomes $\ln (\mathrm{MORT}_{1970} / \mathrm{MORT}_{1980}) = \alpha + \beta \, (\mathrm{DRUGS}_{1970,1980} / \mathrm{DRUGS}_{1980}) + \varepsilon$. The dependent variable is the percentage reduction in mortality between 1970 and 1980. The independent variable is the fraction of drugs prescribed in 1980 that were approved in 1970 or later.

The intercept (α) is usually not a parameter of much interest in regression analysis, but it is in this particular case, for it is the predicted change in mortality in the absence of any pharmaceutical innovation. Suppose that the pharmaceutical industry performed no R&D during the period 1960–70, and that, as a result, no new drugs were approved between 1970 and 1980 (in this example, we assume a ten-year lag between R&D expenditure and new drug approval). Then $\mathrm{DRUGS}_{1970,1980} = 0$ and $E[\ln (\mathrm{MORT}_{1970}/\mathrm{MORT}_{1980})] = \alpha$. The total benefit of innovation is the difference between the actual change in mortality (the mean of the dependent variable) and α, the predicted change under no innovation. When MORT is defined as life-years lost, this is an *annual* benefit of innovation, since the dependent variable is the percentage change in annual life-years lost.

We will estimate the model for several alternative definitions of MORT and of $DRUGS_{t-k,t}$, and two sample periods. The mortality variables we will analyze are:[13] mean age at death, life-years lost before age sixty-five by all decedents under sixty-five, and by decedents in three age categories: age nought to one, age one to twenty-five, and age twenty-five to sixty-five. The disaggregation of total LYL into three age categories enables us to distinguish the impact of pharmaceutical innovation on infant mortality from its impact on other premature mortality. Each record in the Mortality Detail file includes a single ICD9 code to indicate the cause of death. We used this code to calculate the various mortality statistics by two-digit ICD9 disease, by year.[14]

The ICD9 classification includes three kinds of codes: disease (or natural causes of death) codes (000–799), nature of injury codes (800–999), and external causes of death codes (E800–E999).[15] In the mortality files, only the first and last sets of codes are used, whereas, in the ambulatory care surveys, only the first and second sets of codes are used. We therefore confine our analysis to diseases (natural causes of death).

Estimates of $DRUGS_{t-k,t}$ – the number of drugs prescribed in year t that received FDA approval in year t–k or later – were obtained by combining data from several sources. The first data source was NAMCS, which contains records of 46,081 doctor-office visits in 1980 and of 33,795 visits in 1991. Each record lists the drugs prescribed (if any) by the physician. Up to five drugs may be coded. If a drug is a combination drug, its ingredients (up to five) are coded, so that as many as twenty-five ingredients (molecular entities) could

---

[13] All mortality statistics were computed from the 1970, 1980, and 1991 Vital Statistics – Mortality Detail files, which constitute complete censuses of the roughly 2 million deaths per year.

[14] The calculated distribution of the log change between 1980 and 1991 of life-years lost included two extreme outliers. The mean, median, and highest and lowest five values of this distribution ($N = 79$) were as follows: mean = .079; median = .093; lowest = $(-4.31, -1.21, -1.07, -0.61, -0.50)$; highest = $(0.66, 0.70, 0.81, 0.86, 3.51)$. A log change of -4.31 corresponds to a seventy-fivefold increase in life-years lost, and a log change of 3.51 corresponds to a thirty-threefold decrease. We were suspicious of such extreme magnitudes, and these observations were statistically quite influential, so we excluded them from the sample.

[15] There is no direct correspondence between the nature of injury codes and the external cause of death codes – e.g. 821 does not correspond to E821.

be coded in a single record. The record also lists up to three diagnoses (ICD9 codes), and a "drug weight," which is used to compute population-level estimates of drug utilization from the sample data. Because multiple diagnoses may be cited in a given record, we sometimes confront the problem of "allocating" the mention of a drug across diagnoses. We adopted the simple, feasible, approach of *equal* allocation of the drug mention across the several diagnoses. For example, if two diagnoses were cited and the drug weight was 10,000, we replaced the mention of that drug by two mentions of the same drug, one for each diagnosis, each with a drug weight of 5000; this procedure does not change the population estimates of drug mentions, by molecule. We then calculated estimates of the aggregate number of prescriptions, by molecule and patient diagnosis.

To calculate the fraction of drugs that were approved by the FDA after a certain date, we linked these data to a list of all new drug applications since 1939 provided by the FDA. Both files included the scientific name of the drug, and the FDA file included the date the drug was first approved as a new molecular entity.[16]

Table 2.2 shows the number of NMEs approved by the FDA from 1940 to the present. A total of 1352 NMEs have been approved. With the single exception of the 1965–69 period, the quinquennial number of NMEs approved has increased steadily over time. Figure 2.1 shows the distribution of drugs prescribed in 1994, by year of FDA approval. Almost a half of the drugs prescribed in 1994 were approved after 1979; about a quarter were approved after 1984.

Only patient visits to doctors' offices are covered by NAMCS. In 1992 the National Center for Health Statistics began conducting a similar

---

[16] The FDA data enable us to identify the dates of "minor" innovations – new formulations of existing molecules – as well as the dates of "major" innovations – new molecules. For example, approval dates of different dosage forms of the molecule aminophylline are as follows:

1940   tablet
1979   tablet, sustained action
1982   solution
1983   suppository
1991   injection

Minor as well as major innovations may confer health benefits. Unfortunately, we are unable to monitor the diffusion of minor innovations, since the drug information provided in NAMCS does not include the dosage form or route of administration of the drug prescribed, only the name of the compound.

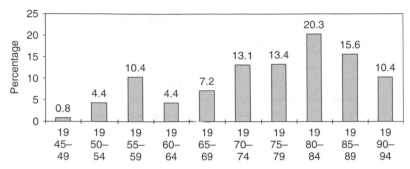

Figure 2.1. The distribution of drugs prescribed in 1994, by year of FDA approval

survey (NHAMCS) of visits to hospital outpatient departments and emergency rooms. The number of hospital outpatient visits is believed to have grown much more rapidly during the last ten to fifteen years than the number of doctor-office visits. Doctor-office visits still accounted for more than 80 percent of total visits in the 1990s,[17] but we sought to improve the precision of our estimates of the new drug proportion by combining information from the NAMCS and NHAMCS surveys. We calculated the share of new drugs in total drugs prescribed in the consolidated sample as well as separately for each of the three outpatient "sectors."[18]

As noted earlier, in the course of the approval process the FDA classifies drugs into two categories: "priority review drugs" – drugs that appear to represent an advance over available therapy – and "standard review drugs" – drugs that appear to have therapeutic qualities similar to those of already marketed drugs.[19] One might reasonably hypothesize that it is only new priority drugs that have a significant

---

[17] Estimated annual number of outpatient visits, by venue, are as follows:
doctors' offices   670 million;
hospital outpatient departments   63 million;
emergency rooms   90 million.

[18] Unfortunately, there is a slight temporal misalignment between the two surveys: the NAMCS data are for 1991 (the same year as the mortality data), but the NHAMCS data are for 1993.

[19] The FDA attempts to review and act on complete applications for "priority" drugs within six months of an NDA's submission date, and to review and act on NDAs for "standard" drugs within twelve months. All AIDS drugs are classified as priority drugs.

Table 2.11a *Percentage distribution of drugs prescribed*
in 1980 that were less than ten years old, by drug class

Drugs used for relief of pain	19.7
Antimicrobial agents	17.1
Respiratory tract drugs	11.2
Psychopharmacologic drugs	9.5
Skin/mucous membrane	8.7
Cardiovascular–renal drugs	8.6
Metabolic and nutrient agents	8.0
Gastrointestinal agents	6.7
Ophthalmic drugs	4.0
Anesthetic drugs	2.3
Neurologic drugs	1.7
Unclassified/miscellaneous	1.2
Oncolytics	0.7
Antidotes	0.2
Hormones and agents affecting hormonal mechanisms	0.2
Antiparasitic agents	0.1

impact on mortality, or at least that their effect is much larger than that of new standard drugs. We estimate versions of equation (2.5) in which $DRUGS_{t-k,t}$ is defined as the number of *priority* drugs prescribed in year t that received FDA approval in year $t-k$ or later, as well as versions in which $DRUGS_{t-k,t}$ is defined as the *total* number of (priority plus standard) drugs prescribed in year t that received FDA approval in year $t-k$ or later. (In both cases $DRUGS_{\cdot,t}$ is defined as the total number of drugs (priority plus standard) prescribed in year t.)

We have mortality data for 1970, 1980, and 1991, and data on the vintage distribution of drugs prescribed in both 1980 and 1991, so we can estimate equation (2.5) for the two periods 1970–80 and 1980–91. The data indicate that the important (in terms of market demand) new drugs introduced in the 1970s and the 1980s were targeted at different diseases and patient populations. Table 2.11a shows the percentage distribution of drugs prescribed in 1980 that were less than ten years old, by drug class. The top three classes of new drugs were drugs used for the relief of pain, as antimicrobial agents, and as respiratory tract drugs; together these accounted for about a half of new drug

Table 2.11b Percentage distribution of drugs prescribed in 1991 that were less than eleven years old, by drug class

Cardiovascular–renal drugs	36.5
Respiratory tract drugs	12.6
Hormones and agents affecting hormonal mechanisms	10.8
Psychopharmacologic drugs	9.9
Antimicrobial agents	7.3
Drugs used for relief of pain	5.9
Gastrointestinal agents	5.1
Skin/mucous membrane	4.7
Metabolic and nutrient agents	4.5
Ophthalmic drugs	2.5
Oncolytics	0.2
Neurologic drugs	0.0
Otologic drugs	0.0

prescriptions. Table 2.11b shows the percentage distribution of drugs prescribed in 1991 that were less than eleven years old, by drug class. Cardiovascular–renal drugs accounted for over a third of the new drugs prescribed in 1991, more than four times their share of new drugs in 1980; the new-drug share of hormones also increased sharply. As Table 2.12 indicates, the average age of patients receiving drugs varies significantly across drug classes. The average age of patients receiving drugs used for the relief of pain – the largest new drug class in 1980 – is forty-eight, while the average age of patients receiving cardiovascular–renal drugs – the largest new drug class in 1991 – is sixty-six. Further calculations show that, consistent with this, the average age of patients receiving any new drug in 1980 was forty-four, whereas the average age of patients receiving any new drug in 1991 was fifty-two. Since the clientele for drugs introduced in the 1980s tended to be older than the clientele for drugs introduced in the 1970s, one would expect the mortality reductions in the 1980s to be more concentrated among older patients. Since we have computed life-years lost by age group, we can test this prediction.

The dependent variable in equation (2.5) is the log-change (growth rate) of mortality, $\ln (MORT_{t-k} / MORT_t)$. The variance of the dependent variable is strongly inversely related to the average size of the

Table 2.12 Average age of patients receiving drugs,
by drug class, 1991

Cardiovascular–renal drugs	66.4
Ophthalmic drugs	60.9
Hematologic agents	57.6
Oncolytics	57.0
Gastrointestinal agents	53.3
Anesthetic drugs	52.2
Psychopharmacologic drugs	50.9
Hormones and agents affecting hormonal mechanisms	49.5
Antiparasitic agents	48.3
Drugs used for relief of pain	48.2
Neurologic drugs	46.4
Unclassified/miscellaneous	43.6
Otologic drugs	40.7
Metabolic and nutrient agents	39.1
Antidotes	36.8
Skin/mucous membrane	36.2
Respiratory tract drugs	32.0
Antimicrobial agents	29.6
Radiopharmaceutical/contrast media	20.4
Immunologic agents	12.4

Source: Author's calculations, based on 1991 NAMCS file.

disease, $(MORT_{t-k} + MORT_t)/2$. The growth rates of diseases affecting a relatively small number of people are likely to be much farther away (in both directions) from the mean than those of major diseases.[20]

---

[20] To illustrate this, we estimated the unweighted regression of the log-change between 1980 and 1991 in LYL on the share of new, priority drugs in 1991, and computed the absolute value of the residuals from this regression. We then divided the diseases into three size categories, based on the average number of LYL in 1980 and 1991. The means of the absolute residuals in each of the three size groups were:
smallest   0.322;
middle   0.288;
largest   0.137.
Also, there is a highly significant (p-value <.01) inverse correlation between the absolute residual and the logarithm of mean LYL.

To correct for heteroskedasticity, the equation is estimated via weighted least squares, where the weight is $(MORT_{t-k} + MORT_t)/2$. Diseases that are responsible for larger average numbers of deaths or life-years lost are given more weight.[21]

There are a number of reasons why estimates of the parameters of equation (2.5) might be biased. Several considerations imply that estimates of β will be biased towards zero, and therefore that our hypothesis tests will be strong tests. Measurement error in the new-drug proportion is perhaps the main reason to suspect downward bias. The variable we calculate $(DRUGS_{t-k,t}/DRUGS_{.t})$ may be a noisy measure of the true share of new drugs, for a number of reasons: (1) sampling error: the NAMCS survey is a random 1 in 10,000 or 1 in 16,000 sample of doctor-office visits; (2) coverage: our drug data refer only to doctor-office visits in 1980, and to outpatient visits in 1991 (no in-patient data); (3) misallocation of drugs to diseases resulting from errors in allocation procedure described above (when there are multiple diagnoses); (4) non-compliance: our data refer to drugs prescribed by physicians, not drugs consumed by patients. It is estimated that only about a half the medicine prescribed is taken correctly. The National Council on Patient Information and Education divides the problem of non-compliance into two categories: acts of omission and acts of commission. Acts of omission include: never filling a prescription; taking less than the prescribed dosage; taking it less frequently than prescribed; taking medicine "holidays"; stopping the regime too soon. Acts of commission include: overuse, sharing medicines, consuming food, drink or other medicines that can interact with the prescribed drug.[22]

It is plausible that reverse causality (endogenous innovation), as well as measurement error in the independent variable, could bias estimates

---

[21] When the dependent variable is the log-change in mean age at death or the coefficient of variation of age at death, the weight used is the average of the number of deaths in years $t-k$ and t.

[22] Compliance rates vary with the disease and setting of the patient group. According to data reported in the *Journal of Clinical Pharmacy and Therapeutics*, patients in homes for the aged had relatively high rates of compliance, as did patients in the first year of antihypertensive treatment. In contrast, patients taking penicillin for rheumatic fever had relatively low rates of compliance.

of β towards zero. Suppose that there is a significant anticipated increase in fatalities from a certain disease (such as AIDS), and that this prospect stimulates a high rate of development and diffusion of new drugs targeted at that disease. Behavior of this sort would reduce the probability of observing a positive relationship across diseases between mortality reduction and new drug utilization.

Estimates of β might conceivably be biased by the omission from equation (2.5) of determinants of mortality change that are correlated with the new drug share. One might postulate that reductions in mortality may be partially attributable to growth in in-patient bed-days and ambulatory visits, and that these might be correlated with our drug measure. However, Skinner and Wennberg's (2000) and Lichtenberg's (2000) findings suggest that these variables do not have a significant effect on mortality; moreover, Lichtenberg's (1996) results suggest that growth in inpatient bed-days is *negatively* correlated across diseases with the extent of pharmaceutical innovation. Nonetheless, we will check the robustness of our estimates by estimating a variant of equation (2.5) in which we include the growth in in-patient bed-days and ambulatory visits.

We can also think of one possible reason for the least squares estimate of β to be biased upwards – i.e. to overestimate the average contribution of pharmaceutical innovation to medical progress. Suppose that the rate of progress (P) against a disease is a deterministic, concave function of research expenditure on the disease (X) and research productivity (π): $P_i = \pi_i X_i^{\theta}$, where i denotes disease i and $0 < \theta < 1$. Taking logarithms of the progress function, $\ln P_i = \ln \pi_i + \theta \ln X_i$. Suppose that disease-specific research productivity (π_i) is unobservable. If π_i were uncorrelated with research expenditure X_i, the least squares estimate of the elasticity of progress with respect to spending θ would be unbiased. However, as argued in Lichtenberg (1997), if decision-makers are efficiently allocating the research budget across diseases (i.e. to maximize the total number of people cured of all diseases), they should devote more research funds to diseases where research productivity is high. Therefore, π_i and X_i are likely to be positively correlated, and the slope coefficient from the simple regression of $\ln P_i$ on $\ln X_i$ would overestimate θ. More progress tends to be made on diseases with high research funding in part because those are the diseases where research productivity is highest. We are examining the relationship between progress and the new drug proportion, not research expenditure, but the latter two variables are likely to be positively correlated.[23]

2.6 Empirical results

Weighted least squares estimates of equation (2.5), based on two alternative measures of mortality change, two measures of pharmaceutical innovation, and two sample periods are presented in table 2.13. The first mortality measure we examine is mean age at death. There is not a statistically significant relationship between the change in mean age at death and the overall new drug share in either period. However, there is a highly significant positive relationship between the increase in mean age and the new priority drug share in both periods. The point estimate of β is four times as large in the first period as it is in the second, but the fraction of cross-disease variance in mortality change explained (R^2) is the same – 7 percent – in the two periods. In the first period the (weighted) mean of the dependent variable – the actual increase in mean age at death – was .101,[24] and α – the predicted increase in the absence of any pharmaceutical innovation – equals -0.01 (and is insignificantly different from zero). This suggests that, in the absence of pharmaceutical innovation, there would have been no increase in mean age at death; new (priority) drugs were responsible for the entire observed increase. It also implies that innovation increased life expectancy, and lifetime income, by just about 1 percent per annum, which appears to be a very sizable contribution to economic growth. In the second period the actual increase in age is much smaller (.036), and the intercept is negative (-0.04) and significantly less than zero, implying that mean age at death would have declined in the absence of innovation. The estimated contribution of innovation to increased longevity

---

[23] We attempted to explore the relationship across drug classes between unpublished PhRMA data on R&D intensity (the ratio of R&D expenditure to sales) and our data on the share of new drugs in total prescriptions. Unfortunately, the PhRMA data are available for only eight highly aggregated drug classes (e.g. cardiovascular drugs), and we failed to find a statistically significant relationship based on such a small sample.

[24] This is a little more than twice as large as the log increase in the mean age at death figures reported in table 2.1, which are means for the entire population of decedents.

Table 2.13 *Weighted least squares estimates of the model* $Y = \alpha + \beta X + \varepsilon$, *for alternative measures of mortality change (Y), pharmaceutical innovation (X), and sample periods*

Dependent variable (Y)	X = New drug share of total number of drugs prescribed at end of period		X = New "priority" drug share of total number of drugs prescribed at end of period	
	1970–80	1980–91	1970–80	1980–91
Percentage increase in mean age at death	$\beta = 0.50$ (1.18) $\alpha = .028$ (0.37) $Y = .101$ $R^2 = .02$	$\beta = -.035$ (0.49) $\alpha = .051$ (1.63) $Y = .036$ $R^2 = .00$	$\beta = 1.19$ (2.28) $\alpha = -0.01$ (0.18) $Y = .101$ $R^2 = .07$	$\beta = 0.30$ (2.34) $\alpha = -0.04$ (5.31) $Y = .036$ $R^2 = .07$
Percentage reduction in age nought to sixty-five life-years lost	$\beta = 3.17$ (5.78) $\alpha = -0.22$ (1.88) $Y = .343$ $R^2 = .32$	$\beta = 0.44$ (2.84) $\alpha = -.081$ (1.28) $Y = .088$ $R^2 = .10$	$\beta = 4.27$ (8.39) $\alpha = -0.21$ (2.40) $Y = .343$ $R^2 = .50$	$\beta = 0.55$ (3.10) $\alpha = -.044$ (0.92) $Y = .088$ $R^2 = .12$

during 1980–91 is .076 (=.036 − (−0.04)), about three-fourths as large as it was during 1970–80.

The second mortality variable we analyze is the percentage reduction in life-years lost before age sixty-five by decedents of all ages. In both periods the LYL reduction is very strongly positively related to both drug innovation measures. The relationship with new priority drugs is slightly stronger than the relationship with all new drugs, so we will focus on the priority drug estimates. The new priority drug share explains a *half* of the cross-disease variation in life-years lost (LYL) reduction during the 1970s. The estimates imply that, in the absence of innovation, LYL would have increased significantly, not declined sharply, as it actually did. The implied contribution of innovation to premature mortality reduction during 1970–80 is enormous: .55 (= .343 − (−.21)).

In the second period the implied contribution is much more modest (the difference between actual and predicted (under no innovation) LYL reduction is .13), although it is still extremely large in economic terms. During the 1980s the average number of LYL per year was 12.6 million. Pharmaceutical innovation reduced annual LYL by about 13 percent, or about 1.6 million. If we value a life-year at 1985 GDP per capita ($16,217) − as discussed earlier, this appears to be conservative − the value of life-years saved each year is $26.6 billion. Of course, innovation costs money. Table 2.14 provides data on total R&D spending by US pharmaceutical companies. Assuming that there is a ten-year lag from research to FDA approval and market introduction of new drugs, R&D expenditures during 1970–81 constitute the additional cost of new drugs (as opposed to existing drugs) consumed by patients during 1980–91. Total (undiscounted, current dollar) industry R&D expenditures during the period 1970–81 were $14.6 billion. Thus, as a first approximation, a "one-time" expenditure of $14.6 billion leads to an annual gain of $26.6 billion. This should be regarded as a very rough estimate; for a number of reasons it may be either too high or too low. Industry R&D expenditure may understate the true social cost of drug development. Toole (1998) presents evidence consistent with the view that the number of new molecular entities approved in a given year is positively related to *government*-funded biomedical research expenditure many (e.g. twenty-five) years earlier, as well as to industry-funded R&D. It may therefore be appropriate to include some government-funded

Table 2.14 *Domestic US R&D and R&D abroad, ethical pharmaceuticals, research-based pharmaceutical companies, 1970–1997*

Year	Domestic US R&D ($ million)	Annual percentage change	R&D abroad ($ million)	Annual percentage change	Total R&D ($ million)	Annual percentage change
1970	566.2	–	52.3	–	618.5	–
1971	626.7	10.7	57.1	9.2	683.8	10.6
1972	654.8	4.5	71.3	24.9	726.1	6.2
1973	708.1	8.1	116.9	64.0	825.0	13.6
1974	793.1	12.0	147.7	26.3	940.8	14.0
1975	903.5	13.9	158.0	7.0	1061.5	12.8
1976	983.4	8.8	180.3	14.1	1163.7	9.6
1977	1063.0	8.1	213.1	18.2	1276.1	9.7
1978	1166.1	9.7	237.9	11.6	1404.0	10.0
1979	1327.4	13.8	299.4	25.9	1626.8	15.9
1980	1549.2	16.7	427.5	42.8	1976.7	21.5
1981	1870.4	20.7	469.1	9.7	2339.5	18.4
1982	2268.7	21.3	505.0	7.7	2773.7	18.6
1983	2671.3	17.7	546.3	8.2	3217.6	16.0
1984	2982.4	11.6	596.4	9.2	3578.8	11.2
1985	3378.7	13.3	698.9	17.2	4077.6	13.9
1986	3875.0	14.7	865.1	23.8	4740.1	16.2
1987	4504.1	16.2	998.1	15.4	5502.2	16.1

Year	Domestic US R&D ($ million)	Annual percentage change	R&D abroad ($ million)	Annual percentage change	Total R&D ($ million)	Annual percentage change
1988	5233.9	16.2	1303.6	30.6	6537.5	18.8
1989	6021.4	15.0	1308.6	0.4	7330.0	12.1
1990	6802.9	13.0	1617.4	23.6	8420.3	14.9
1991	7928.6	16.5	1776.8	9.9	9705.4	15.3
1992	9312.1	17.4	2155.8	21.3	11,467.9	18.2
1993	10,477.1	12.5	2262.9	5.0	12,740.0	11.1
1994	11,101.6	6.0	2347.8	3.8	13,449.4	5.6
1995	11,874.0	7.0	3333.5	[a]	15,207.4	[a]
1996[b]	13,378.5	12.7	3539.6	6.2	16,918.1	11.2
1997[b]	15,045.1	12.5	3816.0	7.8	18,861.1	11.5

[a] R&D abroad affected by merger and acquisition activity.
[b] Estimated.

Notes: 1 – Ethical pharmaceuticals only. Domestic US R&D includes expenditures within the United States by research-based pharmaceutical companies. R&D abroad includes expenditures outside the United States by US-owned research-based pharmaceutical companies. 2 – Totals may not add exactly due to rounding.

Source: PhRMA (1997).

R&D in our cost estimate.[25] On the other hand, pharmaceutical innovation probably confers benefits other than reduced mortality, such as reduced hospitalization and surgical expenditures (see Lichtenberg, 1996), reduced workdays and schooldays lost, and improved quality of life; the above calculation does not account for these.

As discussed earlier, the (simple regression) coefficients on pharmaceutical innovation presented in table 2.13 could be biased due to the omission of covariates such as the growth in in-patient and ambulatory care utilization. We have data on these covariates for the 1980–91 period, and we estimated a generalized version of equation (2.5) that includes them. The estimates are presented in table 2.15. In the mean age at death equation, neither of these variables is significant, and their inclusion has virtually no effect on the drug coefficient. In the LYL equation, both the in-patient and ambulatory care coefficients are highly significant, but they have a perverse (negative) sign: diseases with the highest growth in in-patient and ambulatory utilization had the lowest rates of LYL reduction. Perhaps this is because the growth in both utilization and LYL are driven by exogenous, unobserved changes in the severity or incidence of diseases. In any case, note that the inclusion of these covariates *doubles* both the coefficient on the drug variable and its t-statistic. These estimates suggest that the estimates presented in table 2.13 do not overestimate the effect of new drugs on mortality change, and may indeed understate it.

The last set of estimates we wish to consider are estimates of the model of LYL reduction, by age class of decedent, for three age classes: ages nought to one, one to twenty-five, and twenty-five to sixty-five. The results reported above indicate that pharmaceutical innovation substantially reduces total LYL before age sixty-five, but the reduction in mortality could conceivably be confined to two or even just one of the age classes (e.g. infants). The estimates presented in table 2.16 strongly suggest that this is not the case: all three age groups benefited (in terms of reduced mortality) from the arrival of new drugs in at least one of the two periods. It is true that infants are the only age class for

[25] In 1992 about a half of total health R&D expenditure in the United States was publicly supported, but during the last few decades private funding has grown much more rapidly than public funding. In 1980 government R&D accounted for almost 80 percent of US health R&D. Source: National Science Board (1993, p. 365).

Table 2.15 The effect of controlling for growth in hospital bed-days and doctor-office visits on the priority drug share coefficient

Dependent variable	Percentage increase in mean age at death, 1980–91		Percentage reduction in life-years lost, 1980–91	
Column	(1)	(2)	(3)	(4)
Regressor:				
New "priority" drug share of total number of drugs prescribed in 1991	0.30 (2.34)	0.28 (2.08)	0.55 (3.10)	1.114 (6.05)
Percentage increase in number of hospital bed-days, 1980–91		.029 (1.31)		−0.171 (3.48)
Percentage increase in number of doctor-office visits, 1980–91		.039 (1.48)		−0.307 (5.01)
R^2	.07	.12	.12	.39

Note: t-statistics in parentheses.

which there is a highly significant relationship between LYL reduction and the share of new priority drugs in both periods. (This relationship is remarkably strong in the first period; the priority drug variable explains 85 percent of the variation across diseases in the 1970–80 decline in infant mortality.) However, in both periods innovation also makes a significant contribution to mortality reduction in one of the other two age classes. Consumption of new drugs reduced mortality among persons aged one to twenty-five between 1970 and 1980, and among persons aged twenty-five to sixty-five between 1980 and 1991. In tables 2.11 and 2.12 we showed that the drugs introduced in the 1980s tended to be targeted at older patients than the drugs introduced in the 1970s, so the "vintage-age interaction" evident in table 2.16 is not at all surprising.

Table 2.16 *Weighted least squares estimates of model of life-years lost reduction, by age of decedent*

Dependent variable (Y): percentage reduction in life-years lost, by age of decedent	X = New drug share of total number of drugs prescribed at end of period		X = New "priority" drug share of total number of drugs prescribed at end of period	
	1970–80	1980–91	1970–80	1980–91
Age nought to one	$\beta = 7.02$ (15.6) $\alpha = -0.91$ (8.21) $Y = .585$ $R^2 = .80$	$\beta = 0.21$ (1.44) $\alpha = .105$ (1.69) $Y = .185$ $R^2 = .03$	$\beta = 6.87$ (17.5) $\alpha = -0.67$ (7.62) $Y = .585$ $R^2 = .83$	$\beta = 0.39$ (3.15) $\alpha = .071$ (1.59) $Y = .185$ $R^2 = .14$
Age one to twenty-five	$\beta = 1.66$ (2.68) $\alpha = .075$ (0.60) $Y = .366$ $R^2 = .09$	$\beta = -.068$ (0.28) $\alpha = .268$ (3.17) $Y = .246$ $R^2 = .00$	$\beta = 1.92$ (2.93) $\alpha = 0.15$ (1.59) $Y = .366$ $R^2 = .11$	$\beta = -0.28$ (0.75) $\alpha = .030$ (3.72) $Y = .246$ $R^2 = .01$
Age twenty-five to sixty-five	$\beta = -.153$ (0.36) $\alpha = .194$ (2.42) $Y = .170$ $R^2 = .00$	$\beta = .986$ (4.61) $\alpha = -0.37$ (4.17) $Y = .021$ $R^2 = .24$	$\beta = 0.27$ (0.45) $\alpha = .145$ (1.96) $Y = .170$ $R^2 = .00$	$\beta = 1.47$ (3.16) $\alpha = -0.31$ (2.86) $Y = .021$ $R^2 = .12$

2.7 The impact of government policy on the rate of pharmaceutical innovation

There have been many important government policy events affecting the pharmaceutical industry in the last fifteen years. These have been in a number of areas of policy, including the regulation of drug prices, domestic and foreign intellectual property (patent) protection, procedures for new drug approval, product liability, trade policy, the regulation of generic drugs, and tax policy. Table 2.17 provides examples of some government policy events affecting the pharmaceutical industry during the period 1985–93. In this section we outline an econometric framework for examining the impact of certain government policy changes on innovation (R&D investment) in the pharmaceutical industry.

Failure to account for the effects of policies on innovation may lead to seriously distorted perceptions of the long-term consequences of these policies for social and consumer welfare. For example, several years ago some policy-makers advocated the imposition of price controls on pharmaceuticals. In the short run, consumers might have benefited from price controls. But the imposition of price controls would probably reduce the expected returns to investment in new drugs, hence (optimal) R&D investment.[26] Jensen (1987) provided evidence that the rate of discovery of new drugs is positively related to past R&D investment, and we have provided evidence that mortality and hospitalization rates are inversely related to the rate of introduction of new drugs. It is quite possible that the long-term harm to consumers of fewer new drugs would more than offset the benefit of lower prices on existing drugs.[27]

---

[26] During the period in which price controls were under active discussion, the market value of pharmaceutical firms declined substantially, and (as we discuss below), according to Tobin's widely accepted q theory of investment, declines in market value should cause declines in investment.

[27] A similar argument may be applied to the patent system. A patent is a temporary monopoly granted by the government. It is well known that there are (static) social welfare losses associated with monopolies (the monopoly price is higher than socially efficient). If the patent system were abolished today, there would be an immediate increase in social welfare, as lowest-cost processes and highest-quality products would be more widely utilized. But, without patent protection, there would be much less inventive activity in the future, and the rate of economic growth would almost certainly decrease sharply. The existence of the patent system suggests that society recognizes that the (dynamic) welfare gains from more innovation outweigh the (static) welfare losses from monopoly.

Table 2.17 Examples of government policy events affecting the pharmaceutical industry, 1985–1993

Policy area	Event	Date event reported
New drug approval process	Changes in the new drug approval regulations are expected to cut 20 percent of FDA's average review time	1/7/85
Trade policy	Senate to approve export of non-approved US drugs to countries that have rigorous approval systems	5/5/86
Product liability	Commerce Committee approved product liability protection to pharmaceutical manufacturers	8/86
Financing of regulation	House Appropriations Committee increased FDA budget, exceeding current budget by $33.4 million	9/86
Trade policy/ intellectual property	Government to impose sanctions on Brazil, which refuses to protect patent rights of US drug manufacturers	7/23/88
Regulation of generic drugs	FDA officials conduct special inspections at plants of eleven generic drug manufacturers	7/31/89
Regulation of drug prices	US Senate rejects measure to punish drug price increases	3/16/92
Intellectual property	House approves drug patent extension bill	8/10/92
Regulation of drug prices	Senators threaten action to control prices of prescription drugs to prevent price gouging	9/28/92
Regulation of drug prices/tax policy	Senator says retail drug prices rising too rapidly; wants to end Section 936 tax break	2/15/93
Regulation of drug prices/antitrust	Drug industry plan to curb price increases via inflation tie-in is rejected as illegally anti-competitive by Justice Department	10/2/93
Trade policy/ intellectual property	United States reaches intellectual property agreement for pharmaceuticals with Hungary	8/9/93

Source: *F&S Index, United States Annual* (various years).

Although one could try to examine the effect of policy on pharmaceutical innovation directly, we believe that a more fruitful approach is to analyze the relationships between each of these variables and a third, "intervening" variable: the market value of pharmaceutical firms. The relationships among these three variables may be represented schematically as follows:

$$
\begin{array}{ccc}
& (a) & (b) \\
\text{government policy} & ===> \text{market value of} & ===> \text{R\&D investment} \\
& \text{pharmaceutical firms} &
\end{array}
$$

The first relationship can be estimated using high-frequency (daily) data. (The timing of policy events can be established accurately using the *F&S Index* and other sources.) The second relationship can be estimated using low-frequency (annual) data (from Compustat and the Center for Research in Security Prices (CRSP)). Though estimated at different frequencies, parameter estimates from the two models can be combined to determine the effect of policy changes on R&D. Examination of these relationships within a unified framework should enable us to make precise inferences about the impact of changes in the environment on innovation in the pharmaceutical industry.[28]

a. Government policy ===> market value of pharmaceutical firms

We hypothesize that some government policy events affect the expected future net cash flows of pharmaceutical firms. If the stock market is efficient, the value of the firm at time t is the expected present discounted value (PDV) of its future net cash flows, conditional on the information available at time t. Hence, policy events affect the value of the firm.

There has been some debate about the relationship between market value and the expected PDV of future net cash flows (or between *changes* in market value and *changes* in the expected PDV of future net cash flows), and about "market efficiency" in general. Several recent empirical studies have found a strong relationship between market value and subsequent cash flows (actual or forecast).[29]

---

[28] In a recent study, Jaffe and Palmer (1997) examined the effect of *environmental* regulation on innovation.

[29] Cornett and Tehranian (1992) examined the performance of bank mergers using pre-merger and post-merger accounting data for fifteen large interstate and fifteen large intrastate bank acquisitions completed between 1982 and 1987.

Malkiel (1990) reviews "all the recent research proclaiming the demise of the efficient market theory.[30] [He concludes that,] while the stock market may not be perfect in its assimilation of knowledge, it does seem to do quite a creditable job" (pp. 187–8). He also notes that "one has to be impressed with the substantial volume of evidence suggesting that stock prices display a remarkable degree of efficiency. Information contained in past prices or any publicly available fundamental information is rapidly assimilated into market prices."

There have been at least two previous studies of the effect of government policy on the market value of pharmaceutical firms. Ellison and Mullin (1997) examined the effect of the evolution of the Clinton health care reform proposal on the path of market-adjusted pharmaceutical firms' stock prices during the period from September 1992 to October 1993. Conroy, Harris, and Massaro (1992) provided evidence that four major legislative and regulatory initiatives directed towards the industry during the 1970s and 1980s – the Maximum Allowable Cost Program (1975), the Prospective Payment Plan (1982/3), the Drug Price Competition and Patent Term Restoration Act (1984), and the Catastrophic Protection Act (1987/8) – had significant, often negative, effects on share returns (market value). Using methodologies similar to

They found "significant correlations ... between stock market announcement-period abnormal returns and the cash flow and accounting performance measures. The results suggest that, for large bank mergers, expectations of improved bank performance underlie the equity revaluations of the merging banks." Healy, Palepu, and Ruback (1992) analyzed accounting and stock return data for a sample of the largest fifty US mergers between 1979 and mid-1984 to examine post-acquisition performance, as well as the sources of merger-induced changes in cash flow performance. They reported that a "strong positive relationship exists between postmerger increases in operating cash flows and abnormal stock returns at merger announcements. This indicates that expectations of economic improvements underlie the equity revaluations of the merging firms." Kaplan and Ruback (1995) compared the market value of highly leveraged transactions (HLTs) to the discounted value of their corresponding cash flow forecasts. "For a sample of 51 HLTs completed over the period 1983–1989, the valuation of discounted cash flow forecasts are within 10%, on average, of the market values of the completed transactions. These valuations perform at least as well as valuation methods using comparable companies and transactions."

[30] Malkiel distinguishes between three forms of the efficient market hypothesis: the *weak form* (random walk hypothesis), which states that investment returns are serially independent; the *semi-strong form*, which states that all public information about a company is already reflected in the stock's price; and the strong form, which states that no technique of selecting a portfolio can consistently outperform a strategy of simply buying and holding a diversified group of securities that make up the popular market averages.

theirs, one could also examine the impact of a number of other potentially significant events on market value, such as:

- the George H. W. Bush administration's "Best Price" proposal for Medicaid (1990);
- the Vaccines for Children Program (1993);
- the Drug User Fee initiative; and
- the approval of phase-out of Section 936 tax benefits.

To determine the appropriate length of the event window – the set of days on which the announcement information was incorporated into the stock price – one can follow the approach of Rogerson (1989), who examined the effect of Defense Department awards of procurement contracts for major weapons systems on the stock prices of aerospace firms. He argued that "some general 'turbulence' in the returns might be expected for a few days before and after the announcement" (p. 1299), and estimated models based on alternative definitions of the event window (u, v), where u is the first day of the window (relative to the event day, day O), and v is the last day. The fourteen event windows considered by Rogerson were $v = 1, 2, \ldots, 14$ and $u = 1 - v$. Rogerson decided that the most suitable event window for this sample was $(-2, 3)$. Similarly, rather than imposing a specific a priori definition of the event window, one can experiment with different definitions, and choose the one that provides the best fit to the data.

b. Market value of pharmaceutical firms ===> R&D investment

The second relationship to be examined is the effect of a firm's market value on its rate of R&D investment. John Maynard Keynes (1936) may have been the first economist to hypothesize that the incentive to invest depends on the market value of capital relative to its replacement cost. James Tobin (1969) provided a rigorous theoretical foundation for this hypothesis. Below we show that, under certain plausible assumptions, a value-maximizing firm's investment (relative to its capital stock) is a (linear) function of Tobin's q – the ratio of the market value of the firm to the replacement cost of its capital.[31]

In each period t, firm i's real net cash flow X is given by

$$X_{it} = F(K_{i,t-1}, N_{it}) - w_t N_{it} - p_t I_{it} + C(I_{it}, K_{i,t\ 1})$$

---

[31] This derivation is adapted from the one provided by Hassett and Hubbard (1996). For simplicity, we ignore taxes.

Where K is the capital stock, F( ) is the real revenue (production) function of the firm, N is employment, w is the wage rate, p is the real price of investment goods, and C( ) is the function determining the cost of adjusting the capital stock. The marginal cost of newly installed capital is therefore $p_t + C_I(I_{it}, K_{i,t-1})$.

Under the assumption of value maximization, the firm maximizes the present value of its future net cash flows. Letting β_{is} be the discount factor appropriate for the ith firm at time s, the firm's value at time t is

$$V_{it} = \max E_{it} \Sigma_{s=t}^{\infty} \left(\Pi_{j=t}^{s}\right) \beta_{is} X_{is}$$

where E_{it} is the expectations operator for firm i conditional on information available at time t. The firm chooses the past of investment and employment, given the initial capital stock, to maximize firm value. The change in the capital stock – net investment – is given by $I_{it} - \delta K_{i,t-1}$, where δ is the (assumed constant) proportional rate of depreciation.

For investment, the solution to the problem requires that the marginal value of an additional unit of investment (denoted by q_{it}) equal its marginal cost:

$$q_{it} = p_t + C_I\left(I_{it}, K_{i,t-1}\right) \qquad (2.6)$$

Assume that the adjustment cost function is quadratic,

$$C\left(I_{it}, K_{i,t-1}\right) = (\omega/2)\left[\left(I_{it}/K_{i,t-1}\right) - \mu_i\right]^2 K_{i,t-1}$$

where μ is the steady-state rate of investment and ω is the adjustment cost parameter. Then equation (2.6) can be rewritten as an investment equation:

$$\left(I_{it}/K_{i,t-1}\right) = \mu_I + (1/\omega)[q_{it} - p_t]$$

This equation cannot be estimated directly, in general, because (marginal) q is unobservable. However, Hayashi (1982) showed that, if the firm is a price-taker in input and output markets, and the production function exhibits constant returns to scale, marginal q equals average q (denoted Q), defined as

$$Q_{it} = (V_{it} + B_{it})/K_{i,t-1}^{R}$$

where V is the market value of the firm's equity, B is the market value of the firm's debt, and K^R is the replacement value of the firm's capital stock. This formulation stresses the relationship between investment and the net profitability of investing, as measured by the difference between the value of an incremental unit of capital and the cost of purchasing capital. The hypothesis that investment in general is positively related to Tobin's q – the ratio of the stock market value of the firm to replacement costs – is now widely accepted.[32] Dornbusch and Fischer (1994) argue that "[W]hen [q] is high, firms will want to produce more assets, so that investment will be rapid," and therefore that "a booming stock market is good for investment" (p. 341). "The managers of the company can ... be thought of as responding to the price of the stock by producing more new capital – that is, investing – when the price of shares is high and producing less capital – or not investing at all – when the price of shares is low" (p. 355). Similarly, Hall and Taylor (1991) state that "investment should be positively related to q. Tobin's q provides a very useful way to formulate investment functions because it is relatively easy to measure" (p. 312).[33]

When the firm has some market power (as pharmaceutical firms do, at least on their patented products), average Q is no longer exactly equal to (a perfect indicator of) marginal q, but it is still highly positively correlated with (a good indicator of) marginal q. Under these conditions, the estimated coefficient on Q will be biased towards zero (the magnitude of the bias depends on the "noise-to-signal ratio" in measured Q), and tests of the hypothesis that market value affects investment are "strong tests." Moreover, many economists believe that there are many industries in which firms exercise some market power, and there is much evidence at both the macro- and micro-level that is consistent with the q theory of investment. Hall and Taylor note that, in 1983, q was quite high, and investment was booming in the United States, even though the real interest rate and the rental

[32] According to the theory, Tobin's q is equal to the ratio of the marginal benefit of capital to the marginal (user) cost of capital. When Tobin's q is high (greater than 1), the firm should invest. We hypothesize that changes in the environment affect the (marginal) expected benefits and costs of "knowledge capital," hence the incentives to invest in R&D.

[33] Hassett and Hubbard (1996, p. 31) argue that "determining the response of investment to [q] is easiest during periods in which large exogenous changes in the distribution of structural determinants occur, as during tax reforms" – or, in our context, changes in the regulatory or legal environment.

price of capital were also high. Eisner (1978, p. 112) presented micro-econometric evidence that supports the theory; he found that, "even given past sales changes, the rate of investment tends to be positively related to the market's evaluation of the firm both for the current year and the past year." Investment is positively related to Q under imperfect as well as perfect competition.

This evidence relates to fixed investment in the business sector as a whole, not specifically to R&D investment in the pharmaceutical industry. But Griliches and others have argued that many of the tools and models developed to analyze conventional investment can also fruitfully be applied to R&D investment. For example, there is a stock of "knowledge capital" (resulting from past R&D investment) analogous to the stock of physical capital. Therefore, one would expect to observe a strong positive relationship between the market value of pharmaceutical firms and their rate of R&D investment. If this is the case, then government policy events that significantly reduce market value also tend to reduce R&D investment.

A study along these lines should complement and extend previous studies of the effect of (lagged) cash flow on R&D investment (Grabowski, 1968; Grabowski and Vernon, 1981; Jensen, 1990). We believe it is desirable to examine the relationship between R&D investment and market value (in addition to cash flow) for two reasons (even though cash flow and market value may be correlated): (1) the neoclassical theory of investment outlined above implies that R&D should depend primarily on expected future cash flows, which are incorporated into market value;[34] (2) market value can be linked with "news" about government policy events.

2.8 Summary and conclusions

Schumpeter argued that innovation plays a central role in capitalist economies, and that innovation can be characterized as a process of

---

[34] The main reason why previous investigators have argued that R&D investment depends on cash flow is that high cash flows reduce the *cost* of capital. But R&D investment should depend on the expected *benefits* of "research capital" as well as the cost of capital. In principle, Tobin's q reflects *both* the expected benefit and cost of capital. Presumably, certain legal and regulatory developments have no effect on current (or lagged) cash flows, but have important effects on the profitability of investment and on market value.

"creative destruction." The development of new products and processes creates enormous value for society, but may also undermine or
destroy existing organizations, markets, and institutions. We have
performed an econometric study of creative destruction in the pharmaceutical industry, a sector in which innovation is both pervasive and
very well documented (due to FDA regulation).

First, we examined the impact of the introduction of new drugs on
the demand for old drugs and on the total size of the market. We found
that drugs that represent an advance over available therapy expand the
total market for drugs in that class, whereas drugs that have therapeutic
qualities similar to those of already marketed drugs mainly reduce the
sales of already marketed drugs. We also found that sales of old drugs
are reduced much more by the introduction of new drugs by firms that
have previously invented drugs in the same class than they are by the
introduction of new drugs by firms that have not previously invented
drugs in the same class.

There is considerable anecdotal and case study evidence indicating
that pharmaceutical innovation has played an important role in the
long-term increase in life expectancy of Americans. We have attempted
to provide some systematic econometric evidence on this issue, by
estimating the relationship, across diseases, between the extent of
pharmaceutical innovation and changes in mortality (e.g. the increase
in mean age at death or the reduction in life-years lost). We found that,
in both of the two periods we studied (1970–80 and 1980–91), there
was a highly significant positive relationship across diseases between
the increase in mean age at death (which is closely related to life
expectancy) and the share of new "priority" drugs in total drugs prescribed by doctors. (Priority drugs are drugs that appear to the FDA to
represent an advance over available therapy.) Mean age at death
increased in both the 1970s and the 1980s (albeit more slowly in the
1980s). Our estimates imply that in the absence of pharmaceutical
innovation, there would have been no increase and perhaps even a
small decrease in mean age at death; new (priority) drugs were responsible for the entire observed increase. They also imply that innovation
has increased life expectancy, and lifetime income, by about 0.75 to
1.0 percent per annum, which represents a substantial contribution
to economic growth.

We also examined the effect of new drug introductions on the
reduction in life-years lost before age sixty-five. In both periods the

LYL reduction is very strongly positively related to drug innovation measures. The new priority drug share explains a *half* of the cross-disease variation in LYL reduction during the 1970s. The estimates imply that, in the absence of innovation, LYL would have increased significantly, not declined sharply, as it actually did. The implied contribution of innovation to premature mortality reduction during 1970–80 is enormous. In the second period the implied contribution is much more modest, although it is still extremely large in economic terms. If we value a life-year at 1985 GDP per capita ($16,217), which appears to be conservative, the value of life-years saved each year is $26.6 billion. The "one-time" R&D expenditure required to achieve this saving is on the order of $14.6 billion. These estimates are admittedly very rough, but they do suggest that the social rate of return to pharmaceutical R&D investment has been quite high.

The innovation-induced mortality reduction is not confined to a single age group; all age groups benefited from the arrival of new drugs in at least one of the two periods. Controlling for growth in in-patient and ambulatory care utilization either has no effect on the drug coefficient or significantly increases it.

We also argued that government policies may have important effects on the pharmaceutical industry. Failure to account for the effects of policies on innovation may lead to seriously distorted perceptions of the long-term consequences of these policies for social and consumer welfare. We outlined an econometric framework for examining the impact of government policy changes on R&D investment in the pharmaceutical industry. The implementation of this approach is a task for future research.

References

Conroy, R., R. Harris, and T. Massaro (1992), *Assessment of Risk and Capital Costs in the Pharmaceutical Industry*, unpublished paper, Darden School of Business, University of Virginia, Charlottesville.

Cornett, M. M., and H. Tehranian (1992), "Changes in corporate performance associated with bank acquisitions," *Journal of Financial Economics*, 31 (April), 211–34.

Cutler, D., M. McClellan, J. Newhouse, and D. Remler (1998), "Are medical prices declining?," *Quarterly Journal of Economics*, 113 (4), 991–1024.

Dornbusch, R., and S. Fischer (1994), *Macroeconomics*, 6th edn., McGraw-Hill, New York.

Dustan, H. P., E. J. Roccella, and H. H. Garrison (1996), "Controlling hypertension: a research success story," *Archives of Internal Medicine*, 56 (23 September), 1926–35.

Eisner, R. (1978), *Factors in Business Investment*, Ballinger (for the National Bureau of Economic Research), Cambridge, MA.

Ellison, S., and W. Mullin (1997), *Gradual Incorporation of Information into Stock Prices: Empirical Strategies*, Working Paper no. 6218, National Bureau of Economic Research, Cambridge, MA.

F&S Index, United States Annual (various years), Information Access Co., Foster City, CA.

Grabowski, H. (1968), "The determinants of industrial research and development: a study of the chemical, drug, and petroleum industries," *Journal of Political Economy*, 76, 292–306.

Grabowski, H. G., and J. Vernon (1981), "The determinants of research and development: expenditures in the pharmaceutical industry," in R. Helms (ed.), *Drugs and Health*, American Enterprise Institute, Washington, DC.

Griliches, Z., and F. Lichtenberg (1984), "R&D and productivity at the industry level: is there still a relationship?," in Z. Griliches (ed.), *R&D, Patents, and Productivity*, University of Chicago Press, Chicago, 465–96.

Hall, R., and J. Taylor (1991), *Macroeconomics*, 3rd edn., Norton, New York.

Hassett, K., and R. G. Hubbard (1996), *Tax Policy and Investment*, unpublished paper, Federal Reserve Board of Governors and Columbia Business School, New York.

Hayashi, F. (1982), "Tobin's marginal q and average q: a neoclassical interpretation," *Econometrica*, 50, 213–24.

Healy, P., K. Palepu, and R. Ruback (1992), "Does corporate performance improve after mergers?," *Journal of Financial Economics*, 31 (April), 135–75.

Jaffe, A., and K. Palmer (1997), "Environmental regulation and innovation: a panel data study," *Review of Economics and Statistics*, 79 (4), 610–19.

Jensen, E. (1987), "Research expenditures and the discovery of new drugs," *Journal of Industrial Economics*, 36, 83–95.

    (1990), *The Determinants of Research and Development Expenditures in the Ethical Pharmaceutical Industry*, unpublished paper, Hamilton College, Clinton, NY.

Kaplan, S., and R. Ruback (1995), "The valuation of cash flow forecasts: an empirical analysis," *Journal of Finance* 50 (September), 1059–93.

Keynes, J. M. (1936), *The General Theory of Employment, Interest and Money*, Macmillan, London.

Lichtenberg, F. (1996), "Do (more and better) drugs keep people out of hospitals?," *American Economic Review*, 86, 384–8.

(1997), "The allocation of publicly funded biomedical research," in E. Bernat and D. Cutler (eds.), *Medical Care Output and Productivity*, Studies in Income and Wealth no. 62, University of Chicago Press, Chicago, 565–89.

(2000), "Comment on paper by Skinner and Wennberg," in D. Cutler (ed.), *The Changing Hospital Industry: Comparing Not-For-Profit and For-Profit Institutions*, University of Chicago Press, Chicago.

Lichtenberg, F. and D. Siegel (1991), "The impact of R&D investment on productivity: new evidence using linked R&D–LRD data," *Economic Inquiry*, 29 (April), 203–28.

Malkiel, B. (1990), *A Random Walk Down Wall Street*, 5th edn., Norton, New York.

National Center for Health Statistics (2004), *Health, United States, 2004: With Chartbook on Trends in the Health of Americans*, available at http://www.cdc.gov/nchs/data/hus/hus04.pdf.

National Science Board (1993), *Science and Engineering Indicators – 1993*, Government Publishing Office, Washington, DC.

New York Times (1995), "Benefit to healthy men is seen from cholesterol-cutting drug," 16 November, A1.

(1998), "AIDS deaths drop 48% in New York," 3 February, A1.

Pharmaceutical Research and Manufacturers Association (1997), *PhRMA Annual Survey*, available at http://www.phrma.org/facts/data/R&D.html.

Rogerson, W. (1989), "Profit regulation of defense contractors and prizes for innovation," *Journal of Political Economy*, 97, 1284–1305.

Schumpeter, J. A. (1942), *Capitalism, Socialism, and Democracy*, Harper and Brothers, New York.

Skinner, J. S., and J. E. Wennberg (2000), "How much is enough? Efficiency and Medicare spending in the last six months of life," in D. Cutler (ed.), *The Changing Hospital Industry: Comparing Not-For-Profit and For-Profit Institutions*, University of Chicago Press, Chicago, 169–93.

Tobin, J. (1969), "A general equilibrium approach to monetary theory," *Journal of Money, Credit, and Banking*, 1 (February), 15–29.

Toole, A. (1998), *The Impact of Public Basic Research on Industrial Innovation: Evidence from the Pharmaceutical Industry*, Discussion Paper no. 98–6, Stanford Institute for Economic Policy Research, Stanford University, CA.

Viscusi, W. K. (1993), "The value of risks to life and health," *Journal of Economic Literature*, 31, 1912–46.

3 | The evolution of pharmaceutical innovation

PAUL NIGHTINGALE AND
SURYA MAHDI

3.1 Introduction

The aim of this chapter is to explain and highlight a consistent pattern of changes within pharmaceutical R&D processes that produce "small molecule" therapeutics. In doing so, it aims to provide an alternative to the traditional "biotechnology revolution" conception of these changes, whereby a search regime focused on chemistry and on large, integrated pharmaceutical firms is being displaced by a new search regime focused on molecular biology and networks of smaller, dedicated biotech firms. It instead suggests that the changes are better understood within a Chandlerian framework in terms of an "industrialization of R&D," whereby firms adapt their technologies and organizational structures to exploit the economies of scale and scope that are made possible by economic interdependencies within knowledge production (see Chandler, 1990; Rothwell, 1992; Houston and Banks, 1997; Nightingale, 1998, 2000). These economically important interdependencies are a consequence of knowledge having value only in context (Nightingale, 2004), and generate scale economies because information that is integrated into a coherent whole can have more economic value than the same information divided into its component parts and distributed between economic agents who fail to realize its full value.[1]

The chapter will suggest that changes within pharmaceutical innovation processes over the last twenty years have been driven largely by a shift from a trial and error approach to drug discovery to one that attempts to use scientific understanding of the biology of disease to find

---

[1] For example, a pharmaceutical firm that knows that a particular chemical modifies a particular drug target, that the drug target is validated for a disease, and that the market for that disease is particularly profitable has considerably more valuable knowledge than three firms each knowing only part of the overall picture.

drugs for particular diseases and markets (see Schwartzman, 1976; McKelvey, 1995). Since this process involves finding molecules that act against disease-causing targets, it has created an incentive to improve the experimental technologies that help generate understanding about drug action. Over the last fifteen years a series of new experimental technologies have been introduced that have industrialized R&D and radically increased the amount of novel experimental information being generated. This process of industrialization has involved a range of traditional hand-crafted, sequential, human-scale experiments being complemented by automated, massively parallel, micro-scale experiments involving the computerized integration of experimental data sets with large quantities of stored and simulated data (Nightingale, 2000). While biologists in the late 1980s may have focused primarily on empirical "wet" biology, today they may spend much of their time in front of computers engaging in more theoretical *in silico* science, experimental design, quality control, or trawling through large data sets. This process has made pharmaceutical R&D more theoretical, more interdisciplinary, and has meant that medicinal biologists may now find themselves working with clinicians in hospitals (Gilbert, 1991; Meldrum, 1995).

These changes have had two important and related implications. First, they have increased the fixed costs of drug discovery in general and of experimentation in particular. As a result, there are substantial economic benefits from spreading these high fixed costs over as much output as possible, with important implications for firm behavior (see Chandler, 1990). In particular, the increase in fixed costs has created an emphasis on organizing the division of labor so that high-cost activities are exploited to the full. This is done by improving R&D throughput, improving capacity utilization and reducing the risks of product failure either in development, in clinical trials, or in the market.

Second, the increased amount and complexity of experimental information being generated, the need to improve the productivity of R&D to spread the higher fixed costs, and the increasingly specialized nature of scientific knowledge have created an emphasis on integrating knowledge (Pavitt, 1998) along the drug discovery process, and particularly between R&D and marketing. Because experimental information has value only in the context of knowledge that is partly project- and firm-specific (Nightingale, 2004), the industrialization of R&D has increased the relative importance of knowledge integration. Genes, targets, and even lead compounds, for example, become valuable only when they are genes involved in disease-related biochemical mechanisms, targets that have been validated for particular diseases, and lead compounds that are effective against such targets yet have good bio-availability and toxicity and can compete in profitable markets. As a consequence, firms that have capabilities along the entire process and can integrate knowledge between each stage can potentially improve their capacity utilization and more successfully direct their R&D towards more profitable markets, thereby spreading the higher fixed costs involved. The potentially greater value of integrated information compared with diffused information is therefore both a cause and a consequence of the economies of scale that industrialization has generated.

It is important to understand the implications of these changes properly, because they have been the subject of a considerable amount of unexamined hype that has the potential to distort government policy, create unrealistic expectations in the health sector, and produce losses for investors (Nightingale and Martin, 2004). Analysis that simply extrapolates from revolutionary changes in scientific research may lead one to expect substantial improvements in R&D productivity (measured by the number of drugs being launched each year), competitive advantage for (biotech) firms that can exploit these technologies, and a shift in industrial structure away from large Chandlerian firms and towards network-based innovation. However, as this chapter will show, these assumptions are problematic. It does not follow that radical improvements in scientific research will lead to revolutionary improvements in productivity, because bottlenecks remain within the drug discovery process, particularly at the target validation stage. As a result, qualitative improvements in research capability do not necessarily translate into quantitative improvements in output. Rather than a revolutionary improvement, the number of new drugs being launched remained static over the 1990s despite increased R&D spending; in fact, approximately $50 billion of industrial R&D in 2000 produced only twenty-four new drugs (Nightingale and Martin, 2004).

This suggests that we should be careful about jumping to conclusions about any implications for changing industrial structure, particularly the notion that networks of small biotechnology firms will take over from large integrated pharmaceutical firms. While there have been some clinical successes with biopharmaceuticals, and a small number of financial successes, most biotech firms lack the financial, marketing, and technical capabilities to compete with the large pharmaceutical firms along all the stages in the innovation process. They do not therefore have the same opportunities to integrate knowledge, direct their R&D programmes, and spread the high fixed costs of both R&D and a global marketing presence. While some biotech firms may be able to build these capabilities by exploiting the profits from blockbuster drugs, their high failure rates and the timescales involved suggest that most will not.

One way in which biotech firms can survive is to form alliances and become part of a distributed knowledge supply chain for large pharmaceutical firms. This involves biotech firms providing targets, leads, and technologies to fill gaps in various stages of pharmaceutical firms' drug discovery pipelines, thereby improving the capacity utilization of both firms' R&D processes. Put crudely, the current economic structure of drug discovery means that biotechnology firms are more likely to be suppliers than superiors to large pharmaceutical firms. This Chandlerian "old economy" analysis suggests that the current network-like industrial structure is being driven by large pharmaceutical firms being prepared to spend a quarter of their R&D budgets externally to ensure a smooth supply of new products, rather than a biotech revolution. This, however, may change in the future, and the interdependencies in knowledge production that provide large firms with the ability to exploit economies of scale and scope may not be maintained.

The following sections explore some of the benefits and risks of modern pharmaceutical R&D and, hopefully, provide a means of understanding how these interdependencies might strengthen or weaken over time. Section 3.2 explores how the indirect relationship between science and technology gives innovation processes an internal structure. Section 3.3 explores the economics of R&D and focuses on the potential Chandlerian economies of scale and scope that are opened up by the interdependencies within the R&D process. Section 3.4 explores the dynamics of the evolution of pharmaceutical R&D and explains how different stages of the drug discovery process have become increasingly industrialized. Section 3.5 draws conclusions.

3.2 Pharmaceutical innovation processes

This section explores how innovation processes are structured by problem-solving choices that create interdependencies between different tasks. Over the last fifty years research on innovation processes has highlighted their diversity (Freeman, 1982; Pavitt, 1984; Nelson and Winter, 1982; Nelson, 1991), the non-linear, indirect interactions between science and technology (Rosenberg, 1974; Pavitt, 2001), and consequently the interdependent functional capabilities required to innovate (Chandler, 1990; Pavitt, 1998). This research highlights how scientific explanations are rarely applied directly to produce technology because the complexity of the real world is so different from the atypical, purified conditions of experiments (Pavitt, 1999). As a result, technology is generally produced by trial and error experimentations that may be guided by scientific explanations, but are definitely not reducible to them (Nightingale, 2004).

Technological knowledge is based on "operational principles" (Polanyi, 1966) that define how a particular technological function can be achieved. Their application generates a hierarchy of increasingly specific problem-solving tasks that have an interrelated structure because any proposed solution generates a new, more specific problem of how to generate the conditions in which it can be applied (Vincenti, 1990; Nightingale, 2000). The extent of this structuring can be seen in the way the major therapeutic classifications in the pharmaceutical industry – small molecule, biopharmaceuticals, vaccines, and gene therapy – are related to problem-solving choices.

Disease symptoms can be caused by a range of abnormal changes in biochemical pathways, which, in turn, may have a range of different causes, such as different pathogens or genetic errors. As a result, there are a variety of potential operational principles that can be applied to treat disease and generate drugs. The main ones of concern here are (a) preventing the pathogen from acting, (b) modifying the biochemistry, or (c) repairing the gene. Choices at this level define problems at the next level. For example, the pathogen can be prevented from acting by either exploiting the immune system using vaccines and antibodies, or by poisoning the pathogen directly using traditional small molecules. Similarly, modifications to the biochemistry of the body can be achieved either through the action of small molecules on proteins or through adding therapeutic proteins. Genes can be repaired, in theory, using various gene therapy approaches. As these problem-solving choices are made, a series of interrelated tasks are generated that define the technological paradigm being followed (Dosi, 1982). These technological paradigms in turn structure the technological capabilities needed within a particular subsection of an industrial sector (see table 3.1).

Table 3.1 *Operational principles and industrial substructure*

| | Stop pathogen | | Stop disease | | |
| | | | Modify biochemistry | | Repair gene |
Main operational principles	Use immune system	Poison directly	Modify protein's action	Adjust protein levels	Change expression
Method	Vaccine	Antibody	Small molecule	Protein therapeutic	Gene therapy
Sub-type	Natural, modified, synthetic	MAb, chimeric, transgenic	Small molecules	Naturally occurring rDNA proteins
Small molecule protein mimics	Gene augmentation, directed killing, anti-sense, targeted expression, immune assisted killing

As table 3.1 shows, there are considerable differences between the subsectors of the pharmaceutical industry, and it is potentially misleading to think of biotechnology as being a new search regime that is displacing the older, chemistry-based search regime for the entire industry. In fact, biotechnology-based therapeutics do not necessarily compete directly with traditional chemistry-based small molecule products, and many biopharmaceuticals, such as insulin, have simply replaced extraction-based production methods with recombinant techniques that are not only safer but allow second-generation products to be modified and improved more easily. In the context of the high expectations surrounding biotechnology, it is important to note that the number of medical conditions that are caused by lack of a protein is limited, and the number of those that can be cured by adding recombinant proteins is further limited because not all such proteins are blood-soluble and tolerated by the body.

Rather than considering the early biopharmaceuticals as the beginning of a new therapeutic class that will displace traditional small molecules, it is perhaps more realistic to suggest that the early successful biopharmaceuticals were the atypical "low-hanging fruit" of a distinct and potentially limited therapeutic class, and were relatively easy to generate. Extending this therapeutic class to produce the next generation of biopharmaceuticals will be substantially more difficult. This is perhaps one of the reasons why only twelve recombinant therapeutic protein products and three molecular antibody products have entered widespread clinical application since 1980 (defined as selling more than $500 million per year in 2002/3), and only sixteen biopharmaceutical products evaluated in the Prescrire data set between January 1982 and April 2004 were considered as better than "minimal improvements" over pre-existing treatments by clinicians (Nightingale and Martin, 2004, p. 566).

Similarly, the number of medical conditions that can be treated by gene therapy is limited in both theory and practice, and currently no gene therapy product is licensed in the European Union or United States. With vaccines and antibody therapies the picture is more complicated. While the introduction of new technologies has opened up new and important therapeutic opportunities, the complexities of science, economics and the regulatory approval processes for vaccines take them beyond the scope of this chapter. Instead, the rest of the chapter will focus on how advances in biotechnology and genomics have changed small molecule innovation processes, which is the traditional and still the most common and profitable form of drug discovery.

3.2.1 Small molecule drug discovery

To understand how the pharmaceutical innovation process has changed, it is necessary to understand the interplay between economic and technical factors. Historically, the majority of drugs were natural products. These suffered from purity problems, which, in turn, made their medical use uncertain and dangerous. The first generation of modern drugs were developed using an operational principle based around purifying natural products to extract their active ingredients. This purification process ensured a greater degree of standardization and therefore certainty about their clinical use. In the nineteenth century chemical firms developed the technological capabilities to analyze the active ingredients in these products and, increasingly, the ability to produce them in synthetic and semi-synthetic forms.

It was a short step from synthesizing natural products to modifying their chemical structure to remove unwanted side effects. However, without an understanding of the relationship between chemical structure and biological activity, innovation was largely unguided and involved an expensive and time-consuming process of testing and modification that often failed to provide alternative molecules when lead compounds failed. The lack of understanding of the relationship between structure and function meant that the economic exploitation of newly discovered chemical products was unfocused, and typically took place in an organizational structure where a centralized R&D department was linked to as broad a marketing potential as possible (Mowery, 1983; Chandler, 1990). This model of finding "markets for products" gradually reversed during the mid-twentieth century towards a model that exploited scientific understanding to focus on finding "products for markets."[2]

A major part of this change occurred in the 1960s, when medicinal chemists, following Sir James Black's work on H_2 antagonists, realized that if they understood the structure of disease targets they could improve their selection and modification of small molecules to use

---

[2] Even today the uncertainties involved in drug discovery are such that this shift is far from complete.

Desired result		Operational principles		Task
Cure disease	→	Modify pathway	→	Find and modify pathway (i.e. kill pathogen)
Modify mechanism	→	Modify protein catalysis	→	Find and modify protein action
Modify protein	→	Block active site	→	Find how it works and modify
Block active site	→	Bind molecule	→	Find molecule that binds
Find molecule	→	Assay (test many)	→	Find lead and optimize

Figure 3.1. Structure-based drug discovery

as drugs. The typical aim was to find an enzymatic reaction within the disease pathway (hopefully the rate-limiting step) and find a molecule that would bind to a protein and modify its behavior. The pathway would typically be characterized from animal tissue, and some information would hopefully be available about the enzyme to allow certain families of small molecules to be targeted, based on an understanding of structure–activity relationships, until a hit was found. Medicinal chemists could then try to improve specificity and bio-availability. This spiraling hierarchy of technological traditions is illustrated in figure 3.1.

These technological traditions turn the initial vague problem of "curing a disease" into a hierarchy of increasingly specific sub-problems involving iterative cycles of testing, understanding, modifying, and retesting potential solutions. In each step, scientific knowledge can be used to guide problem-solving and reduce the number of experimental dead ends that are pursued (Rosenberg, 1974). This increases the efficiency of search and improves the utilization of R&D capacity over a given period of time (Mahdi, 2003; Nightingale, 2000). Moreover, unlike random screening, where trial and error experiments are largely independent, when experiments are guided by scientific understanding each experiment can generate new knowledge that can steer and improve the next round of testing (Nightingale, 2004). As the knowledge produced by experiments is no longer independent, individuals or firms who engage in multiple experiments and learn from them can potentially exploit scale economies and improve the speed and quality of their drug discovery programs.

Unsurprisingly, this change has had organizational implications. As understanding of the relationship between chemical structure and biological activity began to improve, the relative importance of the economies of scope that linked products to a wide range of markets began to decline. The large, diversified chemical firms began to be replaced by a more specialized industrial structure characterized by firms focusing on high-value specialty chemicals and pharmaceuticals. Large chemical firms demerged their bulk chemical and pharmaceutical-specialty chemical businesses, and, as links to national and international regulators became increasingly important, the pharmaceutical businesses increasingly demerged from the specialty chemicals businesses.

This shift towards an R&D-intensive model of finding innovative products for known markets exploited a series of Chandlerian interdependencies, particularly between marketing and R&D. The cost structure of the pharmaceutical industry was – and is – dominated by the high fixed costs of both R&D (approximately $500 million per product) and marketing, the long timescales (typically eight to ten years) before revenue is generated (Scherer, 1993; Henderson, Ossenigo, and Pisano, 1999; Cockburn and Henderson, 1998), the highly skewed distribution of profits, the very high failure rates of new products in clinical trials, and the rapid accumulation of costs as one moves further towards the market (DiMasi, 1995). Firms have to ensure that their R&D programs produce enough profitable drugs to cover their development costs, with the actual number of successful products being a product of the number of projects and their success rates. In Chandlerian terms, firms that can invest in R&D and improve their throughput of new drugs can potentially spread their higher fixed costs further and therefore generate economies of scale and scope (Nightingale, 2000). Firms therefore attempt to increase the speed of innovation, reduce the failure rate of projects (particularly late failures), increase the number of projects, and improve their quality.

In the 1980s a dominant strategic model for the industry emerged, based on heavy investment in marketing and R&D. This model involved investments in global marketing capabilities to maximize profits and generate the knowledge required to direct R&D towards profitable markets. Similarly, substantial R&D investments were made to increase the throughput of profitable drugs and ensure that the high fixed costs of R&D and global marketing were spread over a regular throughput of new drugs.

Within the United Kingdom, and to a lesser extent the United States, this model was reinforced by the financial links between large pharmaceutical firms and institutional investors acting for pension funds or insurance houses. Within the United Kingdom these institutional investors may own 70 to 80 percent of the shares in large pharmaceutical firms. The substantial size of the funds that institutional investors manage means that their standard unit of funds is large enough to expose them to substantial liquidity risk if the firms they invest in fail to achieve their expected profits and they are forced to sell stock in illiquid markets. As a result, institutional investors take an active role in ensuring that the senior managers they appoint reach their expected profits and keep profit volatility low. This is done mainly by ensuring that the pharmaceutical firms have a sufficiently high throughput of new drugs to spread the high fixed costs and reduce the impact of late regulatory failures on profit volatility. This has involved managers engaging in mergers to increase market capitalization, divesting non-core businesses, investing in marketing to improve sales, and investing in R&D to improve product performance and catch product failures *early on*, before they eat away at the productive utilization of R&D capacity. Institutional investors' concerns about liquidity risk are important enough that they will encourage mergers that increase market capitalization even if the mergers have no impact on internal productivity.

During the 1980s the growth of a number of large pharmaceutical firms was based on this integrated R&D–marketing model, which generated huge sales from a small number of blockbuster drugs. However, despite this success, the 1980s and 1990s were a time of declining R&D productivity, and managerial concerns focused on bottlenecks within the drug discovery process where throughput might be improved. The economics were fairly simple: drugs need to generate about $400 million a year to cover their development costs, and during the 1990s a typical large firm needed to produce three new drugs a year to deliver on its expected profit growth. This, in turn, required approximately sixty targets moving into discovery and about twenty drug candidates moving into clinical trials each year (Drews, 2000). During the late 1980s it became clear that few firms could possibly approach this level of output, and in the 1990s the relative strategic focus shifted towards a complementary innovative strategy involving the exploitation of economies of scale and scope in experimentation to address these productivity problems (Nightingale, 2000). This process was intended to improve the quality of products (and therefore their profitability) and increase the number of successful programmes by reducing the number of late failures.

3.3 The industrialization of R&D

The industrialization of R&D involved firms investing in a range of technologies that were expected to improve the productivity of experimentation by changing the size, scale, speed, scope, and accuracy of experiments. These improvements are possible because the amount of knowledge generated by an experiment is largely independent of its size. Simplifying hugely, the amount of knowledge produced is a function of the difficulty of the problem, the intelligence and training of the experimenters, and the number and accuracy of the experiments undertaken. The cost of experiments, on the other hand, is a function of fixed and variable costs; and variable costs, in turn, depend on the costs of the reagents, sample and assay, the cost of disposal (potentially very high), the cost of storage, and the cost of maintaining a safe environment, *all of which are functions of size*. As a consequence, if you reduce the size of experiments and run more of them over a given period of time (either in parallel or by speeding up), you can potentially achieve more experimental output per unit cost. This is the basic idea behind the industrialization of R&D – running and analyzing experiments in parallel, reducing their unit size and increasing their speed. As all real-world experiments are constrained by cost, waste, space, accuracy, speed, and the physical ability to "see" results at biochemical scales, the ultimate reduction in size and costs involves performing experiments *in silico* on a computer. Consequently, the industrialization of "wet" science has been complemented by the analysis of stored and simulated data, which has followed the same pattern of shifting towards massively parallel analysis.

A consistent pattern of industrialization emerged across a range of technologies during the 1990s that has shifted experimentation from a process of fishing with single hooks to one where expensive capital goods are used to trawl through experimental space. In doing so, they generate large amounts of data that can be analyzed offline. However, the ability to exploit scale economies will depend on the technical characteristics of the experiments, with some experiments being more amenable to industrialization than others. As a result, the rate of industrialization has been different for different stages within the innovation process. Because pharmaceutical managers are concerned with the overall performance of their innovation processes, there is a managerial emphasis on ensuring that the various stages are in balance (see

Penrose, 1959; Hughes, 1983; Nightingale et al., 2003). For example, there is no point using resources to develop 400 targets if the firm can validate no more than twenty per year. Consequently, productivity improvements in one step can be wasted if the rest of the system is "out of sync" and lacks the capacity to deal with them. When this happens, the bottlenecks act as focusing devices (Rosenberg, 1976) or reverse salients (Hughes, 1983) for innovative attention. As a result, the last twenty years have seen a series of shifting bottlenecks within the drug discovery process, with innovations in experimental technologies solving one bottleneck only to generate another elsewhere in the process.

For example, one of the main bottlenecks historically has been the screening of molecules. This was overcome by the introduction of high-throughput screening (HTS) in the early 1990s, which produced orders of magnitude improvements in the ability of firms to test compounds (Nightingale, 2000). This created a new bottleneck, as firms would quickly run out of chemicals to test. This new bottleneck was solved in the mid-1990s by combinatorial chemistry, a series of technologies that allowed the automated, parallel synthesis of large numbers of chemicals. Within Pfizer, for example, the production of molecules increased 2000-fold between 1986 and 1996, and the number of compounds tested by GSK increased more than tenfold in the following eight years. This created a new bottleneck related to the limited ability of firms to deal with the vast quantities of data coming from these automated experiments. This new bottleneck was partly overcome by improved use of QSAR (quantitative structure activity relationship), databases, and simulations. Perhaps the most important implication of these changes has been the improvements in the quality of the new leads. Between 1996 and 2004 the average potency of lead compounds going into lead optimization improved by more than an order of magnitude (from 3000 nM to 10 nM) (GSK, 2004). Moreover, in 1996, when Glaxo was testing 100,000 compounds annually, its success rate was only 20 percent, but by 2004, when GSK was testing 1,050,000 compounds, its success rate was up to 65 percent and the number of leads per target had doubled from one to two (GSK, 2004). These improvements clearly have the potential to have a drastic effect on product throughput, R&D risk, and – given their potential ability to spread high fixed costs further – firm profitability.

However, the information generated by these industrial experiments has value only when it has been validated and placed in context. This process remains time-consuming and difficult, and is why these changes are complements rather than alternatives to the traditional skills of medicinal chemistry. Part of the bottleneck preventing the full exploitation of combinatorial chemistry has been related to the organizational difficulties involved in integrating it with traditional medicinal chemistry, so that results can be validated and contextualized and combinatorial synthesis directed towards profitable areas (Bradshaw, 1995). The value of combinatorial chemistry without this knowledge integration is minimal. Combinatorial chemistry is, after all, a wonderful technology for producing large numbers of molecules that aren't blood-soluble and therefore have no value as drugs.

By the early 1990s combinatorial chemistry had solved one bottleneck and created another. The new constraint on the throughput of R&D was the lack of drug targets that development programs could be directed towards. As the next sections will show, the bottleneck in target discovery has largely been solved. However, the industrialization of this stage of drug discovery has only had the effect of shifting the bottleneck along to target validation, where the process of industrialization has been far slower and the productivity increases far more modest.

3.4 The industrialization of target discovery and validation

Diseases are caused by very complex interactions between proteins, gene expression products, the complex biochemical machinery of our cells, various metabolites, and our environments, making biology significantly more complex than chemistry. The shift towards an R&D model based on finding products for specific markets is based on exploiting scientific understanding of the biology of disease to select specific targets that small molecule drugs can be targeted against. It is, therefore, dependent on understanding the biology of disease, as products that successfully modify the behavior of targets that are irrelevant to disease processes are not only valueless in themselves, they also eat up R&D resources and reduce the throughput of drug output. As a consequence, attempts to improve the number of drugs being produced have focused on improving scientists' understanding of disease.

While the biology of diseases is extremely complex, patterns fortunately exist within this biological complexity, because of evolutionary redundancy (whereby evolution adapts pre-existing features to new functions) and the conserved relationships between genes and proteins.

This allows a scientist trying to understand a cellular process in a human to work on a similar process in a yeast cell, if it is conserved across evolutionary time. Similarly, the conserved symmetries between DNA, RNA, and proteins allow proteins to be investigated by exploring genes and their expression products.

This ability to shift between levels, species, and *in silico* is useful, because experiments may be (a) substantially easier or (b) substantially easier to industrialize because they are more amenable to miniaturization, analysis, amplification, detection, synthesis, sequencing, or automation. For example, the genetic analysis of fast-breeding model organisms such as nematode worms, yeast, mice, or zebra fish is easier than for human subjects, while transcribing DNA sequence into RNA sequence is much easier with a computer (which can automatically change "T"s to "U"s) than with "wet biology," involving a whole host of extremely complex cellular machinery.

3.4.1 Target discovery: from "single-gene" disorders *to* in silico *science*

Because there is a strongly conserved symmetry between proteins and genes, knowledge about biochemical processes involving proteins can be generated by analyzing their corresponding genes and their expression products. This shift is important, as the DNA level of analysis is more amenable to industrialization than the protein level. As a consequence, there has been a gradual shift when finding genes from (a) traditional biochemical analyses at the protein level to (b) positional cloning methods at the gene level of single-gene disorders, which allowed the generation of (c) increasingly high-resolution genome maps and databases, which, with the completion of the Human Genome Project, provided the foundations for (d) the *in silico* and SNP-based (single-nucleotide polymorphism) analysis of the types of complex disease of interest to the pharmaceutical industry.

It is important to note that the earliest stages of this shift were undertaken in the public rather than private sector, reinforcing the point made in the introduction that knowledge is valuable only in context. The earliest work was not valuable to the pharmaceutical industry, because the knowledge being generated about the genetic causes of diseases was not linked to knowledge about potential cures or potential profitable markets. Moreover, before the Human Genome

Project there was no infrastructure in place to link genetic information more widely to other aspects of pharmaceutical problem-solving.

Early work in medicinal genetics focused on inherited genetic disorders such as cystic fibrosis (CF), hemophilia and Huntington's disease, which were initially thought to involve one gene mutation producing one dysfunctional protein, which produces one disease. Research proceeded at the protein level and relied on the elucidation of biochemical mechanisms and the laborious gel separation of protein fragments. The gene for hemophilia A, for example, was found by purifying large quantities of pigs' blood to isolate small quantities of factor VIII, which were used to generate oligonucleotide probes that were then used to screen human DNA libraries. This protein-based approach is extremely difficult, and was applicable only to a small minority of medical conditions where the protein of interest is expressed in large quantities, is easy to access, and is amenable to this type of analysis.

The shift to the second stage of gene hunting followed David Botstein's realization that genetic variation would influence the ability of restriction enzymes, which act only at specific points, to cut up DNA (Botstein et al., 1980). The resulting fragments of different length can be separated on gels so that genetic variations act as markers that can be used to follow statistically the inheritance of genetic diseases and find the position of genes. The idea behind this "positional cloning" method is that, because DNA is randomly recombined at meiosis, there is a relationship between the recombination frequencies of DNA markers and their proximity. In theory, therefore, genes can be positioned by statistically correlating genome marker and disease states (based on good clinical characterization, which is typical of single-gene-defect diseases) through their occurrence in populations (association studies) or in families (linkage analysis). The approximated region is then used to generate a series of overlapping DNA clones, and the gene is identified, sequenced, confirmed, and used to produce diagnostics.[3]

In practice, however, collecting large numbers of families and densely covering candidate regions with polymorphic markers can be

---

[3] The candidate genes can be confirmed by (i) confirming mutations in unrelated affected individuals, (ii) restoring normal function in vitro, and (iii) then producing a transgenic model (typically mouse) of the disease. The CF gene, for example, was confirmed because the 3-bp deletion (ΔF508) was found in 145/214 CF chromosomes (68 percent), and 0/198 normal chromosomes – producing a rather convincing P value of $10^{-57.5}$ (Rommens, Januzzi, and Kerem, 1989).

extremely time-consuming and laborious. By 1995 only about fifty genes had been discovered using this approach (Collins and Guyer, 1995). The method is problematic because linkage analysis is statistically very sensitive to the quality of the data, and the number of experimental dead ends increases dramatically if the initial gene localization is larger than about 1 Mb (typically, the candidate region was about 10 Mb, or even 20 Mb – enough to contain 500 genes).

The main way around this problem was not university-based research or biotechnology firms but links to hospital-based clinicians, who were alert to unfortunate patients with chromosome abnormalities that reproduced "genetic" diseases. These abnormalities allowed the location of the gene to be approximated more easily. Data from clinicians guided the discovery of a range of genetic diseases: Duchenne muscular dystrophy (DMD) in 1986, neurofibromatosis in 1990, Wilms' tumour in 1990, and Fragile X in 1991 (Strachan and Read, 2000, p. 356). Cystic fibrosis, by contrast, lacked chromosomal pointers, and a laborious four-year process of chromosome walking was needed to generate and map a region of about 500 kb (Rommens, Januzzi, and Kerem, 1989). These early gene hunts were hard won, time-consuming, and very expensive. The hunt for the cystic fibrosis gene cost some $200 million, and the process from linking Huntington's disease to a marker to isolating the gene took a large international research community about ten years (Sudbury, 1999).

Given these high costs, a series of technologies have been developed that exploit scale and scope economies. These have involved reducing the distance between markers to improve their statistical discrimination (which reduces the number of experimental dead ends) and spreading the high fixed costs involved in generating genetic markers, clones, and experimental data over a larger number of experiments by putting them in accessible stores, databases, and *in silico* environments where they can be reused at low marginal cost. Interestingly, one of the major innovations for spreading the benefits of investment was institutional and involved setting up a global sequencing consortium, supported by charities and public funds.

3.4.2 Economies of scale and the resolution of genetic maps

The search for genes involves statistically linking them to markers, and so can be improved by reducing the average distance between markers

Table 3.2 The increasing resolution of genetic markers

Name	Date	Number of markers	Distance	Technology
Blood groups	1910–60	20 loci	Large	Blood typing
Enzyme polymorphisms	1960–75	~30	Large	Measure protein serum mobility
Micro-satellites: restriction fragment length polymorphisms (RFLPs)	1987	Potentially 10^5	>10 cM (clustered near ends)	Southern blots, restriction enzymes, radio-labelled probes
Mini-satellites: variable-number tandem repeats (VNTRs)	1992	5,000+	1.6 cM (1994 1.0 cM)	Gels and PCR
Single nucleotide polymorphisms (SNPs)	2000	Millions	Very short	Micro-arrays

Source: Strachan and Read (2000, p. 273).

and the amount of information each marker provides about family pedigree. However, increasing the number of markers increases the experimental workload, so increases in the resolution of genetic maps have required complementary improvements in the ease of experimentation (see table 3.2).

The first map with enough markers to make searching for statistical links between markers (genes) and diseases (phenotypes) a practical possibility was produced in 1987 (Donis-Keller et al., 1987). Initially, this RFLP map was used with very expensive, difficult, and time-consuming techniques because the markers provided limited information about which parent the marker was inherited from. The next generation of "mini-satellite" markers measured the variation in the number of repeats in small sections of DNA, and were therefore more informative about which parent a marker was inherited from, but they tended to be unevenly distributed and were difficult to automate.

The third generation of markers – "micro-satellites" – were based on $(CA)_n$ repeats and were more common and evenly distributed along the genome and, importantly, small enough to allow part of the process to be automated with polymerase chain reaction (PCR) (although they still required time-consuming gel separation) (Weissenbach et al., 1992; Murray et al., 1994; Hudson et al., 1995). The latest technology involves single nucleotide polymorphisms, which make up for their lack of informativeness by offering both better resolution (with potentially 2 million markers) and, most importantly, the possibility of being typed on solid-state micro-arrays, removing the need for expensive and time-consuming gel electrophoresis (Cargill, Altshuler, and Ireland, 1999). This potentially offers a shift towards the massively parallel automated analysis of an entire human genome on a single, mass-produced chip (Wang, Fan, and Siao, 1998; Chakravarti, 1999).

The automated use of micro-satellites allowed practically all the mutations for single-gene genetic disorders to be found. However, the problem for the pharmaceutical industry was to find and validate targets for low-penetrance diseases associated with common, non-Mendelian disorders such as heart disease. Finding the susceptibility loci for these diseases is extremely difficult, because very large data sets, which make consistently good clinical characterization very hard, are required to generate decent statistics. To understand how this problem is being solved, we need to look at the Human Genome Project, and understand its role in industrializing sequencing and placing information on reusable databases to allow a shift towards *in silico* science and massively parallel experimentation (Venter et al., 2003, p. 220).

3.4.3 Scale economies and the industrialization of sequencing

The ability to find and sequence DNA is a core capability in molecular biology. Because sequenced genetic data, markers, and clones are expensive to obtain, but can be reused at minimal marginal cost, there is an incentive to share genetic resources, and the Human Genome Project was set up to generate a shared public data set and improve the technologies involved. The project was highly ambitious, as sequencing the 3.2×10^9 base pairs of the human genome would require at least 6 million gel runs *even if* the sequences don't overlap.[4] Unfortunately, because sequences are selected randomly, ensuring adequate coverage requires the genome to be sequenced fourfold to produce a draft, and ninefold to produce a finished version. The original sequencing operation was, therefore, an immense international effort, planned to run for fifteen years, cost $3 billion, and conditional on improvements in sequencing technology that improved capacity from 90 Mb per year to 500 Mb per year (Collins et al., 1998; Collins and McKusick, 2001; Collins and Galas, 1993). As this section will show, these targets were reached easily, because sequencing technologies are amenable to industrialization. However, in order to mobilize the financial and organizational resources to undertake the project, wide-ranging expectations about the benefits of the project had to be created and modified over the lifetime of the project as new technical opportunities opened up or closed down (Martin, 1998).

The basic process of sequencing involves taking a single-stranded template DNA and adding a site-specific primer that starts the synthesis of a complementary DNA strand. Four reactions are carried out, involving the addition of normal A, C, G, and T with the addition of a low concentration of four modified dideoxynucleotides that have been changed so that they terminate the replication of DNA. This generates four mixtures of different-length DNA strands that terminate at

---

[4] Each run typically sequences about 500 bp. Overall, the integrated physical and genetic maps required 15 million PCR reactions, and the final results took up 900 printed pages (Sudbury, 1999).

different points, which correspond to the position of the modified A, C, G, and T. These can be separated out on polyacrylamide gel to produce a read-out of the sequence of bases within the original DNA.

Traditional manual sequencing techniques required technically demanding and dangerous autoradiography-based gel reads, and could generate only about three to four hundred base pairs per day. The dangerous and complex steps were removed with the introduction of fluorescence technologies, whereby a fluorophore would be attached to the dideoxynucleotides so that a monitor would detect its signal as the DNA passed through a fixed point on the gel (Smith et al., 1985, 1986). This innovation made sequencing more predictable and allowed machines to undertake tasks previously undertaken by human hands. This, in turn, generated a series of productivity improvements. For example, it removed the need for autoradiography, which allowed continuous rather than batch processing; it automated detection, which reduced errors in transcribing results; and it increased through-put by having four different indicators for each of the four letters of the sequence on one gel line, rather than the previous situation of four gels and one indicator.

The second key innovation that increased sequencing capacity involved enhancing the speed of samples through gels. The speed is proportional to the applied voltage, but too high a voltage will melt the gel. To get around heat diffusion problems, capillaries were developed with a 1 μm internal diameter that allowed better heat dispersal and higher voltages, and produced a fourteenfold increase in speed. This technology was amenable to automation, and sequencers that used it were able to sequence up to 400,000 bp per day (Venter et al., 2003, p. 223).

The third key innovation involved the production of clones for sequencing, which are used to generate the physical maps that plot the location of DNA sequences and are essential for ordering sequences. The most important physical maps are clone maps made up of libraries of overlapping and ordered clones (contigs). During the production and ordering of clones there are complex trade-offs between the size of the DNA inserts, the number of clones, the difficulties of ordering them, and their quality. For example, yeast artificial chromosomes (YACs) carry about 1 Mb of DNA (i.e. about 10,000 cover the genome), which potentially makes them easy to order, but they are often unstable and can produce a high rate (up to 60 percent) of chimeric clones (Chumakov et al., 1992). As a consequence, second-generation clone maps rely on higher-quality, smaller bacterial artificial chromosomes (BACs) and P1 artificial chromosomes (PACs) (about 300 kb and 100 kb, respectively). Their small size requires more clones and makes them harder to order, but, as the amount of *in silico* genome data increased, ordering smaller inserts became less of a problem.

The fourth main innovation was the shift towards using expressed sequence tags (ESTs) to identify genes (Adams, Kelly, and Gocayne, 1991). These are small sections of DNA that are generated from mRNA, and therefore code only for specific genes within the genome. They can be used to focus sequencing and have allowed the "shotgun sequencing" of genomes, where, rather than going through a long, laborious process of ordering individual clones before breaking them up and sequencing them, an entire section of DNA is broken up into small sections and sequenced repeatedly to generate overlaps that are ordered using sophisticated algorithms and massively parallel computer systems. This process allowed a team headed by Craig Venter at the Institute for Genomics Research to sequence the 1.8 Mb *H. influenzae* genome by assembling around 24,000 DNA fragments (Venter et al., 2003).

These innovations were all part of a steady industrialization of sequencing. The productivity of automated sequencing was further improved during the 1990s by a shift towards parallel sequencing involving multiple gel runs per machine and multiple machines per organization. The emergence of "sequencing farms" generated further productivity improvements and contributed to the exponential rise in the amount of sequencing undertaken. Table 3.3 shows the jump following the introduction of automated sequencing in 1994, after little change in the previous decade, and the forty-fourfold increase in the four years before 2000 (Meldrum, 2000a, 2000b). The improvements in the speed and scale of sequencing illustrate the implications of industrialized research methods. While the earliest methods barely managed to sequence about three to four hundred base pairs a day, at the height of International Human Genome Sequencing Consortium operations in 1999 they covered the genome every six weeks and were producing about a thousand base pairs per second (Sudbury, 1999).[5]

---

[5] At the same time, Celera was running 175,000 gel reads per day on 300 ABI PRISM3700 DNA analyzers working on a 24/7 basis to sequence more than a billion bases every month (Sudbury, 1999).

Table 3.3 Advances in sequencing throughput

Date	Organism	Sequence length
1978	SV40	5.2 kb
1979	HPV	4.9 kb
1981	MtDNA	16.6 kb
1984	Epstein Barr virus	172 kb
1992	Chromosome III of *S. cerevisiae*	315 kb
1994[a]	Human chromosome 19q13.3	106 kb
1994	Human cytomegalovirus	230 kb
1994[a]	Chromosome XI of *S. cerevisiae*	666 kb
1995[a]	*H. influenzae*	1.83 Mb
1996[a]	*S. cerevisiae* genome	12 Mb
1998[a]	*C. elegans* genome	97 Mb
2000[a]	Drosophila	120 Mb
2003[a]	Human	3300 Mb

[a]Automated sequencers.
Sources: Strachan and Read (2000, p. 303); Venter et al. (2003).

3.4.4 Finding genes for complex diseases

As the Human Genome Project databases and clone stores have matured, finding genes has become much easier, as biologists no longer have to go through a complex and lengthy process of cloning and sequencing, but can instead look up information on databases. As a result, the interesting parts of the human genome are now sequenced and most of the genes have been discovered and annotated (Deloukas et al., 1998). However, while the genes may be on data bases, linking them to economically significant diseases is still not easy (Bork and Bairoch, 1996), and finding cures once the gene is discovered is substantially harder still (Cocket, Dracopoli, and Sigal, 2000). To put the difficulties in perspective, in the late 1990s finding a particular gene might take a year of a Ph.D. project, but finding and validating a disease-causing biochemical mechanism would take a large team of researchers engaged in international collaboration decades of work, and would normally result in a major academic prize.

This is because, in contrast to the rare Mendelian genetic diseases caused by single, highly penetrant genes (i.e. having the gene almost always means having the disease), most common diseases have a low penetrance, as gene polymorphisms that confer increased disease risk tend to be naturally selected against (Lander and Schork, 1994). This makes finding genes involved in common diseases *much* more complicated, because of the weak correlation between any genetic effects and the disease state, and because several genes tend to be involved.[6] Even with large numbers of markers, getting good enough statistics to avoid chance association between genes and phenotypes can require prohibitively large sample sizes (Terwilliger et al., 2002).

One way around this problem has been to focus on disease sub-types that have a higher genetic influence, in order to help understand the mechanisms and pathways involved in the more common forms. Successful disease susceptibility genotypes have been found for subtypes of breast cancer, colon cancer, and rare types of early onset Alzheimer's and type-2 diabetes. In each case, the genotypes found account for about 3 percent of all occurrences of the disease, but offer illuminating insights into the mechanisms of the more common forms (Holzman and Marteau, 2000).

Even in these cases, where the genetic link is stronger, the target selection process is often guided using biochemical characteristics to select candidate genes. For example, when linkage analysis suggested that susceptibility to Alzheimer's disease was linked to chromosome 19, the ApoE locus was a candidate because it codes for apolipoprotein, which is found in amyloid plaques, which are characteristic of the disease (Roses, 2001). While linkage analysis was used successfully to test models of inheritance and discover the BRCA1 gene for familial breast cancer, it has had problems when applied to other diseases and has suffered from replication failures, because large numbers of related individuals have to be analyzed to find the influence of weakly acting genes (Terwilliger et al., 2002).

Similarly, association studies can be problematic if the association between the disease and the marker is caused by shared common ancestors (Risch, 2000; Risch and Merikangas, 1996; Roses, 2000). This linkage disequilibrium effect is a particular problem in countries

---

[6] As Terwilliger et al. (2002) note, "There must be a strong correlation between marker and trait locus genotypes (i.e. strong linkage and or LD) and there must be a strong predictive relationship between the ascertained phenotypes and the underlying genotypes for a mapping study to prove successful."

such as the United Kingdom and the United States, where ethnic sub-groups cause population stratification. As a consequence, there has been much attention on isolated populations in Utah, Iceland and Finland – although the "purity" of these samples has been questioned.

Most pharmaceutical companies use a variety of these methods but attention is currently focused on employing SNP markers, which can be used as internal controls to correct for some of the problems of linkage and association studies (Risch, 2000). However, even SNP-based methods have been criticized (Weiss, 1993; Pennisi, 1998; Weiss and Terwilliger, 2000).

The outcome of most of these studies has, so far, been modest. The basic problem seems to be that, for weakly acting genes, a lot of statistical power is needed, which requires large data sets even with densely packed SNP markers. Generating such data sets is time-consuming and costly, but the fact that studies are being undertaken suggests that they generate valuable information. However, the problem is not finding genes but validating targets and finding cures. The gene involved in cystic fibrosis, for example, has been known for decades yet an effective therapy has not been forthcoming. The problems involved in moving from a gene to a therapy take the discussion to the next stage of the innovation process, where the process of industrialization continues. While a decade ago firms suffered from a lack of targets, today they suffer from too many targets and not enough understanding of which targets should be prioritized.

3.4.5 Genomics technologies in target validation

As noted previously, finding and validating disease-causing mechanisms involves substantial scientific resources over long periods of time. Although many technologies and experimental methods have been used, four main technologies play a central role. These are: model organisms, differential expression including micro-array technologies, various gene "knock-out" or "-down" regulation technologies, and complementary database research, which is more commonly known as bioinformatics (Murray Rust, 1994; Gelbart, 1998; Hopkins, 1998). These technologies allow experimental investigation to jump between proteins, RNA, and DNA, between organisms, and between "wet" and *in silico* environments in similar ways to the industrialization of gene sequencing. However, the biology involved is much more difficult, and, while these technologies have had a major impact on bioscience, the complexity of the problems that biologists face has meant that this stage remains a major bottleneck in drug discovery. The next section will highlight changes in model organisms and hybridization technologies to illustrate the common pattern of industrialization in this stage of the innovation process.

3.4.6 Model organisms

The industrialization of genomics has created a self-reinforcing focus onto the "super six" organisms –*Arabidopsis*, yeast (*Saccharomyces cerevisae*), nematode worm (*Carnorhabditis elegans*), fruit fly (*Drosophila*), mouse, and human genomes – and consequently away from more traditional objects of study in biology. The increased research focus on the "super six" is driven, in part, by the ability to share clones, link genetic data in databases, and exploit the larger, more specialized market for products, experimental kits, and services that increase the speed of experimentation.

These model organisms have long histories as subjects of genetic research because their short life cycles, simple structures, and short genomes make them easy to study. In general, reductions in scale (and genome size) are associated with reductions in complexity (which makes extrapolation to humans more difficult) but provide speedier life cycles and easier automation, which make experimentation easier.

For example, about 20 percent of the human genes associated with disease have yeast homologs (Andrade and Sander, 1997) and their function can be studied much more easily in yeast, where the relatively small size of the genome has allowed the entire genome to be put on gene chips for automated, massively parallel expression studies (Wodicka et al., 1997). Similarly, because the nematode worm is transparent, the development of its 959 cells and primitive nervous system can be studied by linking green fluorescent protein to genes in order to follow their expression in time and space. Worms, like fruit flies, are also particularly useful for RNA interference (RNAi) technology, which allows specific genes to be knocked out in the early stages of development, the resulting mutants then being analyzed for clues about gene function.

Mice, on the other hand, are much more similar to humans in the complexity of their biology and their short breeding cycles, small size, and rapid development, and the relative ease with which their germ lines can be engineered using stem cells to produce sophisticated animal models for target validation make them extremely useful models for large-scale studies. All these organisms make studying biology substantially easier, but their complexity makes validating and contextualizing the results of experiments very difficult, and extrapolating the results to humans, and particularly diseased human subjects, more difficult still.

3.4.7 Differential expression

The post-genome-sequencing era has shifted genomics research away from looking for genes towards exploring the diversity of gene expression over time (particularly during development), across species, between cell types, and between diseased and non-diseased states (Lander, 1996; Gelbert and Gregg, 1997). If a gene is expressed only in a diseased tissue, or in a particular cell type, then it is more likely to be involved in the mechanism – either as a cause, in which case it can be targeted for prevention, or as an effect, which may lead to drugs that reduce symptoms. Specificity is also a useful indicator of a potential lack of side effects. Expression patterns can be found for known genes using a variety of *in situ* hybridization methodologies, which have followed a pattern of increasing experimental density, automation, increases in speed, and reductions in size (Duggan et al., 1999).

Nucleic acid hybridization is a core tool in molecular biology and takes advantage of the very specific way that RNA and DNA bind to their complementary strands. The earliest technologies used radioactively labelled DNA, which required technically demanding autoradiograms and long development times (days and, in some cases, weeks). In a similar way to improvements in sequencing, the process was dramatically simplified and speeded up by the use of fluorophore labels attached to the probe. As a result, instrumentation methodologies such as FisH (fluorescence *in situ* hybridization) allow the position of any piece of DNA to be found easily using a microscope to see where a fluorescent probe, made of a complementary strand of DNA, binds onto a chromosome.

The shift towards the industrialization of hybridization technologies was started with Ed Southern's development of the Southern blot, whereby labelled probes could be used to interrogate nucleic acid molecules that have been broken up with restriction enzymes, separated on gels, and immobilized onto a nylon membrane, where they are hybridized to complementary DNA probes (Southern, 1975). A similar process, called the Northern blot, is used for RNA.

The separation and immobilization steps make possible increasingly parallel experimentation, greater experimental standardization, and therefore greater data reuse. One of the first applications of this technology was in "colony blotting," where large DNA clone libraries are screened for particular sequences of DNA (see Grunstein and Hogness, 1975). This was vital to the industrialization of sequencing, which depended on finding clones containing particular fragments of DNA within extremely large clone libraries. The process involved transferring individual colonies onto very large, gridded membranes that were then hybridized against a specific probe, which linked particular strands of DNA to particular colonies. This process was amenable to automation, and today robots automatically pipette clones onto membranes containing tens of thousands of points. This automated process allows individual clones to be identified rapidly and then copied and distributed around the world (Southern, Mir, and Shchepinov, 1999), further spreading the benefits of high-fixed-cost research.

Micro-arrays have further exploited the potential economies of scale available from miniaturization, automation, and massively parallel experimentation. Their development was based on two key innovations (Duggan et al., 1999; Southern, Mir, and Shchepinov, 1999). First, Pat Brown's group at Stanford developed non-porous solid supports that improved hybridization by stopping the sample diffusing into the gel, allowing better detection and further miniaturization (Duggan et al., 1999; Southern, Mir, and Shchepinov, 1999). The second innovation involved Steve Fodor's group developing methods for synthesizing oligonucleotides at very high densities, which reduced scale, increased the number of tests, and allowed more automated and much cheaper production (Lipshutz et al., 1999; Southern, Mir, and Shchepinov, 1999).[7]

These technologies have allowed molecular genetics to move from studying one gene, or possibly a few genes, to studying several

---

[7] The commonalities within the industrialization process are particularly apparent here, as a similar technology, using similar methods and equipment, was used by the *same* research group to industrialize solid-phase chemical synthesis in combinatorial chemistry (Nightingale, 2000) as was used to industrialize solid-phase hybridization in genomics.

thousand genes. This opens up the possibility of whole genome analyses through time or between organs to study development, or between diseased and non-diseased states to study their differences, or between organisms to study evolution, or between individuals to study how genetic variation increases the risk of disease or adverse drug reactions, potentially allowing genetic profiles to screen patients and provide more individualized medicine.

The word "potentially" in the last sentence is important, as these technologies are currently very expensive, and experimental variation makes experiments difficult to design, validate, and interpret (Butte, 2002; *Nature Genetics*, 1999, 2002). Micro-array technologies may have much higher throughput, but the amount of information they generate has made data validation and analysis difficult and time-consuming. Moreover, at present they are non-quantitative. As a result, complementary "wet" biological experimentation has to be undertaken to confirm initial hypotheses. While the technology of experimentation has undergone technical change, the complementary problem of understanding what experiments mean remains constrained by human cognition. Processes such as sequencing are readily amenable to almost full automation, but processes that still require human problem-solving, while improvable by technology, remain bottlenecks. Interestingly, since sequencing is now an industrialized and standardized process, it offers limited opportunities for competitive advantage. As a consequence, by the early 2000s some large firms were outsourcing sequencing.

3.5 Discussion and conclusions

This chapter has attempted to explain how pharmaceutical innovation processes have changed over recent years. It has highlighted how the desire to generate economies of scale and scope in experimentation has driven a consistent pattern of technical change in a range of technologies, resulting in a gradual transformation of drug discovery. In technologies based around biology, such as genetic markers, transgenic organisms, and micro-arrays, and in technologies that have little to do with biology, such as combinatorial chemistry and informatics, a similar pattern of reductions in size, a shift towards massively parallel experimentation, and increased integration between "wet" experimentation and *in silico* science has been observed. This change has made pharmaceutical research increasingly multidisciplinary, increased the importance of large data sets and the skills needed to explore them, increased the relative importance of data-driven research compared to traditional hypothesis-driven research, and made R&D increasingly reliant on expensive, high-fixed-cost capital goods.

Conceptualizing empirical changes in pharmaceutical innovation this way is important, because it emphasizes important distinctions between therapeutic classes, technologies, and steps in the innovation process that often get lost in policy discourses framed around a biotech revolution model of technical change. These distinctions, in turn, highlight the empirical diversity at work, the systemic nature of the methodologies, and the fact that substantial bottlenecks persist, particularly in target validation and other areas dependent on human problem-solving.

These bottlenecks mean that advances in scientific understanding do not necessarily produce advances in product development. As noted in the chapter, over two decades after the discovery of the CF gene we remain a long way from a cure. Similarly, despite the increases in R&D investment, there is still no unambiguous evidence of any improvement in the throughput of drugs. While absence of evidence is not evidence of absence, it seems clear that many expectations have not yet been realized. The introduction of new chemical entities is currently lower than at any time since 1945, and many of the high-tech methods may be decades away from routine commercial application (see Lahana, 1999; Horrobin, 2001, 2003; Pollack, 2002; Terwilliger et al., 2002; Beattie and Ghazal, 2003; and the original – Lewontin, 2000), although improvements are starting to be detected in patenting.

This is, however, exactly what one would expect based on previous historical experience of major changes in technology (Nightingale and Martin, 2004). As Rosenberg (1979), David (1990) and Freeman and Louca (2001) have shown, major technical change is a slow and cumulative process, because new technologies exist within wider socio-technical systems that need to adjust to accommodate them, and this process of adjustment can take a long time to occur. David (1990), for example, draws on the historical experiences of the introduction of electricity to suggest that the lack of evidence of productivity improvements from IT investments is to be expected. It could be argued that these lessons have more general application to genetics in pharmaceutical R&D.

While there have been some major changes in a range of technologies that underpin drug discovery, the process of moving from changes in technology to increases in productivity is an incremental one. This is because the shift from craft to industrialized experimentation has produced a trade-off between, on the one hand, increasingly accurate experimental information and, on the other, the need to contextualize that information in order to derive value. This process is localized and dependent on the organization of a division of specialized labor (Nightingale, 2004), suggesting substantial variation in how well firms can exploit these technologies.

The slow, incremental, and systemic view of technical change proposed by economic historians, and supported by this chapter, can be contrasted with the biotech revolution model of technical change, in which molecular biology in general and genomics in particular are revolutionizing drug discovery. The considerable amount of hype in the public domain about genomics might lead one to suppose that it is part of a new economy, where the lessons of the past don't apply. However, this chapter suggests that the technical problems that are being addressed by genomics technologies are extremely complicated and finding technical solutions requires a large amount of time and money. As a result, mobilizing the financial and social resources required for the many years of trial and error "tinkering" that are needed to get experimental technologies to work reliably involves creating widely shared, social expectations about what the developing technologies will be able to do (Nightingale and Martin, 2004). The creation and diffusion of these shared expectations is, therefore, a fundamental feature of innovation in this sector, and the high expectations surrounding genomics technologies are not consequences of their being revolutionary technologies. Instead, it is precisely because these technologies are not revolutionary, and take many years and substantial time and financial resources to produce productivity improvements, that high expectations are needed in the first place.

Similarly, the Chandlerian model proposed in this chapter offers an alternative to the networked model of industrial structure associated with the biotech revolution. There is no doubt that there has been a substantial increase in the outsourcing of R&D within the sector, with large firms typically spending about 25 percent of their R&D budgets through external networks. However, this industrial structure is driven by the Chandlerian interdependencies along the drug discovery process and the need to spread the high fixed costs of drug development as widely as possible. The productivity improvements in different stages produced by changes in experimental technologies have produced a series of shifting bottlenecks and excess capacity within the system. Large firms have therefore relied on external alliances to bring their R&D process into sync. Given the interdependencies outlined in this chapter, rather than selling off the excess capacity, as Penrose (1960) suggested in her theory of the growth of the firm, firms bring in resources from outside (see Nightingale et al., 2003). The pattern of alliances is not, therefore, a radical change in industrial organization, but a Toyota-style knowledge supply chain, where a range of diverse technologies and leads are generated in small biotechnology firms, focusing on very uncertain and potentially problematic technologies. When the uncertainties are reduced, large pharmaceutical firms bring them in-house to expand their capacity utilization. If this hypothesis is correct – and it is at this stage only a hypothesis – then much of the current policy focus on biotechnology firms as the vanguard of the new economy may be misplaced, as their current organizational structure as small firms may not be an early stage of their growth into large firms but a fundamental feature of their technological regime (Pavitt, 1984).

Relatedly, the analysis in this chapter provides support for the institutional success of open science in promoting innovation and raises concerns about the ongoing shift towards an increasingly privatized science system focusing on intellectual property rights and "markets in technology" (see also Geuna, Salter, and Steinmueller, 2003; Gibbons et al., 1994; and Dosi, Llerena, and Sylos Labini, 2005). The development of all the major technologies reviewed in this chapter drew heavily on academic research in the open literature, and it is doubtful if any of them would be available today without it. Where private money has been invested in infrastructure, such as the SNP consortia, it has typically been as a complement rather than an alternative to public projects. The complexity of the medical problems being addressed by the pharmaceutical industry, the high fixed costs of capital-intensive experimentation, the uncertainties involved in R&D, and the knowledge interdependencies that generate value all suggest that an open science system is a more efficient and effective means of spreading the benefits of large-scale investments in research (David, 1997, 2004; Nelson, 2004). By contrast, a privatized system would have limited ability to spread high fixed costs, would impose substantial transaction costs on the spread of scientific knowledge, and would require firms to invest in so much more R&D that their current structures would be economically untenable (David, 1997, 2004; Nelson, 2004). Given that even the largest firms only publish 1 to 2 percent of all the scientific papers in their fields of specialization, the "creative accumulation" of technological resources within large firms is an investment that allows them to access the wider open science system (Pavitt, 1998, 1999) and is viable only because that system is in place.

In conclusion, the Chandlerian reconceptualization of current changes within the pharmaceutical industry presented in this chapter suggests that much of the hype surrounding genomics is questionable. Undoubtedly, there are major changes at work, but the complexity of the subject matter and the problem-solving processes of medicinal biology should caution us against expecting any revolutionary changes. The industrialization of R&D has potential in addressing these complexities, but does not provide all the answers. These changes raise a range of theoretical, empirical, and policy issues that may lack the excitement of a "biotechnology revolution," with all the promise that the phrase evokes, but lack of excitement doesn't detract from their importance.

References

Adams, M. D., J. M. Kelly, and J. D. Gocayne (1991), "Complementary DNA sequencing: expressed sequence tags and the Human Genome Project," *Science*, 252, 1651–6.

Andrade, M. A., and C. Sander (1997), "Bioinformatics: from genome data to biological knowledge," *Current Opinion in Biotechnology*, 8, 675–83.

Beattie, J., and P. Ghazal (2003), "Post-genomics technologies: thinking beyond the hype," *Drug Discovery Today*, 8 (20), 909–10.

Bork, P., and A. Bairoch (1996), "Go hunting in sequence databases but watch out for the traps," *Trends in Genetics*, 12 (10), 425–7.

Botstein, D., R. L. Ehite, M. Skolnick, and R. W. Davis (1980), "Construction of a genetic linkage map in man using restriction fragment length polymorphism," *American Journal of Human Genetics*, 32, 314–31.

Bradshaw, J. (1995), *Pitfalls in Creating a Chemically Diverse Compound Screening Library*, unpublished document, Glaxo, London.

Butte, A. (2002), "The use and analysis of micro-array data," *Nature Reviews Drug Discovery*, 1, 951–60.

Cargill, M., D. Altshuler, and J. Ireland (1999), "Characterization of single nucleotide polymorphisms in coding regions of human genes," *Nature Genetics*, 22 (3), 231–8.

Chakravarti, A. (1999), "Population genetics – making sense out of sequence," *Nature Genetics*, 21 (1 Supplement), 56–60.

Chandler, A. D., Jr. (1990), *Scale and Scope: The Dynamics of Industrial Capitalism*, Belknap Press, Cambridge, MA.

Chumakov, I., P. Rigault, S. Guillou, P. Ougen, A. Billaut, G. Guasconi, P. Gervy, I. LeGall, P. Soularue, L. Grinas, L. Bougueleret, C. B. Chantelot, B. Lacroix, E. Barillot, P. Gesnouin, S. Pook, G. Vaysseix, G. Frélat, A. Schmitz, J. L. Sambucy, A. Bosch, X. Estivill, J. Weissenbach, A. Vignal, H. Riethman, D. Cox, D. Patterson, K. Gardinar, M. Hattori, Y. Sakaki, H. Ichikawa, M. Ohki, D. L. Paslier, R. Heilig, S. Antonarakis, and D. Cohen (1992), "Continuum of overlapping clones spanning the entire human chromosome 21q," *Nature*, 359, 380–7.

Cockburn, I. M., and R. Henderson (1998), "Absorptive capacity, co-authoring behaviour, and the organization of research in drug discovery," *Journal of Industrial Economics*, 46 (2), 157–82.

Cocket, M., N. Dracopoli, and E. Sigal (2000), "Applied genomics: integration of the technology within pharmaceutical research and development," *Current Opinion in Biotechnology*, 11, 602–9.

Collins, F. S., and Galas D. (1993), "A new five-year plan for the U.S. Human Genome Project," *Science*, 262, 11.

Collins, F. S., and M. S. Guyer (1995), "How is the Human Genome Project doing, and what have we learned so far?," *Proceedings of the National Academy of Sciences*, 92 (24), 10841–8.

Collins, F. S., and V. A. McKusick (2001), "Implications of the Human Genome Project for medical science," *Journal of the American Medical Association*, 285, 540–4.

Collins, F. S., A. Patrinos, E. Jordan, R. Gesteland, L. Walters, and members of DOE and NIH planning groups (1998), "New goals for the US Human Genome Project: 1998–2003," *Science*, 282, 682–9.

David, P. A. (1990), "The dynamo and the computer: a historical perspective on a modern productivity paradox," *American Economic Review*, 80, 355–61.

(1997), "Knowledge, property and the system dynamics of technological change," in L. Summers and S. Shah (eds.), *Proceedings of the World Bank Conference on Development Economics*, World Bank, New York, 215–48.

(2004), "Understanding the emergence of open science institutions: functionalist economics in historical context," *Industrial and Corporate Change*, 13 (3), 571–89.

Deloukas, P., G. D. Schuler, G. Gyapay, E. M. Beasley, C. Soderlund, and P. Rodriguez-Tome (1998), "A physical map of 30,000 human genes," *Science*, 282, 744–6.

DiMasi, J. A. (1995), "Trends in drug development costs, times and risks," *Drug Information Journal*, 29, 375–84.

Donis-Keller, H., P. Green, C. Helms, S. Cartinhour, B. Weiffenbach, K. Stephens, T. P. Keith, D. W. Bowden, D. R. Smith, and E. S. Lander (1987), "A genetic linkage map of the human genome," *Cell*, 23 (51), 319–37.

Dosi, G., P. Llerena, and M. Sylos Labini (2005), *Science–Technology–Industry Links and the "European Paradox": Some Notes on the Dynamics of Scientific and Technological Research in Europe*, Working Paper no. 2005:2, Laboratory of Economics and Management, Sant'Anna School for Advanced Studies, Pisa.

Dosi, G. (1982), "Technological paradigms and technological trajectories: a suggested interpretation," *Research Policy*, 11, 147–62.

Drews, J. (2000), *The Impact of Cost Containment Initiatives on Pharmaceutical R&D*, Annual Lecture, Centre for Medicines Research, London.

Duggan, D. J., M. Bittner, Y. Chen, P. Meltzer, and J. M. Trent (1999), "Expression profiling using cDNA microarrays," *Nature Genetics*, 21 (1 Supplement), 10–14.

Freeman, C. (1982), *The Economics of Innovation*, 3rd edn., Pinter, Cheltenham.

Freeman, C., and F. Louca (2001), *As Time Goes By: The Information and Industrial Revolutions in Historical Perspective*, Oxford University Press, Oxford.

Gelbart, W. M. (1998), "Databases in genomic research," *Science*, 282, 659–661.

Gelbert, L. M., and R. E. Gregg (1997), "Will genetics really revolutionise the drug discovery process?," *Current Opinion in Biotechnology*, 8, 669–74.

Geuna, A., A. Salter, and W. E. Steinmueller (2003), *Science and Innovation: Rethinking the Rationale for Funding and Governance*, Edward Elgar, Cheltenham.

Gibbons, M., C. Limoges, H. Nowotny, S. Schwartzman, P. Scott, and M. Trow (1994), *The New Production of Knowledge: The Dynamics of Science and Research in Contemporary Societies*, Sage Publications, London.

Gilbert, W. (1991), "Towards a paradigm shift in biology," *Science*, 249, 99.

Grunstein, M., and D. S. Hogness (1975), "Colony hybridisation: a method for the isolation of cloned DNAs that contain a specific gene," *Proceedings of the National Academy of Sciences*, 72, 3961–5.

GSK (2004), "Briefing to industrial analysts," Glaxo SmithKline, London.

Henderson, R. M., L. Orsenigo, and G. Pisano (1999), "The pharmaceutical industry and the revolution in molecular biology: exploring the

interactions between scientific, institutional, and organizational change," in D.C. Mowery and R.R. Nelson (eds.), *Sources of Industrial Leadership: Studies of Seven Industries*, Cambridge University Press, Cambridge, 267–311.

Holzman, N.A., and T.M. Marteau (2000), "Will genetics revolutionize medicine?," *New England Journal of Medicine*, 343, 141–4.

Hopkins, M. (1998), *An Examination of Technology Strategies for the Integration of Bioinformatics in Pharmaceutical R&D Processes*, Masters dissertation, Science and Technology Policy Research Unit, University of Sussex, Brighton.

Horrobin, D.F. (2001), "Realism in drug discovery – could Cassandra be right?," *Nature Biotech*, 19, 1099–100.

   (2003), "Modern biomedical research: an internally self-consistent universe with little contact with medical reality?," *Nature Review Drug Discovery*, 2, 151–4.

Houston, J.G., and M. Banks (1997), "The chemical-biological interface: developments in automated and miniaturised screening technology," *Current Opinion in Biotechnology*, 8, 734–40.

Hudson, T.J., L.D. Stein, S.S. Gerety, J. Ma, A.B. Castle, J. Silva, D.K. Slonim, R. Baptista, L. Kruglyak, and S.H. Xu (1995), "An STS-based map of the human genome," *Science*, 270, 1919–20.

Hughes, T. (1983), *Networks of Power: Electrification in Western Society, 1880–1930*, Johns Hopkins Press, Baltimore.

Lahana, R. (1999), "How many leads from HTS?," *Drug Discovery Today*, 4 (10), 447–8.

Lander, E.S. (1996), "The new genomics: global views of biology," *Science*, 274, 536–9.

Lander, E.S., and N. Schork (1994), "Genetic dissection of complex diseases," *Science*, 265, 2037–48.

Lewontin, R.C. (2000), *It Ain't Necessarily So: The Dream of the Human Genome and Other Illusions*, New York Review of Books, New York.

Lipshutz, R.J., S.P.A. Fodor, T.R. Gingeras, and D.J. Lockhart (1999), "High-density synthetic oligo-nucleotide arrays," *Nature Genetics*, 21 (1 Supplement), 20–4.

Mahdi, S. (2003), "Search strategy in product innovation process: theory and evidence from the evolution of agrochemical lead discovery process," *Industrial and Corporate Change*, 12 (2), 235–70.

Martin, P. (1998), *From Eugenics to Therapeutics: Science and the Social Shaping of Gene Therapy*, D.Phil. thesis, Science and Technology Policy Research Unit, University of Sussex, Brighton.

McKelvey, M. (1995), *Evolutionary Innovation: The Business of Biotechnology*, Oxford University Press, Oxford.

Meldrum, D. R. (1995), "The interdisciplinary nature of genomics," *IEEE Engineering in Medicine and Biology*, 14, 443–8.

(2000a), "Automation for genomics: part 1, preparation for sequencing," *Genome Research*, 10 (8), 1081–92.

(2000b), "Automation for genomics: part 2, sequencers, micro-arrays, and future trends," *Genome Research*, 10 (9), 1288–1303.

Mowery, D. C. (1983), "The relationship between intrafirm and contractual forms of industrial research in American manufacturing, 1900–1940," *Explorations in Economic History*, 20, 351–74.

Murray Rust, P. (1994), "Bioinformatics and drug discovery," *Current Opinion in Biotechnology*, 5, 648–53.

Murray, J. C., K. H. Buetow, J. L. Weber, S. Ludwigsen, T. Shirpbier-Meddem, and F. Manion (1994), "A comprehensive human linkage map with centimorgan density," *Science*, 265, 2049–54.

Nature Genetics (1999), "The chipping forecast," *Nature Genetics*, 21 (1 Supplement), 1–60.

(2002), "The chipping forecast II," *Nature Genetics*, 32 (4 Supplement), 461–552.

Nelson, R. R. (1991), "Why do firms differ, and how does it matter?," *Strategic Management Journal*, 12 (1), 61–74.

(2004), "The market economy, and the scientific commons," *Research Policy*, 33 (3), 455–71.

Nelson, R. R., and S. G. Winter (1982), *An Evolutionary Theory of Economic Change*, Belknap Press, Cambridge, MA.

Nightingale, P. (1998), "A cognitive theory of innovation," *Research Policy*, 27 (7), 689–709.

(2000), "Economies of scale in experimentation: knowledge and technology in pharmaceutical R&D," *Industrial and Corporate Change*, 9 (2), 315–59.

(2004), "Technological capabilities, invisible infrastructure and the un-social construction of predictability: the overlooked fixed costs of useful research," *Research Policy*, 33 (9), 1259–84.

Nightingale, P., T. Brady, A. Davies, and J. Hall (2003), "Capacity utilisation revisited: software, control & the growth of large technical systems," *Industrial and Corporate Change*, 12 (3), 477–517.

Nightingale, P., and P. Martin (2004), "The myth of the biotech revolution," *Trends in Biotechnology*, 22 (11), 564–9.

Pavitt, K. (1984), "Sectoral patterns of technological change: towards a taxonomy and a theory," *Research Policy*, 13, 343–74.

(1998), "Technologies, products and organisation in the innovating firms: what Adam Smith tells us and Schumpeter doesn't," *Industrial and Corporate Change*, 7 (3), 433–52.

(1999), *Technology, Management and Systems of Innovation*, Edward Elgar, Cheltenham.

(2001), "Public policies to support basic research: what can the rest of the world learn from US theory and practice? (And what they should not learn)," *Industrial and Corporate Change*, 10 (3), 761–79.

Pennisi, E. (1998), "A closer look at SNPs suggests difficulties," *Science*, 281, 1787–9.

Penrose, E. (1960), "The growth of the firm: a case study of the Hercules Powder Company," *Business History Review*, 43, 1–23.

1959, *The Theory of the Growth of the Firm*, Basil Blackwell, Oxford.

Polanyi, M. (1966), *The Tacit Dimension*, Doubleday, New York.

Pollack, A. (2002), "Despite billions for discoveries, pipeline of drugs is far from full," *New York Times*, C1.

Risch, N. J. (2000), "Searching for genetic determinants in the new millennium," *Nature*, 405, 847–56.

Risch, N. J., and K. Merikangas (1996), "The future of genetic studies of complex human diseases," *Science*, 273, 1516–17.

Rommens, J. M., M. C. Januzzi, and B.-S. Kerem (1989), "Identification of the cystic fibrosis gene: chromosome walking and jumping," *Science*, 245, 1059–65.

Rosenberg, N. (1974), "Science innovation and economic growth," *Economic Journal*, 84, 90–108.

(1976), *Perspectives on Technology*, Cambridge University Press, Cambridge.

(1979), "Technological interdependence in the American economy," *Technology and Culture*, 20 (1), 25–40.

Roses, A. D. (2000), "Pharmacogenetics and the practice of medicine," *Nature*, 405, 857–65.

(2001), "Pharmacogenetics," *Human Molecular Genetics*, 10 (20), 2261–7.

Rothwell, R. (1992), "Successful industrial innovation: critical factors for the 1990s," *R&D Management*, 22 (3), 221–39.

Scherer, M. F. (1993), "Pricing, profits and technological progress in the pharmaceutical industry," *Journal of Economic Perspective*, 7, 97–115.

Schwartzman, D. (1976), *Innovation in the Pharmaceutical Industry*, Johns Hopkins University Press, Baltimore.

Smith, L. M., S. Fung, M. W. Hunkapiller, T. J. Hunkapiller, and L. E. Hood (1985), "The synthesis of oligonucleotides containing an aliphatic amino group at the 5' terminus: synthesis of fluorescent DNA primers for use in DNA sequence analysis," *Nucleic Acids Research*, 13, 2399–412.

Smith, L. M., J. Z. Sanders, R. J. Kaiser, P. Hughes, C. Dodd, C. R. Connell, C. Heiner, S. B. H. Kent, and L. E. Hood (1986), "Fluorescence detection in automated DNA sequence analysis," *Nature*, 321, 674–9.

Southern, E. M. (1975), "Detection of specific sequences among DNA fragments separated by gel electrophoresis," *Journal of Molecular Biology*, 98, 503–17.

Southern, E. M., K. Mir, and M. Shchepinov (1999), "Molecular interactions on microarrays," *Nature Genetics*, 21 (1 Supplement), 5–9.

Strachan, T., and A. P. Read (2000), *Human Molecular Genetics 3*, Garland Press, Oxford.

Sudbury, P. (1999), *Human Molecular Genetics*, Cell and Molecular Biology in Action Series, Addison-Wesley, London.

Terwilliger, J. D., F. Haghighi, T. S. Hiekkalinna, and H. H. Goring (2002), "A *bias*-ed assessment of the use of SNPs in human complex traits," *Current Opinion in Genetics and Development*, 12, 726–34.

Venter, C. J., S. Levy, T. Stockwell, K. Remington, and A. Halpern (2003), "Massive parallelism, randomness and genomic advances," *Nature Genetics*, 33 (3 Supplement), 219–27.

Vincenti, W. G. (1990), *What Engineers Know and How They Know It*, Johns Hopkins University Press, Baltimore.

Wang, D. G., J. B. Fan, and C. J. Siao (1998), "Large-scale identification, mapping and genotyping of single nucleotide polymorphisms in the human genome," *Science*, 280, 1077–82.

Weiss, K. M. (1993), *Genetic Variation and Human Disease: Principles and Evolutionary Approaches*, Studies in Biological Anthropology no. 11, Cambridge University Press, Cambridge.

Weiss, K. M., and J. D. Terwilliger (2000), "How many diseases does it take to map a gene with SNPs?," *Nature Genetics*, 26 (2), 151–7.

Weissenbach, J., G. Gyapay, C. Dib, A. Vignal, J. Morissette, P. Millas-Seav, G. Vaysseix, and M. Lathrop (1992), "A second-generation linkage map of the human genome," *Nature*, 359, 794–801.

Wodicka, L., H. Dong, M. Mittmann, M. H. Ho, and D. J. Lockhart (1997), "Genome-wide expression monitoring in *Saccharomyces cerevisiae*," *Nature Biotechnology*, 15 (13), 1359–67.

4 | *Firm and regional determinants in innovation models: evidence from biotechnology and traditional chemicals*

MYRIAM MARIANI

4.1 Introduction

Firm competencies are discussed in the literature as important sources of firms' competitive advantage. They develop over time and, together with specific routines and communication mechanisms internal to the organization, affect the direction and the outcome of firms' R&D activities (see, for example, Nelson and Winter, 1982, and Dosi et al., 1988).

Recently, however, the economic literature has highlighted the importance of an alternative model for organizing production and innovative activities: the geographical proximity among researchers and institutions in a technological cluster. The contributions on regional clustering and geographically localized knowledge spillovers argue that the cost of transferring knowledge increases with the geographical distance among the parties involved in the exchange. This explains the tendency of innovative activities to locate together (see, for example, Jaffe, 1986; Jaffe, Trajtenberg, and Henderson, 1993; Audretsch and Feldman, 1996; and Swann, Prevezer, and Stout 1998).

So far the literature has analyzed these issues separately. This chapter recombines these two streams of the literature and compares their relative importance in explaining a research output: the value of innovations. It estimates how much of the value of an innovation is affected by the affiliation of the inventors to the same organization, as opposed to spillovers that arise when the inventors are geographically close to each other and to external sources of knowledge. In so doing it expands upon a previous paper by Mariani (2004).

The empirical investigation uses a data set of 4262 European Patent Office (EPO) patents applied for in the chemical industry by 693 firms between 1987 and 1996. It merges patent information with other data concerning the characteristics of the firms that apply for the patents, and the regions in which the innovations originate. The multiple

112

correlation analysis in the empirical sections uses the numbers of citations received by the patents in the five years after the application date,
excluding self-citations, as a proxy for the value of the innovations.
This is regressed on three sets of other variables: (1) firms' characteristics: characteristics at the level of the firm and characteristics at the
level of the R&D project that led to the innovation; (2) regional
characteristics, and in particular the scientific and technological characteristics of the regions in which the patents are invented; and (3) a set
of controls.

The finding of this research is that the factors leading to high-value
research output in the chemical industry differ across sectors. In the
traditional chemical sectors (i.e. materials, organic chemistry, pharmaceuticals, and polymers), firm competencies, and specifically the R&D
intensity of the firm and the scale of the research projects, are positively
correlated with the probability of producing valuable innovations,
with no role for external knowledge spillovers. In biotechnology, not
only are firm competencies important (i.e. firms' technological specialization), but also the geographical proximity of the inventors to external sources of knowledge positively affects the probability of
developing a "technological hit." These results, which are robust to
different models, specifications, and controls, are consistent with other
studies in the literature. Klepper (1996), for example, demonstrates
that new industries benefit from local knowledge spillovers more than
industries in the mature stages of the industry life cycle. Along the same
lines, Audretsch and Feldman (1996) show that innovative activities
cluster more in skilled and R&D-intensive industries.

This chapter is organized as follows. After a brief review of previous
studies on firm competencies, geographical spillovers, and the use of
patent citations (section 4.2), section 4.3 describes the database.
Sections 4.4 and 4.5 discuss the empirical specifications and present
the estimated results. Section 4.6 summarizes and concludes the chapter.

4.2 Firm competencies, geographical spillovers, and patent indicators

How much of the value of an innovation depends on firms' competencies and characteristics vis-à-vis the characteristics of the location in
which the innovation is produced? There is a long literature on firm
competencies. The capabilities to coordinate and integrate different

people and different parts of the firm, and the peculiar learning processes and routines that a firm develops over time, are important factors
in explaining firms' innovativeness and long-term profitability (Nelson
and Winter, 1982; Dosi et al., 1988; Klepper and Sleeper, 2002; Teece,
Pisano, and Shuen, 1997. See also Henderson and Cockburn, 1996).

In recent years, however, an alternative candidate to the firm as a
means to coordinate people and activities has emerged in the literature:
the geographical cluster. Quite a few contributions describe the importance of locations and geographical spillovers. They argue that knowledge spillovers arise when people and organizations are geographically
close to each other, and to external sources of scientific and technological knowledge (see, for example, Jaffe, 1986; Jaffe, Trajtenberg, and
Henderson, 1993; Swann, Prevezer, and Stout, 1998; and Verspagen,
1999). The empirical evidence highlights that there are sectoral differences in the extent to which firms take advantage of knowledge spillovers (see, for example, Audretsch and Feldman, 1996, and Klepper,
1996). For example, Powell, Kaput, and Smith-Doerr (1996) show that
the locus of innovation in the biotechnology sector is a network of
different organizations, rather than the individual firm. To make this
point, they describe two important biotechnology discoveries in the
mid-1990s that have been co-authored by more than thirty researchers
located in different organizations. They also map the network structure
of the US biotechnology industry by using data on the formal agreements set up by 225 biotechnology firms, and argue that firms colla-
borate in order to expand their competences. Zucker, Darby, and
Brewer (1998) show that the growth and location of basic scientific
research was the main driving force behind the growth and location
of biotech firms in the United States. Finally, Zucker, Darby, and
Armstrong (1998) associate the growth of biotech firms in California
with the fact that they established "formal" relationships with university scientists.

This chapter combines these two strands of the literature and estimates the relative contribution of firm competencies and geographical
proximity to the production of technological hits. The value of the
innovations is proxied by the number of citations received by the patent
in the five years after its application date. The use of patent citations is
now fairly standard in the literature. Several contributions demonstrate
that there is a positive relationship between patent indicators that
appear after the innovation has been discovered and the actual

ex post value of the innovation as given by traditional accounting evaluation (see, for example, Hall, Jaffe, and Trajtenberg, 2005 and 2001, for a survey). Trajtenberg (1990) shows that there is a non-linear and close association between patent counts weighted by forward citations and the social value of innovations, in the computer tomography scanner industry. Harhoff, Scherer, and Vopel (2003) demonstrate that the numbers of citations to past patent literature, as well as the numbers of citations to the scientific literature and the numbers of citations received by the patent after its publication date, are positively correlated with the market value of the innovation. They also show that oppositions and annulment procedures are positively correlated with the probability of a patent being of high value. The same applies for patents granted in a large number of countries. Griliches, Pakes, and Hall (1987) demonstrate that there is a positive correlation between patents' renewal rates and fees on the one hand, and the private value of patent rights on the other hand. Finally, Lanjouw and Schankerman (2004) construct a composite measure of the quality of patents, and show that forward citations are one of the least noisy indicators of the value of the innovations.

Of course, patent indicators also have a number of limitations (Griliches, 1990). For example, citations cannot be made to or by innovations that have not been patented. Second, as the time series move closer to the latest date in the data set, patent citations increasingly suffer from missing observations (i.e. "truncation" problem). Third, patents applied for in different years and technological classes might differ in their propensity to be cited, suggesting that changes in the numbers of citations per patent might stem from factors other than the actual changes in the technological impact of the innovations. This research will not ignore these problems, and it will use remedies to limit them as proposed by Hall, Jaffe, and Trajtenberg (2001).

4.3 Data

This work uses a large and detailed data set that links patent information drawn from the European Patent Office to other data concerning the characteristics of the firms that apply for the patents, and the characteristics of the locations in which the innovations are invented. From a universe of 201,531 biotechnology and chemical patents applied for during 1987–96 I selected a random sample of 10,000 patents. By reading the abstracts of the patents and the description of the three-digit International Patent Classification (IPC) codes of the main/obligatory technological class, an expert pharmacologist helped me assign these patents to one of the following five technological classes: biotechnology, materials, organic chemistry, pharmaceuticals, and polymers.[1]

4.3.1 Patent data

From the patent document I collected information on the name of the applicants, the name of the inventors, the application year, and the numbers of citations received by the patent after the application date up to 2000.

4.3.2 Regional data

I used the zip code contained in the address of the inventors of the patents to assign them to the specific NUTS region (Nomenclature des Unités Territoriales Statistiques) in which the patents were invented.[2] While doing this, however, I found out that more than a half of the 10,000 patents had been invented in countries outside the European Union. Since I had complementary regional data only for the European regions, I decided to drop these patents from the sample. This left me with a final sample of 4607 patents for which at least one inventor is located in Europe. The Regio database (released in 1999) provides data about the characteristics of the NUTS regions in which the inventors are located, including the total number of patents invented in the area in the period 1987–96. The European R&D database published by Reed Elsevier in 1996 collects the number and type of R&D laboratories in Europe. I downloaded from this database about 20,000 laboratories as at December 1995, and classified them as university

---

[1] I thank Rossana Pammolli for doing this.

[2] When the inventors of a patent are all located in one region, there is a clear identification of the NUTS region in which the innovation is invented. When the inventors of a patent are located in different regions, I assigned the patent to the region in which the largest share of inventors is located (this applied to 24 percent of the total sample). Finally, when the inventors are located 50 percent in one region and 50 percent in another region, I assigned the patent to the region in which the first inventor of the list is located (this applied to 9.5 percent of the cases). See Mariani (2004) for more details. The appendix lists the NUTS regions used in this chapter.

laboratories, government laboratories, and chemical laboratories if they focused on chemical research. Each laboratory was assigned to the NUTS region in which it was located.

4.3.3 Firm data

The name of the applicant of the patent was used to collect information on the organization to which the inventors were affiliated. For a small share of patents with more than one applicant, I collected information on the first applicant on the list (7.6 percent of the patent sample). There are 166 patents applied for by public organizations, excluding universities; 45 patents owned by universities; 134 patents applied for by individual inventors; and 4262 patents applied for by 693 private companies. This research focuses on these 4262 patents applied for by companies, which cover 92.6 percent of the patents in the sample. The Who Owns Whom database (1995) was used to merge parent and affiliate firms under the same name. To collect company data, I first consulted Aftalion (1991), who lists the top 250 chemical companies worldwide in 1988, and provides firm-level information on R&D spending and firm turnover, not just in chemicals but in all sectors. The Compustat database (1999) complemented the data on R&D and sales that I could not get from Aftalion (1991).[3] In the end, I could not find information on a tail of applicants covering 852 patents in the sample, the distribution of which is not biased in any particular direction across regions and technological classes. Nevertheless, a dummy variable is included in the regressions to take into account those observations with missing data on sales and R&D.

The final data set is composed of 4262 observations with information on the characteristics of the innovation, the characteristics of the firms that apply for the patent, and the characteristics of the region in which the inventors were located at the time of the invention. Table 4.1 shows the definition of the variables.

The empirical analysis performs a multiple correlation exercise by means of negative binomial regressions. The dependent variable in my regressions is the value of the individual innovations, which I proxy by the

---

[3] I also looked at the websites of some companies for which data were not provided by other sources. I thank Fabio Pammolli for providing me with data on R&D and sales for an additional group of companies.

Table 4.1 List of variables for regressions

Firm and project characteristics	
CITS	Dependent variable; number of citations received by the patent in the five years after the application date, excluding self-citations
SALES	Company sales in 1988 (1988 $ millions)
R&D/SALES	Company R&D spending over sales in 1988
TECHSPEC	Number of EPO patents in the same technological sector of each patent in the sample (biotechnology, materials, organic chemistry, pharmaceuticals, and polymers) filed by the firm during the five years before the application date
TECHCOMP	Number of EPO patents filed by the firm during the five years before the date of each patent in the sample in the other four technological sectors
INVENTORS	Number of inventors who collaborate to develop the innovation
DLOC	Dummy; it takes the value 1 if all the inventors listed in the patent are located in the same region, 0 if at least one inventor is located in a different NUTS region
MAPPL	Dummy; it takes the value 1 if there are multiple applicants, 0 otherwise
NOCHEM	Dummy for non-chemical companies
MISSING	Dummy for missing values on SALES and R&D

Regional characteristics	
REGHLABS	Number of higher education laboratories located in the region (stock in 1995)
REGPATS	Number of patent applications in all sectors invented in the regions (units – average 1987–96)
GDP	Regional per capita gross domestic product ($ millions, purchasing power parity basis, corrected for inflation – average 1987–96)
POP	Population density of the region (thousands – average 1987–96)
AREA	Area of the region (km^2)

Other controls	
SCIENCE	Number of citations made by the patent to past scientific literature
CITSSEC	Citation intensity of patents, computed as the average number of citations received by the patents applied for in the same year and technological class as the patent application (three-digit IPC classes)

INVCY	Dummy for the country of the inventors[a]
APPLCY	Dummy for the country of the applicant firms[b]
YEAR	Dummy for the application date (1987–96)
SECTOR	Dummy for the sector in which the patent is classified: biotechnology, materials, organic chemistry, pharmaceuticals, and polymers

---

[a] Countries covered are Austria, Belgium, Denmark, Finland, France, Germany, Greece, Ireland, Italy, Luxembourg, the Netherlands, Spain, Sweden, Switzerland, and the United Kingdom.

[b] Countries covered include all of the above plus Japan, the United States, and others.

number of citations received by the patents in the five years after the application date up to 2000, excluding self-citations (*CITS*). Self-citations – i.e. citations from patents applied for by the same applicant – are excluded from *CITS* because they may not have the same role in measuring the value of the innovation as compared to "external" cites (Hall, Jaffe, and Trajtenberg, 2005).[4] Consistently with other contributions that use patent citations, the distribution of *CITS* is skewed. The number of patent citations excluding self-citations ranges between nought and thirteen, with mean 0.74 and standard deviation 1.34. When self-citations are included in *CITS*, the number of patent citations ranges between nought and nineteen, with mean 1.20 and standard deviation 1.91.

4.4 Empirical analysis: variables and negative binomial regressions

This section compares the effect of firm and regional characteristics on the probability of producing highly cited patents in the chemical industry. Because of the count nature of the dependent variable – i.e. *CITS* – I perform negative binomial regressions.[5]

---

[4] Self-citations in small and specialized companies could be an indicator of internal spillovers and cumulative processes of knowledge creation rather than value per se. Large firms might cite themselves because they have large patent portfolios to cite, irrespective of the actual value of the cited patents. In any event, when I performed the regressions by including self-citations in *CITS* there were no relevant changes in the results.

[5] Since there is over-dispersion in the data, as suggested by the value of α and by the LR-chi-square test of the null hypothesis that $\alpha = 0$, the negative binomial model fits the data better than the Poisson model.

The first set of factors included in the regressors is composed of firm characteristics. Company size is proxied by firms' *SALES* in 1988 dollars. *SALES* is expected to affect positively the probability of inventing valuable innovations. Large firms may have an advantage in the financial markets compared to smaller firms. Moreover, large firms have the opportunity to benefit from economies of scale in research by spreading the fixed cost of R&D over a large number of projects and over a large sales base (Henderson and Cockburn, 1996). Moreover, by centrally coordinating different research projects, they can avoid duplication in research. *R&D/SALES* – i.e. the R&D intensity of the companies – is also expected to have a positive impact on *CITS*: R&D-intensive firms are expected to initiate a portfolio of different research projects the complementarities of which can generate research spillovers internal to the firm. Moreover, if such complementarities exist, knowledge produced in one project can be used in other projects, therefore enhancing the whole firm's efficiency in doing research. Research-intensive companies are also likely to hire many researchers and to gain from their specialization. Finally, a firm's R&D investment also enhances the capacity to absorb knowledge produced by external sources.

The other two firm characteristics included in the regressions are *TECHCOMP* and *TECHSPEC*, respectively the technological competencies and the technological specialization of the firms. These variables are constructed by using the five technological sectors in which the patents are classified (biotechnology, materials, organic chemistry, pharmaceuticals, and polymers) and the count of the patents applied for by the firms in these sectors. Specifically, *TECHSPEC* is the number of EPO patents applied for by the firm during the five years before the date of each sample patent, in the same technological sector. *TECHCOMP* is the same as the previous variable but it uses the number of EPO patents in the other four technological sectors. A dummy for the missing values of *SALES* and *R&D*, and a dummy for the large non-chemical companies, are also included in the regressions.

At the project level, the number of inventors listed in the patent (*INVENTORS*) is a proxy for the scale of the research project, which is expected to affect positively the probability of developing major innovations. If many researchers are involved in a common R&D project, there may be benefits from their specialization in specific tasks and competencies that they embody. There may also be gains from the scale of the project if the cost of the research is partially fixed. Another project level variable is *MAPPL*, a dummy that indicates if the patent is the output of a collaboration among different institutions. The expectation is that a joint research effort that goes beyond the firm's boundaries increases the probability of developing technological hits. Finally, I included *DLOC*, a dummy for the co-location of the inventors in the same NUTS region. The idea is that a complex and interdisciplinary research project may need competencies that are mastered internationally by pulling together a wide range of competencies from different firms' units, organizations, and locations. This would produce a positive correlation between *DLOC* and *CITS*. This dummy also controls for the arbitrary geographical assignment of patents when the inventors are located in different NUTS regions.

The second group of variables used in the empirical analysis is the technological characteristics of the regions. This work uses the number of higher education laboratories (*REGHLABS*) as a proxy for the location of scientific institutions in the region. The average number of patents invented in each region in all sectors (*REGPATS*) in 1987–96 is an indicator of the regional technological capabilities. The area of the regions (*AREA*), the population density (*POP*), and the per capita GDP (*GDP*) are exogenous controls. Consistently with the idea of agglomeration economies, the population density and the per capita *GDP* are expected to correlate positively with *CITS*, while *AREA* is expected to have a negative effect.[6]

Finally, the third set of data is composed of controls. An important one is *CITSSEC*, a citation benchmark value that controls for the truncation problem and for differences in the citation intensities over time and across sectors that are unrelated with the value of the patents (Hall, Jaffe, and Trajtenberg, 2001). *CITSSEC* weights the number of citations of each patent by the average citation intensity of a group of patents with similar characteristics in terms of application year and technological class. It controls for technology-specific citation intensities by using a three-digit IPC classification. I have also controlled for the extent to which a patent is related to basic research. This is done by including the number of citations listed in each patent to

---

[6] The Regio database (1999) does not provide information on the regional characteristics of Switzerland, Finland, Norway, and Sweden.

Table 4.2 Descriptive statistics of variables

Variable	Mean	Standard deviation	Minimum	Maximum
		Firm characteristics		
CITS	0.61	1.21	0	13
SALES	16,974	14,902	4	87,542
R&D/SALES	0.06	0.11	0.002	3.12
TECHSPEC	275	340	0	1230
TECHCOMP	1003	1166	0	6264
INVENTORS	3.19	1.85	1	16
DLOC	0.62	0.49	0	1
MAPPL	0.06	0.24	0	1
		Regional characteristics		
REGHLABS	46	70	0	461
REGPATS	593	610	0	2263
GDP	16.7	4.4	7.4	27.5
POP	0.70	0.88	0.005	5.93
AREA	6243	4211	97	55,401
		Other controls		
SCIENCE	1.40	2.04	0	24
CITSSEC	1.47	0.60	0.67	4.02

Note: Number of observations: 4262 chemical patents.

non-patent literature such as scientific journals, books, and proceedings (*SCIENCE*). Since more "scientific" work may be used in a large number of different applications, *SCIENCE* is expected to correlate positively with the probability of receiving forward citations. All the regressions also include: dummies for the country of the inventors and the country of the applicants, to capture the effect of regional characteristics that are independent of the variation across countries; dummies for the year in which the patent was applied for; and dummies for the five technological sectors in which the patents are classified, to control for technology-specific effects.

Table 4.3 Estimates of negative binomial regressions

	Firm and project characteristics	
SALES	−0.012 (0.025)	0.005 (0.028)
R&D/SALES	0.124[a] (0.040)	0.135[a] (0.040)
TECHSPEC	–	−0.035 (0.037)
TECHCOMP	–	−0.011 (0.034)
INVENTORS	0.181[a] (0.056)	0.205[a] (0.060)
DLOC	–	0.060 (0.066)
MAPPL	0.297[a] (0.115)	0.319[a] (0.115)
	Regional characteristics	
REGHLABS	0.037 (0.037)	0.033 (0.037)
REGPATS	0.014 (0.055)	0.016 (0.055)
GDP	−0.275 (0.218)	−0.277 (0.219)
POP	0.023 (0.081)	0.024 (0.081)
AREA	−0.024 (0.91)	−0.022 (0.091)
	Other controls	
SCIENCE	−0.0006 (0.050)	−0.003 (0.051)
CITSSEC	0.836[a] (0.193)	0.811[a] (0.195)
Number of observations	4015	3984
Log-likelihood	−4052.32	−4028.003
Pseudo R^2	0.0388	0.0390
α	1.50 (0.10)	1.50 (0.10)

Note: Sample: 4262 patents. Variables are in logs. Cluster-robust standard errors are in parentheses. All regressions include dummies for non-chemical companies, a missing value for R&D and SALES, inventor country, applicant country, year of application, and technological field.
[a] Significant at 0.01 level.

Table 4.2 shows the descriptive statistics of the variables used in the regressions. Table 4.3 shows the results of the econometric estimates of the negative binomial regressions for the whole chemical industry, the dependent variable being the number of citations received by the patent in the five years following the application date, excluding self-citations (*CITS*). The variables are in logs.

The two specifications in table 4.3 differ for the inclusion of firm and project characteristics: the first specification does not include *TECHSPEC*, *TECHCOMP*, and *DLOC*. The results, however, are robust across the two specifications: economies in R&D internal to the firm are key factors for explaining the probability of developing technological hits. The elasticity of *CITS* with respect to *R&D/SALES* is 0.124 and 0.135 in the two specifications and it is statistically significant. In addition, the number of inventors involved in a common R&D project (*INVENTORS*) has a positive and statistically significant impact on the probability of receiving forward citations, as well as the collaboration with other partners (*MAPPL*). The technological specialization of the firms does not add anything to the expected value of the innovations.

What is really surprising is that the technological characteristics of the regions in which the patents are invented are not correlated with the probability of producing highly cited innovations. The expectation was that in technology-intensive regions, where innovative activities agglomerate, it would be easier to find the specialized and complementary competencies needed in complex R&D projects. Moreover, since people with complementary expertise are located close to one another, the probability of collaborating and of exchanging knowledge increases. However, the coefficients of *REGHLABS* and *REGPATS* are not statistically significant. The regional characteristics are also jointly insignificant. I tested the unrestricted model in table 4.3 via a likelihood ratio test against a restricted model with $REGHLABS = REG\text{-}PATS = GDP = POP = AREA = 0$. The null hypothesis cannot be rejected. Given this further check, and the fact that the regressions are run with a large number of controls, these results suggest that economies in R&D internal to the firm are the key factors that positively affect the probability of producing high-value innovations in the chemical industry.[7]

---

[7] When self-citations are included in the dependent variable, the econometric estimates (not shown here) do not change significantly compared to those in table 4.3.

4.5 Looking for geographical spillovers: biotechnology versus traditional chemicals

The result that localized spillovers do not matter is puzzling, especially given the number of contributions in the literature that emphasized them. In fact, my results indicate that, in chemicals, the model of innovation that leads to patents with high expected value is dominated by the large firm that invests heavily in internal R&D activities and large-scale projects, with no role for spillovers generated by the external technological environment in which the research is performed.

The chemical industry also offers an interesting basis, however, for exploring this issue in greater detail. The industry is composed of different sub-sectors, ranging from bulk chemicals to biotechnology products. The heterogeneity of these sectors might hide differences in the way innovations are produced. Biotechnology is a natural candidate in this respect. As many authors have discussed, not only is it a novel sector, but also the way innovation is organized seems to be different from the traditional chemical sectors. While, in the latter, innovation is centered on large, incumbent chemical manufacturers, biotechnology is populated by smaller technology-intensive companies that are organized in cluster-like forms, and that establish external linkages for exchanging and developing new knowledge (see, for example, Orsenigo, 1989). This section tests if there is evidence of geographical spillovers in biotechnology compared to the more traditional branches of the chemical industry. Given the extensive controls that I use for the features of the firms, then, if even in biotechnology there is little evidence of geographical spillovers, it may well turn out that most of the effects are really at the firm level, with little role for geographical factors.

My sample is composed of 525 biotechnology patents and 3737 traditional chemical patents.[8] Patents in traditional chemicals receive a lower number of citations than in biotechnology. The average

---

[8] I have grouped organic chemistry, materials, pharmaceuticals, and polymers together in the "traditional chemical" sector because, compared to biotechnology, the empirical analysis shows that the model of innovation leading to high-value innovations relies on similar factors.

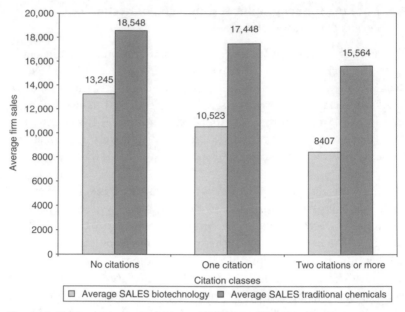

Figure 4.1. The average size (mean of *SALES*) of firms with patents in biotechnology and traditional chemicals

Note: Sample: 2570 patents. Since the values are calculated at the patent level, fifty-nine firms in the sample with patents both in biotechnology and in traditional chemicals enter the calculations in both sectors.

Source: Elaboration from the EPO, Aftalion (1991), and Compustat data, 1987–1993.

number of *CITS* is 0.69 in traditional chemicals; it is 1.03 in biotechnology. This difference is statistically significant.[9]

Figures 4.1 and 4.2 show some insights on the relationship between the characteristic of the applicant firms and the value of the innovations in biotechnology and in traditional chemicals. The horizontal axis in figure 4.1 groups the patents according to the number of *CITS*: patents

[9] To limit the "truncation" problem, figures 4.1 and 4.2 use a sub-sample of 3080 patents applied for between 1987 and 1993. This is so that all these patents had at least seven years of potential citations, as I noted that patents filed in 1987 received about 70 percent of their citations in the first seven years. The econometric analysis uses the whole sample of 4262 patents, and it employs controls for the truncation problem.

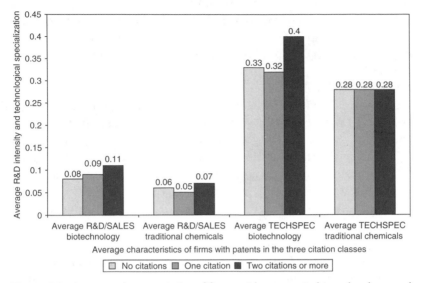

Figure 4.2. Average characteristics of firms with patents in biotechnology and traditional chemicals: R&D intensity (mean of *R&D/SALES*) and technological specialization (mean of *TECHSPEC*) of firms

Note: Sample: 2517 patents for *R&D/SALES*, 3057 patents for *TECHSPEC*. Since the values are calculated at the patent level, fifty-nine firms in the sample with patents both in biotechnology and in traditional chemicals enter the calculations in both sectors.

Source: Elaboration from the EPO, Aftalion (1991), and Compustat data, 1987–1993.

with no citations (i.e. the median number of citations), those with one citation (i.e. patents in the fourth quartile), and those with two citations or more (i.e. patents in the top 10 percent). The vertical axis measures the average size of the firms in biotechnology and traditional chemicals, as measured by company sales in millions of 1988 dollars (*SALES*).

Figure 4.1 confirms the expectation that companies in biotechnology are, on average, smaller than firms in traditional chemicals. Moreover, in both biotechnology and in traditional chemicals there is a negative relationship between the size of the firms and the probability of receiving citations. The average R&D intensity (*R&D/SALES*) and the average technological specialization (*TECHSPEC*) of companies in biotechnology and traditional chemicals are drawn in figure 4.2. As expected, firms in biotechnology are on average more R&D-intensive and technologically specialized than companies in traditional chemicals. Furthermore, there is a positive relationship between the R&D intensity of the firms and the probability of producing highly cited patents in both sectors. In biotechnology, *R&D/SALES* jumps from 8 percent in the no citation class, to 9 percent for companies with patents in the top quartile, to 11 percent for companies with patents in the top 10 percent class. In traditional chemicals, *R&D/SALES* goes from 6 percent, to 5 percent, and 7 percent in the three citation classes. Differently, firm technological specialization is positively correlated with the probability of developing important patents only in biotechnology: *TECHSPEC* goes from 0.33 in the bottom 50 percent class to 0.40 in the top 10 percent citation class.

I also explored the relationship between the technological characteristics of the regions (*REGHLABS* and *REGPATS*) in which the inventors were located, and the value of the innovations in biotechnology and traditional chemicals. The results (not shown here) are inconclusive. This might be due to the fact that the tables do not highlight the net effect of being in a technologically intensive area. Therefore, by means of multiple correlation analysis, table 4.5 shows the results of the negative binomial regressions for biotechnology versus traditional chemicals (the variables are in logs). Table 4.4 presents the descriptive statistics of the variables for the two sectors separately.

The results in table 4.5 confirm that the determinants of technological hits in the traditional branches of the chemical industry reside within the firm. The R&D intensity of the companies and the large scale of the projects that involve collaborations with other institutions affect the probability of inventing highly cited innovations. The technological specialization of traditional chemical companies is negatively correlated with the probability of being cited. The estimated coefficient of *TECHSPEC* is −0.093, and it is statistically significant. Probably, rather than the effect of technological diversification, this is due to the fact that large chemical companies have a higher propensity to patent compared to the smaller biotechnology companies. This is confirmed by the fact that traditional chemical companies have a high share of self-citations per patent, probably because of their larger patent portfolios. It is then plausible that a higher propensity to patent by large firms occurs at the cost of a lower average number of citations per

Table 4.4 Descriptive statistics: traditional chemicals and biotechnology

Variable	Mean	Standard deviation	Minimum	Maximum
		Firm characteristics		
CITS	0.58	1.16	0	13
	0.84	*1.54*	*0*	*11*
SALES	17,574	14,895	4	87,542
	10,889	*13,575*	*4*	*79,643*
R&D/SALES	0.06	0.10	0.002	3.12
	0.09	*0.11*	*0.002*	*0.85*
TECHSPEC	307	351	0	1230
	44	*59*	*0*	*252*
TECHCOMP	1073	1174	0	6264
	506	*971*	*0*	*4502*
INVENTORS	3.19	1.85	1	16
	3.15	*1.84*	*1*	*14*
DLOC	0.62	0.49	0	1
	0.59	*0.49*	*0*	*1*
MAPPL	0.06	0.23	0	1
	0.10	*0.31*	*0*	*1*
		Regional characteristics		
REGHLABS	46	71	0	461
	45	*69*	*0*	*461*
REGPATS	610	609	0.8	2263
	467	*605*	*0*	*2263*
GDP	16.9	4.4	7.4	27.5
	17.1	*4.3*	*7.6*	*27.5*
POP	0.70	0.87	0.01	5.93
	0.67	*0.92*	*0.005*	*5.73*
AREA	6207	3981	97	35,291
	6499	*5574*	*97*	*55,401*
		Other controls		
SCIENCE	1.16	1.8	0	24
	3.14	*2.7*	*0*	*17*
CITSSEC	1.44	0.58	0.13	4.02
	1.70	*0.67*	*0.07*	*3.49*

Note: Number of observations: 525 biotechnology patents, 3737 traditional chemical patents. Biotechnology figures are in italics.

Table 4.5 Estimates of negative binomial regressions: traditional chemicals and biotechnology

	Traditional chemicals	Biotechnology	Traditional chemicals	Biotechnology
	Firm and project characteristics			
SALES	−0.002	−0.035	0.019	0.037
	(0.027)	(0.056)	(0.031)	(0.065)
R&D/SALES	0.134[a]	−0.018	0.151[a]	−0.079
	(0.043)	(0.116)	(0.043)	(0.124)
TECHSPEC	–	–	−0.093[b]	0.264[b]
			(0.040)	(0.108)
TECHCOMP	–	–	0.031	−0.196[a]
			(0.037)	(0.074)
INVENTORS	0.183[a]	0.125	0.200[a]	0.154
	(0.062)	(0.119)	(0.066)	(0.129)
DLOC	–	–	0.033	0.169
			(0.072)	(0.168)
MAPPL	0.271[b]	0.504[b]	0.278[b]	0.539[b]
	(0.133)	(0.229)	(0.131)	(0.232)
	Regional characteristics			
REGHLABS	0.054	−0.061	0.052	−0.076
	(0.041)	(0.081)	(0.042)	(0.081)
REGPATS	−0.025	0.303[b]	−0.021	0.323[b]
	(0.058)	(0.151)	(0.059)	(0.154)
GDP	−0.220	−0.474	−0.231	−0.373
	(0.236)	(0.526)	(0.238)	(0.550)
POP	0.036	−0.176	0.041	−0.186
	(0.089)	(0.185)	(0.090)	(0.182)
AREA	0.006	−0.300	0.010	−0.274
	(0.101)	(0.210)	(0.101)	(0.209)
	Other controls			
SCIENCE	−0.055	0.293[b]	−0.063	0.302[a]
	(0.056)	(0.116)	(0.056)	(0.111)
CITSSEC	0.653[a]	1.447[a]	0.676[a]	1.090[a]
	(0.224)	(0.405)	(0.223)	(0.404)

	Traditional chemicals	Biotechnology	Traditional chemicals	Biotechnology
Number of observations	3518	497	3494	490
Log-likelihood	−3460.4	−568.5	−3439.5	−559.9
Pseudo R^2	0.0358	0.0797	0.0366	0.0851
α	1.54 (0.12)	1.04 (0.17)	1.54 (0.11)	0.098 (0.17)

Note: Sample: 4262 patents. Variables are in logs. Cluster-robust standard errors are in parentheses. All regressions include dummies for non-chemical companies, a missing value for *R&D* and *SALES*, inventor country, applicant country, year of application, and technological field.
[a] Significant at 0.01 level.
[b] Significant at 0.05 level.

patent compared to biotechnology companies. The results in table 4.5 also confirm that the regional variables are not correlated with the probability of developing valuable innovations in traditional chemicals.

Valuable innovations in biotechnology are the outcome of a different model of innovation. Firm characteristics are still correlated with the probability of inventing important patents: the extent of technological specialization in producing biotechnology innovations and the collaboration with external institutions positively and significantly affect the probability of developing valuable innovations. The estimated coefficients of *TECHSPEC* and *TECHCOMP* are 0.264 and −0.196 and are statistically significant. The coefficient of *MAPPL* is 0.539, and it is statistically significant. Once controlling for technological specialization, however, firm R&D intensity does not play any role in inventing big biotechnology innovations.

What really distinguishes the biotechnology model from the traditional chemical model, though, is the importance of the technological characteristics of the regions. The technological environment in which research is carried out influences the probability of developing

technological hits in biotechnology: the estimated coefficient of
REGPATS is positive and significant.[10]

The differential effects of the characteristics of the location in which
the research is conducted in biotechnology compared to traditional
chemicals confirms that the model of innovation leading to important
patents is different according to the characteristics of the technologies
being produced. In biotechnology, a new and science-intensive sector,
big innovations are more likely to be produced by firms that are
technologically specialized in that sector. Knowledge spillovers from
being located in a technologically intensive region help in raising the
probability of inventing hits. Differently, in the traditional branches
of the chemical sector, large established companies that invest heavily
in internal R&D activities have a higher probability of producing big
innovations, with no role for external technological spillovers.

4.6 Conclusions

This chapter has addressed the question of how much of the value of a
patent is affected by the characteristics of the organization in which the
innovation is developed, and how much is explained by the technological characteristics of the European region in which the inventors are
located. To analyze this issue, the empirical work uses a new and
detailed data set composed of 4262 European patents applied for in
1987–96 and classified in five chemical sectors. It combines these data
with information on firm and regional characteristics. The geographical units of analysis are the European regions according to the NUTS
classification at the second and third level of disaggregation.

The results indicate that R&D-intensive firms that engage in large-
scale research projects are more likely to produce important chemical
innovations, while spillovers external to the firm do not add anything
to the probability of developing technological hits. However, when
biotechnology is separated from the traditional chemical sectors (i.e.
materials, organic chemistry, pharmaceuticals, and polymers) the
empirical analysis suggests that different models of innovation exist

---

[10] I checked the robustness of the results by using an ordered probit model, and by
controlling for heteroskedasticity with cluster-robust estimators for firms and
regions alternatively, and with dummy variables for firms and regions (not
shown here). The significance of the estimated coefficients is sometimes smaller,
but the results are consistent with those in table 4.5.

in different sub-sectors of the chemical industry. In the traditional chemical sectors spillovers are still bounded within the same R&D-intensive firm: large R&D projects and the R&D intensity of the firm explain the probability of developing valuable innovations. Firm competencies, and specifically the technological specialization of the firms, are also important in biotechnology. But what really distinguishes the biotechnology model of innovation is that the technological characteristics of the location in which R&D is conducted also positively affect the probability of inventing technological hits.

A previous version of this work (Mariani, 2004) provides additional qualitative evidence of these differences. It traces the geographical network of patent citations in biotechnology and in traditional chemicals, and it shows that the share of local citations (i.e. citations made to patents invented in the same region of the citing patent), although small, is higher in biotechnology than in traditional chemicals. This suggests that geographical proximity is somewhat more important in the former sector compared to the latter. Differently, self-citations are much more frequent in traditional chemicals, confirming that the process of innovation in these sectors relies more on knowledge internal to the firm.

One possible explanation for these results is that biotechnology is in a different stage of the industry life cycle compared to the more mature sectors of the chemical industry. This would be consistent with other contributions showing that there are sectoral differences in spatial clustering, with skilled and R&D-intensive industries benefiting more from co-location and knowledge spillovers compared to more mature industries (Audretsch and Feldman, 1996; Breschi, 1999). Klepper (1996) demonstrates that innovative activity in the early phases of an industry life cycle benefits more from locally bounded knowledge spillovers compared to the mature or declining stages. Finally, Wlash (1984) discusses how knowledge-led factors are particularly important for innovation in the early phases of an industry.

The sources of geographically localized spillovers could be generic knowledge spillovers, or they could stem from human capital mobility or from the foundation of new firms. They might be generated by the initial technological characteristics of specific regions where top universities and high-tech companies locate, making these areas better than others at doing certain types of research activities and affecting the regional numbers of patents and their citation intensities. Whatever the sources, this chapter confirms that there are sectors in which such external factors are important, and others in which they are less relevant for the innovation process. In other words, by studying the process of innovation at the level of the individual inventors, companies, and regions, this work contributes to the analysis of the sources of successful innovation, and confirms the key role played by firm characteristics and the evolution of industries.

Appendix: Regional classification used in the chapter

This chapter uses the NUTS regional classification provided by Eurostat. This classification subdivides the European Union in groups of regions (NUTS1), regions (NUTS2), and provinces (NUTS3). In order to have homogeneity in the size of the regions, I used the NUTS3 regions for Austria, Denmark, Spain, Finland, France, Italy, and Sweden, and the NUTS2 classification for Belgium, Germany, Greece, the Netherlands, and the United Kingdom. Luxemburg, Ireland, and Switzerland were considered as a whole.

The table below lists the NUTS regions used in this work.

Austria

at111 Mittelburgenland	at112 Nordburgenland	at113 Südburgenland
at121 Mostviertel-Eisenwur	at122 Niederösterreich-Süd	at123 Sankt Pölten
at124 Waldviertel	at125 Weinviertel	at126 Wiener Umland/Nordteil
at127 Wiener Umland/Südteil	at13 Vienna	at211 Klagenfurt-Villach
at212 Oberkärnten	at213 Unterkärnten	at221 Graz
at222 Liezen	at223 Östliche Obersteiermark	at224 Oststeiermark
at225 West- und Südsteier.	at226 Westliche Obersteierm.	at311 Innviertel
at312 Linz-Wels	at313 Mühlviertel	at314 Steyr-Kirchdorf
at315 Traunviertel	at321 Lungau	at322 Pinzgau-Pongau
at323 Salzburg und Umgebung	at331 Außerfern	at332 Innsbruck
at333 Osttirol	at334 Tiroler Oberland	at335 Tiroler Unterland
at341 Bludenz-Bregenzer W.	at342 Rheintal-Bodenseegebiet	

Belgium

be1 Région Bruxelles	be21 Antwerpen	be22 Limburg (B)
be23 Oost-Vlaanderen	be24 Vlaams Brabant	be25 West-Vlaanderen
be3 Région Wallonne	be31 Brabant Wallon	be32 Hainaut
be33 Liège	be34 Luxembourg (B)	be35 Namur

Denmark

dk001 København og Frederik.	dk002 Københavns amt	dk003 Frederiksborg amt
dk004 Roskilde amt	dk005 Vestsjællands amt	dk006 Storstrøms amt

dk007 Bornholms amt dk008 Fyns amt dk009 Sønderjyllands amt
dk00a Ribe amt dk00b Vejle amt dk00c Ringkøbing amt
dk00d Århus amt dk00e Viborg amt dk00f Nordjyllands amt

Finland

fi131 Etelä-Savo fi132 Pohjois-Savo fi133 Pohjois-Karjala
fi134 Kainuu fi141 Keski-Suomi fi142 Etelä-Pohjanmaa
fi143 Pohjanmaa fi144 Keski-Pohjanmaa fi151 Pohjois-Pohjanmaa
fi152 Lappi fi161 Uusimaa (maakunta) fi162 Itä-Uusimaa
fi171 Varsinais-Suomi fi172 Satakunta fi173 Kanta-Häme
fi174 Pirkanmaa fi175 Päijät-Häme fi176 Kymenlaakso
fi177 Etelä-Karjala fi2 Åland

France

fr1 Île de France fr211 Ardennes fr212 Aube
fr213 Marne fr214 Haute-Marne fr221 Aisne
fr222 Oise fr223 Somme fr231 Eure
fr232 Seine-Maritime fr241 Cher fr242 Eure-et-Loir
fr243 Indre fr244 Indre-et-Loire fr245 Loir-et-Cher
fr246 Loiret fr251 Calvados fr252 Manche
fr253 Orne fr261 Côte-d'Or fr262 Nièvre
fr263 Saône-et-Loire fr264 Yonne fr301 Nord
fr302 Pas-de-Calais fr411 Meurthe-et-Moselle fr412 Meuse
fr413 Moselle fr414 Vosges fr421 Bas-Rhin

fr422 Haut-Rhin
fr433 Haute-Saône
fr512 Maine-et-Loire
fr515 Vendée
fr523 Ille-et-Vilaine
fr532 Charente-Maritime
fr611 Dordogne
fr614 Lot-et-Garonne
fr622 Aveyron
fr625 Lot
fr628 Tarn-et-Garonne
fr633 Haute-Vienne
fr713 Drôme
fr716 Rhône
fr721 Allier
fr724 Puy-de-Dôme
fr813 Hérault
fr821 Alpes-de-Haute-Prov.
fr824 Bouches-du-Rhône
fr831 Corse-du-Sud

fr431 Doubs
fr434 Territoire de Belfort
fr513 Mayenne
fr521 Côte-du-Nord
fr524 Morbihan
fr533 Deux-Sèvres
fr612 Gironde
fr615 Pyrénées-Atlantiques
fr623 Haute-Garonne
fr626 Hautes-Pyrénées
fr631 Corrèze
fr711 Ain
fr714 Isère
fr717 Savoie
fr722 Cantal
fr811 Aude
fr814 Lozère
fr822 Hautes-Alpes
fr825 Var
fr832 Haute-Corse

fr432 Jura
fr511 Loire-Atlantique
fr514 Sarthe
fr522 Finistère
fr531 Charente
fr534 Vienne
fr613 Landes
fr621 Ariège
fr624 Gers
fr627 Tarn
fr632 Creuse
fr712 Ardèche
fr715 Loire
fr718 Haute-Savoie
fr723 Haute-Loire
fr812 Gard
fr815 Pyrénées-Orientales
fr823 Alpes-Maritimes
fr826 Vaucluse
fr9 French overseas depts.

Germany

de11 Stuttgart
de14 Tübingen

de12 Karlsruhe
de21 Oberbayern

de13 Freiburg
de22 Niederbayern

de23 Oberpfalz	de24 Oberfranken	de25 Mittelfranken
de26 Unterfranken	de27 Schwaben	de3 Berlin
de4 Brandenburg	de5 Bremen	de6 Hamburg
de71 Darmstadt	de72 Gießen	de73 Kassel
de8 Mecklenburg-Vorpom.	de91 Braunschweig	de92 Hannover
de93 Lüneburg	de94 Weser-Ems	dea1 Düsseldorf
dea2 Köln	dea3 Münster	dea4 Detmold
dea5 Arnsberg	deb1 Koblenz	deb2 Trier
deb3 Rheinhessen-Pfalz	dec Saarland	ded1 Chemnitz
ded2 Dresden	ded3 Leipzig	dee1 Dessau
dee2 Halle	dee3 Magdeburg	def Schleswig-Holstein
deg Thüringen		

ie Ireland   Ireland

Italy

it111 Torino	it112 Vercelli	it113 Biella
it114 Verbano-Cusio-Ossola	it115 Novara	it116 Cuneo
it117 Asti	it118 Alessandria	it131 Imperia
it132 Savona	it133 Genova	it134 La Spezia
it201 Varese	it202 Como	it203 Lecco
it204 Sondrio	it205 Milano	it206 Bergamo

it207 Brescia
it208 Pavia
it209 Lodi
it20a Cremona
it20b Mantova
it311 Bolzano-Bozen
it312 Trento
it321 Verona
it322 Vicenza
it323 Belluno
it324 Treviso
it325 Venezia
it326 Padova
it327 Rovigo
it331 Pordenone
it332 Udine
it333 Gorizia
it334 Trieste
it401 Piacenza
it402 Parma
it403 Reggio nell'Emilia
it404 Modena
it405 Bologna
it406 Ferrara
it407 Ravenna
it408 Forlì-Cesena
it409 Rimini
it511 Massa-Carrara
it512 Lucca
it513 Pistoia
it514 Firenze
it515 Prato
it516 Livorno
it517 Pisa
it518 Arezzo
it519 Siena
it51a Grosseto
it521 Perugia
it522 Terni
it531 Pesaro e Urbino
it532 Ancona
it533 Macerata
it534 Ascoli Piceno
it601 Viterbo
it602 Rieti
it603 Roma
it604 Latina
it605 Frosinone
it711 L'Aquila
it712 Teramo
it713 Pescara
it714 Chieti
it72 Molise
it721 Isernia
it722 Campobasso
it801 Caserta
it802 Benevento
it803 Napoli
it804 Avellino
it805 Salerno
it911 Foggia
it912 Bari
it913 Taranto
it914 Brindisi
it915 Lecce
it921 Potenza
it922 Matera
it931 Cosenza
it932 Crotone
it933 Catanzaro
it934 Vibo Valentia
it935 Reggio di Calabria
ita01 Trapani
ita02 Palermo
ita03 Messina
ita04 Agrigento
ita05 Caltanissetta
ita06 Enna

ita07 Catania	ita08 Ragusa	ita09 Siracusa
itb01 Sassari	itb02 Nuoro	itb03 Oristano
itb04 Cagliari		

Luxembourg

lu Luxembourg

Netherlands

nl1 Noord-Nederland	nl11 Groningen	nl12 Friesland
nl13 Drenthe	nl21 Overijssel	nl22 Gelderland
nl23 Flevoland	nl31 Utrecht	nl32 Noord-Holland
nl33 Zuid-Holland	nl34 Zeeland	nl41 Noord-Brabant
nl42 Limburg (NL)		

Spain

es111 La Coruña	es112 Lugo	es113 Orense
es114 Pontevedra	es211 Álava	es212 Guipúzcoa
es213 Vizcaya	es22 Comunidad de Navarra	es23 La Rioja
es241 Huesca	es242 Teruel	es243 Zaragoza
es3 Comunidad de Madrid	es411 Avila	es412 Burgos
es413 León	es414 Palencia	es415 Salamanca
es416 Segovia	es417 Soria	es418 Valladolid
es419 Zamora	es421 Albacete	es422 Ciudad Real
es423 Cuenca	es424 Guadalajara	es425 Toledo

es431 Badajoz	es432 Cáceres	es511 Barcelona
es512 Gerona	es513 Lérida	es514 Tarragona
es521 Alicante	es522 Castellón de la Plana	es523 Valencia
es53 Illes Balears	es611 Almería	es612 Cadiz
es613 Córdoba	es614 Granada	es615 Huelva
es616 Jaén	es617 Malaga	es618 Sevilla
es62 Murcia	es631 Ceuta	es632 Melilla
es701 Las Palmas	es702 Santa Cruz de Tenerife	

Sweden

se011 Stockholms län	se021 Uppsala län	se022 Södermanlands län
se023 Östergötlands län	se024 Örebro län	se025 Västmanlands län
se041 Blekinge län	se044 Skåne län	se061 Värmlands län
se062 Dalarnas län	se063 Gävleborgs län	se071 Västernorrlands län
se072 Jämtlands län	se081 Västerbottens län	se082 Norrbottens län
se091 Jönköpings län	se092 Kronobergs län	se093 Kalmar län
se094 Gotlands län	se0a1 Hallands län	se0a2 Västra Götalands län

Switzerland

ch Switzerland

United Kingdom

ukc1 Tees Valley and Durham	ukc2 Northumb., Tyne and Wear	ukd1 Cumbria
ukd2 Cheshire	ukd3 Greater Manchester	ukd4 Lancashire

ukd5 Merseyide	uke1 East Riding, N. Lincolnsh.	uke2 North Yorkshire
uke3 South Yorkshire	uke4 West Yorkshire	ukf1 Derbyshire, Nottinghamshire
ukf2 Leicester, Rutland, North.	ukf3 Lincolnshire	ukg1 Hereford, Worcester, Warks
ukg2 Shropshire and Staffordshire	ukg3 West Midlands	ukh1 East Anglia
ukh2 Bedfordshire, Hertfordshire	ukh3 Essex	uki1 Inner London
uki2 Outer London	ukj1 Berkshire, Bucks, Oxford.	ukj2 Surrey, East and West Sus.
ukj3 Hampshire and Isle Wight	ukj4 Kent	ukk1 Gloucester, Wilts, N. Som.
ukk2 Dorset and Somerset	ukk3 Cornwall and Isles of Scilly	ukk4 Devon
ukl1 West Wales – The Valleys	ukl2 East Wales	ukm1 North Eastern Scotland
ukm2 Eastern Scotland	ukm3 South Western Scotland	ukm4 Highlands and Islands

Source: Regio database, Eurostat (data as at 1999).

References

Aftalion, F. (1991), *A History of the International Chemical Industry*, University of Pennsylvania Press, Philadelphia.

Audretsch, D. B., and M. P. Feldman (1996), "Knowledge spillovers and the geography of innovation and production," *American Economic Review*, 86 (3), 630–40.

Breschi, S. (1999), "Spatial patterns of innovation: evidence from patent data," in A. Gambardella and F. Malerba (eds.), *The Organisation of Economic Innovation in Europe*, Cambridge University Press, Cambridge, 71–102.

Dosi, G., C. Freeman, R. R. Nelson, G. Silverberg, and L. Soete (1988), *Technical Change and Economic Theory*, Francis Pinter, London.

Griliches, Z. (1990), "Patent statistics as economic indicators: a survey," *Journal of Economic Literature*, 28, 1661–707.

Griliches, Z., A. Pakes, and B. Hall (1987), "The value of patents as indicators of inventive activity," in P. Dasgupta and P. Stoneman (eds.), *Economic Policy and Technological Performance*, Cambridge University Press, Cambridge, 97–124.

Hall, B., A. Jaffe, and M. Trajtenberg (2001), *The NBER Patent Citations Data File: Lessons, Insights and Methodological Tools*, Working Paper no. 8498, National Bureau of Economic Research, Cambridge, MA.

(2005), "Market value and patent citations," *RAND Journal of Economics*, 36, 16–38.

Harhoff, D., F. Scherer, and K. Vopel (2003), "Citations, family size, opposition and the value of patent rights: evidence from Germany," *Research Policy*, 32, 1343–63.

Henderson, R. M., and I. Cockburn (1996), "Scale, scope and spillovers: the determinants of research productivity in drug discovery," *RAND Journal of Economics*, 27 (1), 32–59.

Jaffe, A. (1986), "Technological opportunity and spillovers of R&D: evidence from firms' patents profits and market value," *American Economic Review*, 76 (8), 984–1001.

Jaffe, A., M. Trajtenberg, and R. Henderson (1993), "Geographic localization of knowledge spillovers as evidenced by patent citations," *Quarterly Journal of Economics*, 63 (3), 577–98.

Klepper, S. (1996), "Entry, exit, growth, and innovation over the product life cycle," *American Economic Review*, 86 (4), 562–83.

Klepper, S., and S. Sleeper (2002), *Entry by Spinoffs*, Paper on Economics and Evolution no. 0207, Evolutionary Economics Group, Max Planck Institute of Economics, Jena.

Lanjouw, J. O., and M. Schankerman (2004), "Patent quality and research productivity: measuring innovation with multiple indicators," *Economic Journal*, 114, 441–65.

Mariani, M. (2004), "What determines technological hits? Geography vs. firm competencies," *Research Policy*, 33 (10), 1565–82.

Nelson, R. R., and S. G. Winter (1982), *An Evolutionary Theory of Economic Change*, Harvard University Press, Cambridge, MA.

Orsenigo, L. (1989), *The Emergence of Biotechnology: Institutions and Markets in Industrial Innovations*, Francis Pinter, London.

Powell, W., K. Kaput, and L. Smith-Doerr (1996), "Interorganisational collaboration and the locus of innovation: networks of learning in biotechnology," *Administrative Science Quarterly*, 41 (1), 116–45.

Swann, P., M. Prevezer, and D. Stout (eds.) (1998), *The Dynamics of Industrial Clustering: International Comparisons in Computing and Biotechnology*, Oxford University Press, Oxford.

Teece, D. J., G. Pisano, and A. Shuen (1997), "Dynamic capabilities and strategic management," *Strategic Management Journal*, 18, 509–34.

Trajtenberg, M. (1990), "A penny for your quotes: patent citations and the value of innovation," *RAND Journal of Economics*, 21 (1), 172–87.

Verspagen, B. (1999), "European regional clubs: do they exist and where are they heading? On economic and technological differences between European regions," in J. Adams and F. Pigliaru (eds.), *Economic Growth and Change: National and Regional Patterns of Convergence and Divergence*, Edward Elgar, Cheltenham, 236–56.

Wlash, V. (1984), "Invention and innovation in the chemical industry: demand-pull or discovery-push?," *Research Policy*, 13, 211–34.

Zucker, L. G., M. R. Darby, and J. Armstrong (1998), "Geographically localized knowledge: spillovers or markets?," *Economic Inquiry*, 36, 65–86.

Zucker, L. G., M. R. Darby, and M. Brewer (1998), "Intellectual human capital and the birth of U.S. biotechnology enterprises," *American Economic Review*, 88 (1), 290–306.

5 | Innovation and industry evolution: a comment

LOUIS GALAMBOS

JOSEPH Schumpeter long ago provided those seeking an understanding of modern capitalist economic growth with a dynamic model. The central engine that made capitalism run, Schumpeter explained, was the entrepreneur, the innovator, the unusual businessman who combined capital, labor, and natural resources to develop a new product, a new service, a new source of raw materials or a new form of business organization (Schumpeter, 1934; Stolper, 1994). Schumpeter rounded out his theory in 1942, when he published *Capitalism, Socialism, and Democracy*, which added a political and social superstructure to the original, unchanged model of economic growth.[1]

At that time, most governments were so fully engaged with World War II and, later, with the serious problems of rebuilding their economies that they paid no attention to Schumpeter's ideas, and even the economics profession was largely oblivious to his concepts.[2] In the postwar years Keynesian analysis was on the throne, securely established, it appeared, by the remarkable recovery of the American economy from the Great Depression of the 1930s. Forced to use the powerful tool of large-scale deficit financing, the United States had emerged from the war as the most powerful and successful economy in the world (Abramovitz and David, 2000; Maddison, 1995). The American economy was so successful that there seemed to be little need to reflect on the quality of the nation's entrepreneurship during

---

[1] Schumpeter also rounded out his theory with essays on imperialism and social classes, but (for good reasons) neither subject has received much attention from subsequent generations of scholars (Schumpeter, 1951).

[2] Even though Schumpeter was elected president of the American Economic Association (AEA) in 1948, there was no serious, broad-scale engagement with his ideas in the profession. The same could be said for the work of AEA presidents Kenneth Boulding (1968) and John Kenneth Galbraith (1972). Their elevations to professional eminence had more symbolic than intellectual content.

what came to be called "the American century" (Stein, 1969; Collins, 1981; Zelizer, 1998; Galambos and Pratt, 1988).

But then, all too soon for most Americans, their promised century of economic dominance gave way to intense global competition. Other countries – including the recently defeated Japan and Germany – challenged US leadership in major manufacturing industries. Intense competition quickly pushed entrepreneurship back towards the center stage in public and private planning circles. The economic and political defeats of the 1970s prepared the ground for Thomas Peters and Robert Waterman, Jr., who launched a gold rush of management guides in 1982, when they published their best-seller, *In Search of Excellence*.[3] *In Search* sounded a loud bugle call for innovation and for those "excellent companies" that knew how to "encourage the entrepreneurial spirit among their people" (pp. 200–1).

In the rush to understand global competition, Schumpeter's theory came back into style in the economics profession too. Perchance – but not entirely perchance – Richard Nelson and Sidney Winter published their call for a new economics, an evolutionary, neo-Schumpeterian economics, in 1982. As Nelson and Winter acknowledged, "Schumpeter pointed out the right problem – how to understand economic change – and his vision encompassed many of the important elements of the answer" (p. ix). Taking aim at neoclassical theory, they pointed out the serious limitations of the dominant brands of equilibrium analysis and the assumption of profit maximization in explanations of long-term capitalist growth. While most economists in the United States were not ready to rush into the effort to develop new dynamic models of growth, Nelson and Winter successfully launched a sub-discipline that has gradually expanded its theoretical and empirical base in the past quarter-century (Fagerberg, 2003). In Europe, in particular, evolutionary economics has attracted a substantial scholarly following (Dosi and Nelson, 1994; Nelson, 1995; Hodgson, 1998; Murmann, 2003).

Nelson and Winter – the Pied Pipers of this new school – have guided the sub-discipline towards an increasingly time- and place-particular

---

[3] The literature is enormous and the gold rush is still on today. For examples, see Hammer and Champy (1993), Collins and Porras (1994), Christensen (1997, with subsequent editions in 2000 and 2003), and Tichy and Cohen, (1997, with subsequent editions in 1998 and 2002).

version of the growth process. Nelson has sponsored and edited a comparative study of national systems of innovation (Nelson, 1993; Balzat and Hanusch, 2004). He has followed up with an international comparison of selected high-tech industries, each of which has played an important role in the third industrial revolution, the information age revolution, of our time (Mowery and Nelson, 1999).

Meanwhile, one of Nelson's former students, Peter Murmann, has refined the concept of the coevolution of institutions using a comparative case study of the synthetic dye industry as his empirical base (Murmann, 2003). Synthetic dyes and the organic chemistry that provided the industry's intellectual foundation were crucial components of the second industrial revolution and of the modern pharmaceutical industry. As Murmann demonstrates, the coevolution of firms, public policies, and science-based technologies played a vital role in the international competition between Germany, the United Kingdom, and the United States – a struggle in which Germany was the clear winner, at least until World War II. Murmann's primary focus is upon the context of the firm and the manner in which that context shapes the selection process at the heart of economic evolution.

Evolutionary studies along these lines have clearly drawn economics closer to history – to the history of business and to historically oriented management studies. One of the latest and most impressive of the management books is Clayton Christensen's *The Innovator's Dilemma* (1997), which explores the destructive side of Schumpeter's "creative destruction." Using historical sketches of a number of leading industries (the same sort of industries on which the Mowery and Nelson book focuses), Christensen offers an explanation of why successful firms are likely to lose market share to smaller firms that can bring to fruition less expensive products that, in effect, create a new sub-market. Christensen peers into the firm, studying the managerial process in defeat. Neither Christensen nor the other management scholars use an explicitly evolutionary framework, but their work at the firm level complements evolutionary studies, and the analysis of the organizational capabilities associated with successful and unsuccessful innovation, that is the Schumpeterian process.

So too with business historians, including Alfred Chandler, who has looked even deeper within the large-scale modern corporation for the sources of success (in the United States and Germany) and failure (in Britain) in the second industrial revolution (Chandler, 1962, 1977,

1990, 2001, 2005). Chandler's subject is management, and, in particular, the management of successful firms and their entrepreneurial
investment decisions. While Chandler drew heavily upon Schumpeter
in framing his early books, he eschewed the Schumpeterian aversion to
bureaucracy and the large-scale enterprise that gave rise to this structure of authority in the private sector (Galambos, 2000). Instead,
Chandler has analyzed in depth the rise and triumph of the modern
corporation, the organization capable of making vital three-pronged
investments in mass production, mass distribution, and professional
management. These first-movers and the strategies they selected were,
Chandler maintains, the heart of the growth process that characterized
modern capitalism. Chandler's synthesis leaves us anchored in a time
and place, embedded in a particular historical context – uncertain
about the future of high-tech industry but absolutely certain of its
past accomplishments.

Evolutionary studies have thus evolved on these three fronts: in
economic theory, in managerial studies, and in history. Which brings
us to the three chapters comprising part I of this volume, and their
contributions to the ongoing attempt to understand the process of
economic change in capitalism. All three essays fit comfortably in an
evolutionary framework. All three have important implications for
that framework, for the history of pharmaceuticals, and for important
policy debates involving this industry.

Most explicitly Schumpeterian is Frank Lichtenberg's study of creative destruction in the pharmaceutical industry in the United States.
Lichtenberg advances our understanding of evolution with two major
points, both of which he establishes with great care using the abundant
data from pharmaceutical regulation and the customary econometric
tools. Most readers will, I'm certain, find his analysis of the impact of
pharmaceutical innovation on mortality (between 1970 and 1991) and
on growth in lifetime per capita income of great interest – personally
and professionally. How could even the youngest scholar not be interested in this subject? His data and analysis fit comfortably in a neo-
Schumpeterian economic paradigm, reinforcing our confidence that
entrepreneurship is a positive phenomenon that generates substantial
benefits for most of the citizens of capitalist countries. The importance
of innovation in this case far exceeds the contribution of pharmaceuticals to gross domestic product, although that too is not negligible.
New "priority" drugs have increased both life expectancy and lifetime

income by close to 1 percent per annum – an impressive figure. Savings reaching as high as $27 billion a year provide strong support for heavy government and non-profit investments in the basic sciences under-lying pharmaceutical innovation, as well as the balance between public and private authority in the industry.

Serious questions have been asked about those public and non-profit investments. Concern about the prices of patented pharmaceutical products has generated a critique based on the assumption that, since the public funds the basic research, private companies should not profit from the science coming out of the National Institutes of Health and the university laboratories running on government grants (Angell, 2004; Eisenberg, 1996). The critics seriously understate, I believe, the scientific research that is required to bridge the gap between basic science and successful pharmaceutical innovation (Cockburn and Henderson, 1998). The best evidence in this regard is the high failure rate experienced by all the best and most productive laboratories in pharmaceutical companies. If the transfer of knowledge was simple and inexpensive, the percentage of new chemical entities that actually become drugs would be very high. It isn't (DiMasi et al., 1995).

Less risky but still very important are the science and engineering that go into new drug development after an effective and safe new product is discovered. The best place to study this process is by looking closely at pilot plant operations. In the pilot plant, science, engineering, and operations interact to ensure that a final product can be produced economically and safely. The pilot plant normally supplies enough of the drug to conduct the phase one, two, and three clinical tests that provide data to the firm and to regulatory bodies in the United States and abroad. There is, in short, nothing automatic about the develop-ment of a new pharmaceutical product discovered after basic research points the way towards a fruitful area for science to explore.

The development of the statins is a good case in point. The basic research done at NIH and elsewhere in the 1950s and 1960s by scien-tists such as Earl Stadtman raised the possibility that drugs to alleviate elevated levels of serum cholesterol – which at that time was thought probably to be linked to heart disease – could be discovered. That research focused attention on enzymes and enzyme inhibition as the crucial targets for pharmaceutical innovation. Nonetheless, it was the 1980s before these initial scientific discoveries yielded practical results; the leading pharmaceutical companies in this field had to acquire new scientists, organize new laboratories, and adopt new approaches to the process of drug discovery before the new understanding of lipid biosynthesis resulted in the statins so widely used around the world today (Grabowski and Vernon, 2000). Specialization of function facilitated that process, and distributed the risks and investments associated with innovation in a socially acceptable and efficient manner that contributed significantly to the United States' new position as the world's leading innovator in pharmaceuticals.

So too with the early development of molecular genetics and biotechnology in the United States. In this case, American companies in this field acquired world leadership in part because the boundaries between non-profit research universities and the private sector were so loose and porous. There were, of course, many other factors that contributed to American prominence in biotechnology, but the ease with which scientific entrepreneurs moved back and forth between their university laboratories and their business ventures certainly promoted innovation in the early years, when uncertainty and risk were extremely high and the outlook for the industry and its political/legal environment unclear (Galambos and Sturchio, 1998). That changed as universities began to police their frontiers in search of royalties and licensing fees. With each research university doing what served its own interests best, society's interests in the rapid development of research capabilities in the new field seem to have suffered. While large firms could well afford to pay up and cross the non-profit frontiers, the new setting erected a formidable barrier for small biotechs. In this case, Adam Smith's "invisible hand" seemed not to serve the social interest in the exploration of new fields of research in pharmaceuticals – at least, not where small biotechs were concerned.

The small firms had taken on a new importance in the industry's innovative capability. For the first time in the history of the modern US pharmaceutical industry – that is, from the 1930s on – large pharmaceutical firms were not dominating the process of new drug development (Bottazzi et al., 2001). There were three phases of structural change. During the first era, which started with the beginnings of biotech development and extended to the 1980s, "big pharma" for the most part left small biotech firms to develop the field. During the second era, the large pharmaceutical firms moved into the field by buying the shares of small biotechs or by acquiring the small firms; it appeared for a time that the industry was thus returning to its normal historical structure. But the process of concentration stopped short of that structure and left the industry today in the third phase of structural change, with a combination of small and large firms, with many strategic alliances, and thus a higher degree of specialization of function than existed in the last two major cycles of new drug development (Orsenigo, Pammolli, and Riccaboni, 2001; Henderson, Orsenigo, and Pisano, 1999).

While it is not apparent today that the new structure will be more successful than the previous structures, Lichtenberg's findings forcefully demonstrate how important the industry's innovative performance is to society. These findings also force us to refine and extend the Schumpeterian theory of political economy. For one thing, the impact of prescription drugs on mortality and morbidity indicate why this particular industry has been singled out in most developed economies for extensive controls of efficacy and safety. And, in most, for price controls. Prescription drugs have seemed too important to the general population, political leaders, and intellectuals to leave their discovery, production, or distribution entirely to market forces. Neither "creative destruction" nor market failure has been the rationale for extensive regulation, and indeed there is no indication that either has been a problem in the modern pharmaceutical industry.

From the perspective of the global industry, however, the ensuing regulations and price controls have been asymmetrical, and this has produced a three-tiered global system, with important implications for the future of the industry. In the first tier, the United States has long tolerated a system in which most prices are not controlled. Since intellectual property rights have been respected and enforced by the courts, the patent system gives substantial incentive to innovate. Moreover, public and non-profit institutions have created a research establishment that is conducive to the development of basic science and to the training of the scientists essential to pharmaceutical R&D. This expensive national innovation system accounts for the fact that the preponderance of new prescription drugs has, for some decades, been developed in the United States (Gambardella, Orsenigo, and Pammolli, 2000; Lacetera and Orsenigo, 2001; Achilladelis and Antonakis, 2001).

The second tier of pharmaceutical firms has their headquarters in price-controlled settings in the developed world, and although these nations have surrendered leadership in innovation to the United States they still produce many important new medicines. Price control would appear to be one of the most important reasons for Europe's decline as a source of new drugs. There are, of course, other reasons, including the nature of the universities and research centers in Europe and Asia (Gambardella, Orsenigo, and Pammolli, 2000), but, regardless of the causes, the nations in this second tier benefit from the innovations coming from the United States. Their firms draw upon US science, and the large European pharmaceutical companies have established operations in the United States to ensure that they will stay abreast of new developments in medical science and that they will stay close to the large American market.

The third tier is the developing world, where virtually no innovations are produced and where neither national governments nor individuals can afford to buy many of the most advanced pharmaceuticals. Of late, the world's attention has been focused on the developing nations and in particular Africa, because of the devastating HIV/AIDS pandemic. But other diseases as well are going untreated. Lichtenberg's data are all from the United States, but the addition of developing world figures would certainly support his conclusions.[4] With modern prescription drugs, the US population has experienced a significant increase in average health and life expectancy. Without those drugs, some African countries are experiencing unbelievable declines in average health and life expectancy. So, the impact as well as the costs of innovation are distributed asymmetrically – a relationship that is putting tremendous strains on the industry in recent years.

Lichtenberg's second conclusion – about "creative destruction" – also forces us to make some significant adjustments in the Schumpeterian model. While Schumpeter's focus was on entrepreneurs and their firms and industries, Lichtenberg takes the process down to the level of individual products. In pharmaceuticals the process of creative destruction is, the author finds, selective. When drugs duplicate or improve upon the performance of existing therapies, they cut into the sales of already marketed drugs along Schumpeterian lines. But

---

[4] In parts of sub-Saharan Africa, life expectancy has fallen significantly as a result of the HIV/AIDS pandemic. An estimated 38.5 percent of those aged fifteen to forty-nine are HIV-positive in Botswana, and average life expectancy has fallen from a projected seventy years to thirty-nine years (Ministry of State President, 2003; Caines et al., 2003; UNAIDS, 2003; Distlerath and Macdonald, 2004).

innovative drugs increase the total market in their therapeutic class. There is creation without destruction.

Moreover, when a company is already producing for a therapeutic class, its new drugs cut more deeply into the sales of existing therapies. Apparently, the sales and marketing teams find it difficult to push the new drug while sustaining their efforts on the older therapies. This suggests that, in studying firm capabilities, we may need to shift some of our attention from R&D and production to sales and marketing, functions that are seldom studied in any detail apart from in business programs. The phenomenon is well understood in pharmaceutical firms: you can't get the sales/marketing force to concentrate on two similar therapies in the same class at the same time (Vagelos and Galambos, 2004). Theoretically, they should be able to do this. But, in reality, they find it difficult to do. Lichtenberg's analysis indicates that this is a general phenomenon, not linked to the marketing ability or lack of marketing skills in any one company's sales and marketing force.

Despite these and other problems, the industry has, Lichtenberg demonstrates, been very successful over the long term in sustaining innovation. The transition to modern bureaucratic corporations – a concern of Schumpeter – has not slowed the pace of innovation. The innovations have come in great waves, inspired and fostered by developments in the basic sciences: first in medicinal chemistry, then in virology, next in biochemistry and enzymology, and most recently in molecular genetics. Each wave has followed a cyclical path, and some of the cycles in this case have been very long.

Only in the most recent cycle of innovation have the large pharmaceutical firms yielded ground to small firms and to new entrants. This is the subject that Paul Nightingale and Surya Mahdi explore, and they too help us extend and improve the neo-Schumpeterian framework. Their study probes the cyclical behavior of innovation and recent attempts to standardize and mechanize the process of new drug development. They examine aspects of the process that have aroused great interest in the media, particularly among business journalists. Media enthusiasm has recently helped to produce a bubble, with pharmaceutical and biotech stock prices in the United States going up sharply and then falling. Investors eager for capital gains provided a ready audience for the hype that put biotech companies and their science/business entrepreneurial teams on the front covers of publications such as *Business Week*.

Eventually reality settled in. The progress with drug discovery has, in fact, been very slow in recent years, and the authors help prick the bubble of media hype. As they demonstrate, the technical changes associated with "wet" biology and with *in silico* research have shifted the nature of the research process and encouraged large pharmaceutical firms to create links with an array of small, specialized biotech companies. These small "partners" then bear a great deal of the risk involved in the early stages of research. The small firms provide highly specialized platform technologies as well as new chemical entities for clinical study.

Meanwhile, the efforts to achieve economies of scale and scope in pharmaceutical innovation seem to have hit two significant barriers. One involves the nature of the science that is being applied and its relationship to the drug discovery process. While biotech has already had a dramatic impact on drug discovery, genomics technology has had more of an effect on the development of enzyme inhibitors and receptor blockers than it has on the production of new, large molecule therapies. But it is new therapies that drive the industry, and, in pharmaceuticals, small molecule therapies continue to be the heart of the innovation process as they do not have to be injected at a physician's office, clinic, or hospital.

The other factor limiting the impact of standardization is more complex. The drug discovery process puts a large premium upon talented leadership by first-class scientists and engineers. There is no substitute for talented risk-takers in research and development. They ensure that the work being done in the laboratories is on the tip, out on the research front, as was the case with Sir James Black at ICI, Roy Vagelos at Merck, and other leaders in innovation (Zucker and Darby, 1995; Hara, 2003; Gambardella, 1995). Neither standardization nor automation can substitute for that kind of top-level, scientific leadership. And, even with that kind of leadership, the process will remain problematical and flat periods in research output will be hard to avoid (Landau, Achilladelis, and Scriabine, 1999).

So, we are left to ponder whether Nightingale and Mahdi's strong dose of realism about the "biotechnology revolution" addresses a short-term or a long-term phenomenon in the evolution of this industry. Are we looking at an extended competitive process that is selecting out the standardization and automation of research techniques as an approach to innovation? Or are we seeing in this instance one of the downturns characteristic of this industry in its modern phase? In the former case, we would be forced to conclude that the current plateau in new drug development will extend rather far into the future and leave the industry vulnerable to political attacks and severe global competition as older products become narrow-margin generic commodities. In the latter case, one could expect the buildup of new approaches to drug development to pay off magnificently in the future, as they did in the 1980s and early 1990s. Whatever the outcome, Nightingale and Mahdi have performed a great service by taking us close to the evolution of the biotech research process in this industry, by shifting the focus from journalistic hype to scholarly research, by dissecting the efforts to industrialize R&D, and by analyzing a setting in which large pharmaceutical firms and small biotechs have developed alliances and evolved complementary capabilities.

Myriam Mariani has also significantly improved the evolutionary paradigm by demonstrating that geographical factors play a significant role in stimulating innovation in the early stage of an industry's life cycle. Of particular interest is her comparison of the traditional industrial chemical sector in Europe and the biotech sector. While firm capabilities in R&D are never entirely internal to the firm, the links between the firm and the relevant sciences and other firms are not geographically constrained in the traditional sector by the need for spillovers. The innovation process is more dependent upon economies of scale and scope, along Chandlerian lines. This is especially true for the organic chemicals, materials, and polymers, all of which are low-margin commodities in industrial sectors characterized by intense global price competition (Arora and Gambardella, 1998). This is, doubtless, as true for the US and Asian industrial sectors as it is for Europe, but further research is needed to establish this comparison as firmly as Mariani does for Europe.

As Mariani's study indicates, the traditional European and US pharmaceutical industries – that is, so-called "big pharma" – have evolved in different ways in so far as geographical concentration is concerned, but their evolutionary paths in the biotech era have been similar. Big pharma in the United States has been heavily concentrated in two locations in the east of the country, and spillovers have apparently been involved in the location decisions. Spillovers have probably not been "critical" to the process of innovation, but they seem to have played some role in making locations close to other pharmaceutical firms and to productive research/university complexes desirable (Gambardella, Orsenigo, and Pammolli, 2000). These kinds of network effects have reinforced political advantages (in taxation, for instance) to favor a continuation of a high degree of concentration that dates back to the beginnings of the industry in the United States. In Europe, these conditions did not exist, and national frontiers long imposed a constraint on geographical concentration.

Certainly, biotech is different, however, and is much more highly concentrated in both the European and US cases. Geographically localized knowledge is important in biotech, and much of that knowledge seems to be associated with particular complexes of research institutions and the personnel they have supplied to the industry. This work provides an important addition to the Piore and Sabel literature on industrial regions (1984), underlining the importance of industry-specific characteristics in the selection process as industries and firms evolved. Of particular significance in this case is the life cycle analysis, a conclusion that should be of special interest to those now framing industrial policy for the European Union. Since geographical concentration is most important for firms in the early stages of their industry's cycle, EU planners should probably make an effort to shape subsidies with that in mind.

These three studies thus enrich the evolutionary framework by exploring in detail the innovation process in the pharmaceutical industry. They offer a fruitful combination of history and economics in their explorations of entrepreneurial success and failure. We can see here the selection processes at work at several levels: shaping and reshaping industrial structures, the process of drug discovery, and the market shares of particular innovations. The overall result in this industry is an entrepreneurial process that has been, is today, and should be in future years of great economic importance to society and to all of us as individuals. These studies illustrate the strength of an evolutionary approach to the history and analysis of the growth process in capitalist systems. The process in this case is clearly more creative than destructive, and we should all be grateful for that.

References

Abramovitz, M., and P. A. David (2000), "American macroeconomic growth in the era of knowledge-based progress: the long-run perspective," in

S. L. Engerman and R. E. Gallman (eds.), *The Cambridge Economic History of the United States*, Vol. III, *The Twentieth Century*, Cambridge University Press, New York, 66–84.

Achilladelis, B., and N. Antonakis (2001), "The dynamics of technological innovation: the case of the pharmaceutical industry," *Research Policy*, 30, 535–88.

Angell, M. (2004), "The truth about the drug companies," *New York Review of Books*, 15 July.

Arora, A., and A. Gambardella (1998), "Evolution of industry structure in the chemical industry," in A. Arora, R. Landau, and N. Rosenberg (eds.), *Chemicals and Long-Term Economic Growth: Insights from the Chemical Industry*, Wiley, New York, 379–413.

Balzat, M., and H. Hanusch (2004), "Recent trends in the research on national innovation systems," *Journal of Evolutionary Economics*, 14, 197–210.

Bottazzi, G., G. Dosi, M. Lippi, F. Pammolli, and M. Riccaboni (2001), "Innovation and corporate growth in the evolution of the drug industry," *International Journal of Industrial Organization*, 19 (7), 1161–87.

Caines, K., J. Bataringaya, L. Lush, G. Murindwa, and H. Njie (2003), *Impact of Public–Private Partnerships Addressing Access to Pharmaceuticals in Low-Income Countries: Uganda Pilot Study*, Global Forum for Health Research, Geneva.

Chandler, A. D., Jr. (1962), *Strategy and Structure: Chapters in the History of the Industrial Enterprise*, MIT Press, Cambridge, MA.

(1977), *The Visible Hand: The Managerial Revolution in American Business*, Harvard University Press, Cambridge, MA.

(1990), *Scale and Scope: The Dynamics of Industrial Capitalism*, Harvard University Press, Cambridge, MA.

(2001), *Inventing the Electronic Century: The Epic Story of the Consumer Electronics and Computer Industries*, Free Press, New York.

(2005), *Shaping the Industrial Century: The Remarkable Story of the Evolution of the Modern Chemical and Pharmaceutical Industries*, Harvard Business School Press, Boston.

Christensen, C. M. (1997), *The Innovator's Dilemma: When New Technologies Cause Great Firms to Fail*, Harvard Business School Press, Boston.

Cockburn, I. M., and R. M. Henderson (1998), "Absorptive capacity, coauthoring behavior, and the organization of research in drug discovery," *Journal of Industrial Economics*, 46 (2), 157–81.

Collins, J. C., and J. I. Porras (1994), *Built to Last: Successful Habits of Visionary Companies*, Harper Business, New York.

Collins, R. (1981), *The Business Response to Keynes, 1929–1964*, Columbia University Press, New York.

DiMasi, J. A., R. W. Hansen, H. G. Grabowski, and L. Lasagna (1995), "Research and development costs for new drugs by therapeutic category," *PharmacoEconomics*, 7 (2), 152–69.

Distlerath, L., and G. Macdonald (2004), "The African comprehensive HIV/AIDS partnerships – a new role for multinational corporations in global health policy," *Yale Journal of Health Policy, Law, and Ethics*, 4 (1), 147–55.

Dosi, G., and R. R. Nelson (1994), "An introduction to evolutionary theories in economics," *Journal of Evolutionary Economics*, 4, 153–72.

Eisenberg, R. S. (1996), "Public research and private development: patents and technology transfer in government-sponsored research," *Virginia Law Review*, 82 (8), 1663–727.

Fagerberg, J. (2003), "Schumpeter and the revival of evolutionary economics: an appraisal of the literature," *Journal of Evolutionary Economics*, 13, 125–59.

Galambos, L. (2000), "The U.S. corporate economy in the twentieth century," in S. L. Engerman and R. E. Gallman (eds.), *The Cambridge Economic History of the United States*, Vol. III, *The Twentieth Century*, Cambridge University Press, New York, 927–67.

Galambos, L., and J. Pratt (1988), *The Rise of the Corporate Commonwealth: United States Business and Public Policy in the Twentieth Century*, Basic Books, New York.

Galambos, L., and J. L. Sturchio (1998), "Pharmaceutical firms and the transition to biotechnology: a study in strategic innovation," *Business History Review*, 72 (Summer), 250–78.

Gambardella, A. (1995), *Science and Innovation: The US Pharmaceutical Industry During the 1980s*, Cambridge University Press, Cambridge.

Gambardella, A., L. Orsenigo, and F. Pammolli (2000), "Benchmarking the competitiveness of the European pharmaceutical industry," European Commission, Brussels.

Grabowski, H. G., and J. Vernon (2000), "The determinants of pharmaceutical research and development expenditures," *Journal of Evolutionary Economics*, 10, 201–15.

Hammer, M., and J. Champy (1993), *Reengineering the Corporation: A Manifesto for Business Revolution*, Harper Business, New York.

Hara, T. (2003), *Innovation in the Pharmaceutical Industry: The Process of Drug Discovery and Development*, Edward Elgar, Cheltenham.

Henderson, R. M., L. Orsenigo, and G. Pisano (1999), "The pharmaceutical industry and the revolution in molecular biology: exploring the interactions between scientific, institutional, and organizational change," in D. C. Mowery and R. R. Nelson (eds.), *Sources of Industrial Leadership: Studies of Seven Industries*, Cambridge University Press, Cambridge, 267–311.

Hodgson, G. M. (1998), *The Foundations of Evolutionary Economics, 1890–1973*, Edward Elgar, Northhampton, MA.

Lacetera, N., and L. Orsenigo (2001), "*Political Regimes, Technological Regimes and Innovation in the Evolution of the Pharmaceutical Industry in the USA and in Europe*," paper presented at Conference on Evolutionary Economics, Johns Hopkins University, Baltimore, 30–31 March.

Landau, R., B. Achilladelis, and A. Scriabine (1999), *Pharmaceutical Innovation: Revolutionizing Human Health*, Chemical Heritage Press, Philadelphia.

Maddison, A. (1995), *Monitoring the World Economy, 1820–1992*, Development Centre, Organisation for Economic Co-operation and Development, Paris.

Ministry of State President, National AIDS Coordinating Agency (2003), *Status of the 2002 National Response to the UNGASS Declaration of Commitment on HIV/AIDS*, Country Report, Government of Botswana, Gaborone.

Mowery, D. C., and R. R. Nelson (eds.) (1999), *Sources of Industrial Leadership: Studies of Seven Industries*, Cambridge University Press, Cambridge.

Murmann, J. P. (2003), *Knowledge and Competitive Advantage: The Coevolution of Firms, Technology, and National Institutions*, Cambridge University Press, New York.

Nelson, R. R. (ed.) (1993), *National Innovation Systems: A Comparative Analysis*, Oxford University Press, New York.

(1995), "Recent evolutionary theorizing about economic change," *Journal of Economic Literature*, 33, 48–90.

Nelson, R. R., and S. G. Winter (1982), *An Evolutionary Theory of Economic Change*, Harvard University Press, Cambridge, MA.

Orsenigo, L., F. Pammolli, and M. Riccaboni (2001), "Technological change and network dynamics: lessons from the pharmaceutical industry," *Research Policy*, 30, 485–508.

Peters, T. J., and R. H. Waterman, Jr. (1982), *In Search of Excellence: Lessons from America's Best-Run Companies*, Harper, New York.

Piore, M. J., and C. F. Sabel (1984), *The Second Industrial Divide: Possibilities for Prosperity*, Basic Books, New York.

Schumpeter, J. A. (1934; Galaxy edn., 1961), *The Theory of Economic Development: An Inquiry into Profits, Capital, Credit, Interest, and the Business Cycle*, Oxford University Press, New York.

(1942; 3rd edn., 1950) *Capitalism, Socialism, and Democracy*, Harper, New York.

(1951; Meridian edn., 1958), *Imperialism; Social Classes*, Meridian Books, New York.

Stein, H. (1969), *The Fiscal Revolution in America*, University of Chicago Press, Chicago.

Stolper, W. F. (1994), *Joseph Alois Schumpeter: The Public Life of a Private Man*, Princeton University Press, Princeton, NJ.

Tichy, N. M., and E. Cohen (1997), *The Leadership Engine: How Winning Companies Build Leaders at Every Level*, Harper Collins, New York.

UNAIDS (2003), *Accelerating Action Against AIDS in Africa*, Joint United Nations Programme on HIV/AIDS, Geneva.

Vagelos, R., and L. Galambos (2004), *Medicine, Science and Merck*, Cambridge University Press, Cambridge.

Zelizer, J. E. (1998), *Taxing America: Wilbur D. Mills, Congress, and the State, 1945–1975*, Cambridge University Press, New York.

Zucker, L. G., and M. R. Darby (1995), *Virtuous Circles of Productivity: Star Bioscientists and the Institutional Transformation of Industry*, Working Paper no. 5342, National Bureau of Economic Research, Cambridge, MA.

PART II

Firm growth and market structure

6 | Heterogeneity and firm growth in the pharmaceutical industry

ELENA CEFIS, MATTEO CICCARELLI,
AND LUIGI ORSENIGO

6.1 Introduction

In this chapter we investigate some properties of the patterns of firm growth in the pharmaceutical industry. The issue of firm growth is interesting as such, but there are two particular (related) reasons why we believe it to be particularly intriguing in the case of pharmaceuticals.

First, surprisingly, very little detailed statistical evidence on the subject is actually available. Second, it is worth exploring how the peculiar underlying patterns of innovation and competition that characterize pharmaceuticals translate into firms' growth.

The main features of the structure of the pharmaceutical industry can be summarized as follows. It is a highly innovative, science-based, R&D- and marketing-intensive industry. However, the industry is characterized by quite low levels of concentration, both at the aggregate level and in the individual sub-markets, such as, for example, cardiovascular, diuretics, and tranquilizers. Similarly, the international pharmaceutical industry is characterized by a significant heterogeneity in terms of firm size, strategic orientations, and innovative capabilities. The "innovative core" of the industry has traditionally been composed of a relatively small group of large corporations ("big pharma"), which includes many of the early entrants from the turn of the twentieth century. These firms are located in the countries that have dominated the industry ever since

Financial support from the European Union (the ESSY project), from the EPRIS project (the European Pharmaceutical Regulation and Innovation System), and from the Italian Ministry of University and Research (grant no. 9913443984, and ex 60% grant no. 60CEFI03, Department of Economics, University of Bergamo) is gratefully acknowledged. Part of Ciccarelli's research was undertaken when he was at the University of Alicante and benefited from funding by DGIMCYT (BEC2002-03097) and Generalitat Valenciana (CTIDIB/2002/175). The views expressed in this chapter are exclusively those of the authors and not those of the European Central Bank. The usual disclaimer applies.

its inception, namely Germany, Switzerland, the United Kingdom, and the United States. These firms maintained an innovation-oriented strategy over time, with both radical product innovations and incremental product and process innovations. A second group of firms – either located in these countries but more frequently found elsewhere, such as in Continental Europe and Japan – are typically much smaller in size and are specialized instead in imitation, minor innovations, and marketing in domestic markets.

The international industrial structure was rather stable up to the mid-1970s, with very few new entrants and a relatively stable and impenetrable "core" of successful innovative firms – at least, until the advent of biotechnology and, more generally, molecular biology (Henderson, Orsenigo, and Pisano, 1999). Such low levels of concentration, as well as the coexistence of a persistent group of giant corporations with a vast fringe of smaller companies, has traditionally been explained on the ground of the properties of the innovation process and – more generally – of competition in pharmaceuticals.

Extremely high costs for the R&D and marketing of new drugs set a lower limit to the size of the core, innovative firms (Sutton, 1998). The process of mergers and acquisitions (M&As) signals these pressures and tends to increase firms' size and industry concentration. However, these tendencies are partly compensated for by a number of different factors.

First, innovation has to be considered as quite a rare event. Only a very small fraction of the compounds that are tested for potential therapeutic effects actually reaches the market. Moreover, innovative activities have often been described as displaying quite low levels of cumulativeness – i.e. it is difficult to use the knowledge accumulated in a particular research project in subsequent innovative efforts, even at the level of narrowly defined chemical families (Sutton, 1998). In this respect, Sutton suggests that pharmaceutical R&D shares some key features of a "pure lottery" model.[1] In addition, pharmaceuticals is

---

[1] This characterization is likely to apply relatively well to the "random screening" era of pharmaceutical R&D, when – in the absence of a detailed knowledge of the biological bases of specific diseases – natural and chemically derived compounds were randomly screened in test tube experiments and laboratory animals for potential therapeutic activity. Thousands of compounds might be subjected to multiple screens before a promising substance could be identified and serendipity played a key role. Recent studies by Henderson and Cockurn (1996) and, albeit less directly, Bottazzi, Cefis, and Dosi (2002) suggest that biotechnology,

actually constituted by a collection of fragmented, independent markets, where even marketing expenditures do not exhibit pronounced economies of scope.

Second, a new drug generates – after its arrival – extremely high rates of market growth, which are sustained by the temporary monopolies afforded by patents. Indeed, the pharmaceutical industry is known to be one of the few industries where patents are unambiguously rated as key mechanisms for the private appropriability of the returns from innovation (Klevorick et al., 1999). These features of the innovative process entail, in turn, a highly skewed distribution of the returns on innovation and of product market sizes, as well as of the intra-firm distribution of sales across products. As a result, a few "blockbusters" dominated the product range of all major firms (Matraves, 1999; Sutton, 1998).

However, entirely new products are only a part of the relevant mechanisms of competition. "Inventing around" existing molecules, or introducing new combinations among them, or new ways of delivering them, etc. constitute a major component of firms' innovative activities more broadly defined. Thus, while market competition centers around new product introductions, firms also compete through incremental advances over time, as well as imitation and generic competition after patent expiration and adaptation to the regulatory policies and conditions of domestic markets. These activities allow a large "fringe" of firms to thrive through commodity production and the development of – in most cases – licensed products.

How do these features of the industry map into firm growth processes? On the one hand, the existence of a highly skewed distribution of firms' sizes, coupled with the properties of the "technological regime" (Breschi, Malerba, and Orsenigo, 2000), would seem to lend support – at first glance – to the plausibility of the assumption that firms' growth is essentially random, being largely driven by rare and weakly cumulative innovative events.

As is well known, this assumption is the essence of the so-called "law of proportionate effects," otherwise known as Gibrat's law (Gibrat, 1931). The law assumes that a firm's size follows a random walk, and hence that the growth of the firm is erratic, being driven by small idiosyncratic shocks.[2] Thus, no stable or predictable differences in

molecular biology, and science-driven research might have strengthened economies of scope in pharmaceutical R&D.

[2] For a recent discussion of this topic, see Sutton (1997) and Bottazzi et al. (2000).

growth exist either in the short or in the long run, and, as a consequence, there is no convergence in firms' sizes within industries,
thereby generating a skewed distribution of firm sizes.

Gibrat's law has been extensively studied, both at the theoretical and
at the empirical level. Theoretically, it contrasts with most fundamental
theories of firm growth, ranging from standard models of convergence
to an optimal size to models where heterogeneous firms, facing idiosyncratic sources of uncertainty and discrete events, are subject to market
selection, so that the most efficient firms grow while the others shrink
and eventually leave the market (see, for example, Geroski, 1999, for a
discussion). Indeed, many recent theoretical models of firm's growth and
industry evolution imply several violations of standard Gibrat-type
processes (e.g. Jovanovic, 1982; Ericson and Pakes, 1995; Dosi et al.,
1995; Pakes and Ericson, 1998; Winter, Kaniovski, and Dosi, 2000).
Moreover, Gibrat's law is at odds with other observed empirical
phenomena, such as the persistence of heterogeneity in some firms'
characteristics and measures of performance (e.g. profits, productivity,
and – more controversially – innovation) (Baily and Chakrabarty, 1985;
Mueller, 1990; Geroski, Machin, and Van Reenen, 1993; Cefis and
Orsenigo, 2001).

Empirically, several violations have been observed with regard to the
standard version of Gibrat's law. The traditional methodology
employed for testing Gibrat's law starts with an equation having the
following form:

$$\ln S_{it} = \beta_0 + \beta \ln S_{it-1} + u_{it}$$

where S_{it} is the size of firm i at time t, and u_{it} is an identically independently distributed (i.i.d.) shock.

Gibrat's law would be confirmed if the null hypothesis, $H_o : \beta = 1$,
could not be rejected versus the alternative, $H_1 : \beta < 1$. Empirical
results are far from being uncontroversial. Some studies confirm
the view that firm size does indeed follow a random walk ($\beta = 1$), at
least as far as large firms are concerned. However, it is now considered
a "stylized fact" (Klepper and Thompson, 2004; Klette and Kortum,
2002) that the growth of firm size follows a random walk only as
far as large firms are concerned, whilst small and/or young firms
tend to grow more rapidly on average – but with a higher variance –
than large or old ones ($\beta < 1$). In other words, according to these

empirical findings, in the long run firms would tend to converge to an ideal size (a steady state), common to all firms.[3]

With reference to the specific case of pharmaceuticals, a recent series of studies by Bottazzi et al. (2002; chapter 7 of this volume) do indeed find substantial violations to Gibrat's law. However, these investigations – although using sophisticated and non-standard statistical techniques – still confirm the absence of any dependence of growth on firms' size.

In this work we explore further these issues by testing Gibrat's law by means of a Bayesian estimation approach. Our findings are that: (i) on average, firm growth rates are not erratic over time; (ii) estimated steady states differ across units, and firm sizes and growth rates do not converge to a common limiting distribution; (iii) initial conditions are important determinants of the estimated distribution of steady states, but there is weak evidence of mean reversion (i.e. initially larger firms do not grow more slowly than smaller firms – in other words, differences in growth rates and in the steady-state size are firm-specific, rather than size-specific); (iv) differences in firm size and in growth rates are likely to reduce at a very slow rate but they do not seem to disappear over time – i.e. they persist; and (v) the specified model provides an adequate fit to the data, and results do not change under plausible alternative models – in other words, they are robust to more general families of prior information.

These different results are to be explained by considering that previous studies have typically used cross-section regressions or short-panel econometric techniques with homogeneity in the parameters across units and over time. Both approaches are problematic. The former ignores the information contained in unit-specific time variation in growth rates. The latter forces the parameters to be the same across individuals, thus pooling possibly heterogeneous units as if their data were generated by the same process, even if it considers information available for all periods and all cross-sectional units. A fixed-effect bias may emerge as a

---

[3] Differentials in growth rates have been explained with respect to firms' age (Mata, 1994; Dunne, Roberts, and Samuelson, 1989), firms' size (Harhoff, Stahl, and Woywode, 1998; Hart and Oulton, 1996; Hall, 1987; Evans, 1987a, 1987b), or both (Farinas and Moreno, 2000). Further observed violations concern the existence of autocorrelation in growth rates and the fall of the variance of growth rates with size.

consequence of this assumption, as is well known in the panel data literature (see, for example, Hsiao, Pesaran, and Tahmiscioglu, 1999).

Thus, most results obtained previously could be econometrically biased. Even though they focus attention on important aspects of the data, they are based on methodologies that force units to be homogeneous and hence that exploit only one side of the information contained in a panel data set. As we shall discuss, these methodologies are inappropriate for studying Gibrat's law and its implications.[4]

Our work proposes instead a general framework the main characteristic of which is plain heterogeneity – i.e. heterogeneity in the intercept and in the slope of the statistical model, which has the twofold useful feature of exploiting all information contained in a panel data set and nesting previous studies on the same topic. Given that there are too many parameters to estimate relative to the number of time series observations for each cross-sectional unit, our point of view is Bayesian. This means that a flexible prior on the parameters must be combined with information contained in the data (likelihood) to obtain posterior estimates. As will be discussed in section 6.2, the procedure solves the small sample problem by estimating separately using only the observations on unit i, since Bayesian estimates are exact regardless of the sample size, and, at the same time, it does not require the stringent assumption of equality of the coefficients across units. The chosen prior shares features with those of Lindley and Smith (1972), Chib and Greenberg (1995, 1996), and Hsiao, Pesaran, and Tahmiscioglu (1999), and it is specified to have a hierarchical structure, which allows for various degrees of ignorance in the researcher's information about the parameters. Both the econometric argument and the Bayesian technique are also related to the procedure

---

[4] Notice that there is not much difference between the cross-section and the panel data approaches if in the latter we force the parameters to be exactly equal in all units – a case that should be formally tested instead. Moreover, the present study shares part of the view expressed in Goddard, Wilson, and Blandon (2002). The authors compare the properties of the standard cross-sectional test of Gibrat's law with those of three alternative panel unit root tests, by means of simulated and real data on Japanese manufacturing. They conclude that Gibrat's law should be rejected, based on the idea that the cross-sectional procedure produces biased parameter estimates and the test suffers from a loss of power if there are heterogeneous individual firm effects, while suitably designed panel tests avoid these difficulties. We share the general view that a panel data approach is a better vehicle to investigate Gibrat's argument than a cross-section, but we argue that it is not just a matter of testing procedure that makes the difference, if the panel data model is not well suited.

adopted by Canova and Marcet (1995), who studied the issue of convergence of per capita income across economic regions.

The chapter is structured as follows. Section 6.2 discusses the statistical model. Section 6.3 describes the data and comments on the estimation results. In section 6.4 we check the robustness of the results, using a different data set and different prior assumptions. Section 6.5 concludes, while details of the estimation and testing techniques are given in the appendix.

6.2 The econometrics

6.2.1 Model specification

Given that our observations are collected across units and time, the evolution of size for all units is determined by a double-indexed stochastic process $\{S_{it}\}$, where $i \in I$ indexes firms, $t = 0, 1, \ldots$ indexes time, and I is the set of the first n integers. Following Sutton (1997), if ε_{it} is a random variable denoting the proportionate rate of growth between period $t - 1$ and t for firm i, then

$$S_{it} - S_{it-1} = \varepsilon_{it} S_{it-1}$$

and

$$S_{it} = (1 + \varepsilon_{it}) S_{it-1} = S_{i0}(1 + \varepsilon_{i1})(1 + \varepsilon_{i2}) \ldots (1 + \varepsilon_{it})$$

In a short period of time, ε_{it} can be regarded as small and the approximation $\ln(1 + \varepsilon_{it}) = \varepsilon_{it}$ can be justified. Hence, taking logs, we have

$$\ln S_{it} \simeq \ln S_{i0} + \sum_{t=1}^{T} \varepsilon_{it}$$

If the increments ε_{it} are independently distributed with mean β_0 and variance σ^2, then $\ln S_{it}$ follows a random walk and the limiting distribution of S_{it} is log-normal.

Hence, to test Gibrat's law, the vast majority of the previous literature has used the following logarithmic specification:

$$\ln S_{it} = \beta_0 + \beta \ln S_{it-1} + u_{it} \tag{6.1}$$

where S_{it} is the size of firm i at time t, and u_{it} is a random variable that satisfies

$$E(u_{it}|S_{it-s}, s > 0) = 0$$

$$E(u_{it}u_{j\tau}|S_{it-s}, s > 0) = \begin{cases} \sigma^2 & i = j, t = \tau \\ 0 & \text{otherwise} \end{cases}$$

Gibrat's law is confirmed if the null hypothesis $\beta = 1$ is not rejected by the data.

An equivalent specification used by the literature and based directly on corporate growth rates is

$$\ln \frac{S_{it}}{S_{it-1}} = \beta_0 + \beta_1 \ln S_{it-1} + u_{it}$$

where, clearly, $\beta_1 = \beta - 1$. In this case Gibrat's law is confirmed if data do not reject the null, $\beta_1 = 0$.

In this work we follow a similar specification. The main difference is that we study the behavior of the (log of) each unit's size relative to the average – i.e. of the variable $g_{it} = \ln(S_{it}/\bar{S}_t)$, where $\bar{S}_t$ represents the average size over all units at each time t. The use of the proportion of size g_{it} as our basic variable, instead of (the log of) plain size S_{it}, alleviates problems of serial and residual correlation, in that possible common shocks α are removed by the normalization.

Therefore, we specify the following statistical model:

$$g_{it} = \alpha_i + \rho_i g_{it-1} + \eta_{it} \tag{6.2}$$

where the random variables η_{it} are assumed normally and identically distributed, with mean zero and variance σ_i^2, and are uncorrelated across units and over time.

Notice that this specification is more general than either a simple cross-sectional analysis or a homogeneous dynamic panel data model. On the one hand, equation (6.2) allows for a more efficient use of the information contained in the time dimension of the panel, since the parameters of the model are estimated by using the firm sizes for all t's. On the other hand, we are not forcing the parameters to be the same across units, as is usually assumed in the empirical literature on Gibrat's law. The reason for considering different intercepts for each unit is simply to avoid the well-known fixed-effect bias due to lack of consideration of the heterogeneity typically found in *micro*-data. Moreover, we think that, even with a fixed-effect specification, the assumption of a common slope is too restrictive. If units are heterogeneous in the slopes but the statistical model does not take this feature into account, then bias and inconsistency problems arise.[5] It is not difficult to show that the neglect of coefficient heterogeneity in dynamic models creates correlation between the regressors and the error term, and causes serially correlated disturbances (Pesaran and Smith, 1995; Hsiao, Pesaran, and Tahmiscioglu, 1999). Hence, any traditional estimator is biased and inconsistent, the degree of inconsistency being a function of the degree of coefficient heterogeneity and the extent of serial correlation in the regressors.

Our main point in this chapter is that the traditional results on Gibrat's law may be econometrically biased in line with the lack of consideration of possible heterogeneity in the data. This argument motivates the choice of the model specification (6.2), which is flexible enough to test formally the restrictions that previous studies impose on the data – i.e. $\alpha_i = \alpha$, and $\rho_i = \rho$, $\forall i$.[6]

Provided that we can estimate the short-run parameters for each unit, we are also able to estimate the steady states directly. Therefore, we can test Gibrat's law both for each single firm and on average, and we can separately examine three further implications of the law. First, we are able to estimate the speed of adjustment $(1 - \rho_i)$ of each unit to its own steady state $(\alpha_i/(1 - \rho_i))$, a question related to the mean reversion argument and the decrease in the variance of the firm size over time. Second, we can verify whether steady states are all equal across units. Finally, if steady states are not common, the model specification can easily be used to test whether these differences across firms are transitory or permanent – i.e. whether there is persistence in size differences.

Given that there are too many parameters relative to the number of time series observations for each cross-sectional unit, we impose a Bayesian prior on the parameters to be combined with information contained in the data (likelihood) to obtain posterior estimates. The procedure solves the small sample problem encountered by estimating

---

[5] Notice that the opposite is not true, in that a well-specified heterogeneous model nests a model without such heterogeneity.

[6] In fact, other studies (e.g. Bottazzi et al., 2000) sometimes estimate $g_{it} = \rho g_{it-1} + u_{it}$, forgetting the specific effect α_i, as if the expected proportionate rate of growth were zero. Even with the kind of normalization used in our work, this is not a correct approach. If $u_{it} = \alpha_i + \eta_{it}$, then ordinary least squares (OLS) estimates are inconsistent, because $E(g_{it}\alpha_i) \neq 0$ for each t, and inefficient, given that the u_{it}'s are serially correlated.

separately using only the observations on unit i, since Bayesian esti-mates are exact regardless of the sample size, and, at the same time, it does not require the stringent assumption of equality of the coefficients across units.[7]

Let $\theta_i = (\alpha_i, \rho_i)'$. Equation (6.2) can then be written in a more com-pact form as

$$g_{it} = X'_{it}\theta_i + \varepsilon_{it} \tag{6.3}$$

where $X_{it} = (1, g_{it-1})'$. Though allowing heterogeneity, the imposed prior distribution assumes that the intercept and the slope of the model do not differ too much across units. Concretely, the population struc-ture is modeled as

$$\theta_i \sim N(\bar{\theta}, \Sigma_\theta) \tag{6.4}$$

where $\bar{\theta}$ and Σ_θ are common to all individuals. In other words, we assume that the parameters of each cross-sectional unit come from a distribution that is common to all firms. The variance of this distribu-tion then determines the degree of uncertainty that the researcher has about the mean. Notice that this assumption is more general than forcing the parameters to be the same for each unit. This limiting case can be obtained by imposing $\Sigma_\theta = 0$. Our opinion is that these restric-tions might be formally tested instead of simply imposed.

Notice also that (6.4) is just a prior assumption, and must then be combined with the data to obtain posterior estimates. If the data are sufficiently informative, the posterior need not be the same as the prior, as it will be clear from the estimation results.

For the prior information to be complete, we assume a normal–Wishart–gamma structure:

$$\bar{\theta} \sim N(\mu, C) \tag{6.5}$$

$$\Sigma_\theta^{-1} \sim W(s_o, S_o^{-1}) \tag{6.6}$$

$$\sigma_i^2 \sim IG\left(\frac{v^2}{2}, \frac{v^2\delta^2}{2}\right) \tag{6.7}$$

---

[7] See Canova and Marcet (1995) and Hsiao, Pesaran, and Tahmiscioglu (1999) for a more detailed discussion on this point.

where the notation $\Sigma_\theta^{-1} \sim W(s_o, S_o^{-1})$ means that the matrix Σ_θ^{-1} is distributed as a Wishart with scale S_o^{-1} and degrees of freedom s_o, and $\sigma_i^2 \sim IG\left(\frac{\nu^2}{2}, \frac{\nu^2\delta^2}{2}\right)$ denotes an inverse gamma distribution with shape ν and scale $\nu\delta$. The hyperparameters μ, C, s_o, S_o, ν, and δ are assumed known. Independence is also assumed throughout.

The entire specification (6.3) through (6.7) is standard in Bayesian literature (e.g. Gelfand et al., 1990) and has the advantage of being sufficiently flexible to answer the kinds of question posed in this work. In particular, notice that this specification easily nests the pure cross-section, the fixed effect, and the pure time series models. In fact, setting $\Sigma_\theta = 0$ is equivalent to imposing equality of coefficients across firms. If the prior variance–covariance matrix is zero, no cross-sectional heterogeneity is present and the parameter vector θ_i is pooled towards the common cross-sectional mean $\bar{\theta}$. This setting, therefore, would replicate the cross-sectional and the homogeneous panel data analysis, while the fixed-effect specification is obtained by forcing only the variance of ρ_i, σ_ρ^2, to be zero. In discussing the empirical results, we will call these two specifications the *pool OLS* and the *fixed effect* respectively. In contrast, if $\Sigma_\theta \to \infty$, the prior information is diffuse. This means that the degree of uncertainty about the mean is infinite, and hence that estimated parameters for different firms are similar to those obtained by applying OLS to (6.3) equation by equation. In other words, when $\Sigma_\theta \to \infty$ only the time series properties of each g_{it} are used, and the estimation results will resemble the *mean group* estimator proposed by Pesaran and Smith (1995). Finally, when Σ_θ is a finite, positive definite matrix, the coefficients are estimated using information contained both in the cross-section and in the time series dimensions of the panel. In the empirical section we will denote the estimation results relative to this setting as *Bayes*. In a recent study, Hsiao, Pesaran and Tahmiscioglu (1999) establish the asymptotic equivalence of the Bayes estimator and the mean group estimator, and show that the Bayes estimator is asymptotically normal for large n (the number of units) and large T (the number of time periods) as long as $\sqrt{n}/T \to 0$ as both n and $T \to \infty$. They also show that the Bayes estimator has better sampling properties than other consistent estimators for both small and moderate T samples. Concretely, the bias of the Bayes estimator never exceeds 10 percent for $T = 5$, while for $T = 20$ it is always less than 2 percent.

Posterior inference can then be conducted using the posterior distributions of the parameters of interest. Specifically, in discussing the validity of Gibrat's law, we will be interested in examining whether the mean coefficient $\bar{\rho}$ is equal to one, as well as finding out how large the percentage of firms is for which the null ($\rho_i = 1$) is not rejected.

In discussing the implications of the law, we also need to verify the null hypothesis $\alpha_i/(1 - \rho_i) = \alpha_j/(1 - \rho_j) \; \forall i, j$ – i.e. the null that the steady states are the same across units versus the alternative, that they are different. The rejection of the null $\alpha_i/(1 - \rho_i) = \alpha_j/(1 - \rho_j) \; \forall i, j$ provides evidence in favor of a lack of unconditional convergence to a common steady state. The final question, then, would be as to whether the initial differences in size are going to persist as time goes by. This issue is examined by running a cross-sectional regression of the form

$$\widehat{SS}_i = c + bg_{i0} + \omega_i \tag{6.8}$$

where $\widehat{SS}_i$ is the mean of the posterior distribution of the steady state for unit i, and g_{i0} is its initial (scaled) size. A positive b would indicate that the distribution of initial size matters for the cross-sectional distributions of steady states, while the magnitude of this estimate will provide an indication of how persistent these differences are.

6.2.2 Estimation and testing

The posterior distributions of the parameters of interest are obtained, as already remarked and as detailed in the appendix, by combining the prior information with the likelihood. More formally, if $\psi = (\theta_i, \bar{\theta}, \Sigma_\theta, \sigma_i^2)'$ is the vector of unknown parameters, and y represents the data, the Bayes rule

$$p(\psi \,|\, y) \propto p(\psi) \, l(y \,|\, \psi)$$

can be applied to obtain the joint posterior distribution of $\psi = (\theta_i, \bar{\theta}, \Sigma_\theta, \sigma_i^2)$. The marginal distribution of each element of ψ can be derived by integrating out the others. Given the complexity of our specification, this integration is analytically intractable and must rely on a numerical method. We use the Gibbs sampling. The ergodic mean of the marginal posterior distributions obtained from the Gibbs sampler are taken as our point estimate.

The null hypotheses $\bar{\rho} = 1, \rho_i = 1, \rho_i = \rho_j$, and $SS_i = SS_j, \forall i, j$ are verified by calculating the (log of) the posterior odds (PO) ratio, as in Leamer (1978) and Sims (1988). The null is not rejected whenever the computed statistics are positive. We also compute the largest prior probability to attach on the alternative in order for the data not to reject the null. This statistic, which we will call ω^*, represents the degree of confidence the researcher should attach to the null so that the data do not overturn his/her prior beliefs. Small values of this measure are the signal that the researcher should put more weight on the null not to reject it, or, equivalently, that the null is unlikely.

In order to choose a preferred model (e.g. $\Sigma = 0$ vs. $\Sigma > 0$), we compare the posterior predictive power of the two models instead of relying on PO ratios. The reason is that the PO ratio may not provide good inferences in the case of comparing the no pooling ($0 < \Sigma < \infty$) and the complete pooling ($\Sigma = 0$) models. The argument can be sketched in a simple way. Let $y_i = (g_{i1}, \ldots, g_{iT})'$. Assuming known variances σ_i^2, the two models can be summarized as

$$0 < \Sigma < \infty : p(y \mid \theta_1, \ldots \theta_n) = \prod_{i=1}^{n} N(y_i \mid \theta_i, \sigma_i^2), \quad p(\theta_1, \ldots, \theta_n) \propto 1$$

$$\Sigma = 0 : p(y \mid \theta_1, \ldots \theta_n) = \prod_{i=1}^{n} N(y_i \mid \theta_i, \sigma_i^2), \quad \theta_1 = \cdots = \theta_n, \quad p(\theta) \propto 1$$

If we use the PO ratio to choose or average among these models, the ratio is not defined because the prior distributions are improper, and the ratio of density functions is 0/0. Therefore, we should assign either proper prior distributions or improper prior carefully constructed as limits of proper prior. In both cases results are unsatisfactory, as shown by Gelman et al. (1995, pp. 176–7).

The details of the estimation techniques can be found in the appendix.

6.3 Empirical results

In this section we describe the data sets used in the analysis and present the empirical results. The latter are shown in figures 6.1 to 6.9 and tables 6.1 to 6.3.

6.3.1 The data

The issue analyzed in this chapter seems particularly relevant to the specific case of the pharmaceutical industry. The latter can be considered an ideal case where the process of growth behaves in accordance with Gibrat's law, due to the peculiar pattern of innovation in the industry. Innovation in this industry has often been described and conceptualized as a pure "lottery model," in which previous innovation (in a particular sub-market) does not influence in any way current and future innovation in the same or in other sub-markets (Sutton, 1998). Thus, to the extent that firms' growth is driven by erratic innovation, it should also be erratic.

Data come from the PHID (Pharmaceutical Industry Database) data set, developed at the Center for the Economic Analysis of Competitiveness and Market Regulation in collaboration with EPRIS (CERM/EPRIS). The database provides longitudinal data for the sales of 210 firms in the seven largest Western markets (France, Germany, Italy, Spain, the United Kingdom, Canada, and the United States) during the period 1987–98. The values are in thousands of pounds sterling at a constant 1998 exchange rate. The companies included in the data set result from the intersection of the top 100 pharmaceutical and biotech-nology companies (in terms of sales) in each national market, obtaining a total of 210 companies. The PHID database was constructed by aggre-gating the values of the sales of these firms in the different national markets: therefore, sales for each firm stand for the sum of their sales in each of the national markets. It is important to emphasize that the panel is unbalanced, since the processes of entry and exit are explicitly considered.

We use sales as a proxy for firm size because it is the only measure available from the data set, and because sales are usually considered the best available proxy of a firm's size in pharmaceuticals and in some recent studies on firms' growth (Hart and Oulton, 1996; Higson, Holly, and Kattuman, 2002).

In this analysis, we are interested in the process of firms' *internal* growth. For this reason, in order to control for M&As during the period of observation, we constructed "virtual firms." These are firms actually existing at the end of the period, for which we constructed backwards the series of their data to allow for the case that they merged or made an acquisition. Hence, if two firms merged during the period, we consider them merged from the start, summing up their sales from the beginning.

Therefore, sales for each firm stand for the sum of their sales in each of the national markets. The use of a broad geographical definition for the relevant market and the consideration of the international firm – i.e. the sum of the sales in each national market – as the unit of analysis is justified, in our view, by the global nature of competition in the pharmaceutical industry. Many firms operate at the same time in different countries, as the competition involves R&D, production, and marketing. More importantly, successful drugs are sold world-wide, and firms' growth depends crucially on the ability to be present at the same time in different countries.

Figure 6.1 reports the histograms at each time period for these firms. The distributions do not show important departures from stability over time. The only relevant feature worth observing is that firms that had a relatively small size at the beginning (the left-hand tail of the empirical distributions, years 1987 and 1988) have probably moved towards the center of the distribution by the end of the sample. We will come back to this point in section 6.4, where the robustness of the results is checked by specifying differently the distributional assumption both of the error term ε_i and of the population structure θ_i.

6.3.2 *The speed of convergence*

The first set of results that it is worthwhile to comment on is contained in table 6.1 and in figure 6.2.

For four different settings of Σ, table 6.1 reports the following information. The first four lines summarize the posterior distribution of $\bar{\rho}$, showing the mean, the median, and the 68 percent confidence bands. Line 5 reports the value of the posterior density computed at the posterior means of the parameters. Line 6 reports the percentage of firms for which we could not reject the null $\rho_i = 1$, while in lines 7 and 8 the average log (PO) and the ω^* are computed to test for unit root across firms. Lines 9 and 10 report the same statistics for the firms with growth that behaves erratically ($\rho = 1$), and for those this is clearly not the case ($\rho < 1$). Lines 11–12 show the statistics for testing the equality $\rho_i = \rho_j$. Finally, lines 13–14 report the statistics for testing the equality $SS_i = SS_j$, with $i \neq j$. The entire posterior distribution of $\bar{\rho}$ under the four settings is also shown in figure 6.2b, while the histograms of mean individual ρ_i are reported in figure 6.2a. Figure 6.2c plots the scatter relation between the mean convergence rates $(1 - \rho_i)$ and the initial firms' sizes, to examine the mean reversion argument.[8]

---

[8] The histograms and the scatter plots are based on the average of the estimated posterior distribution of ρ_i firm by firm. The information contained in the histogram is therefore different from that contained in the posterior distribution of $\bar{\rho}$,

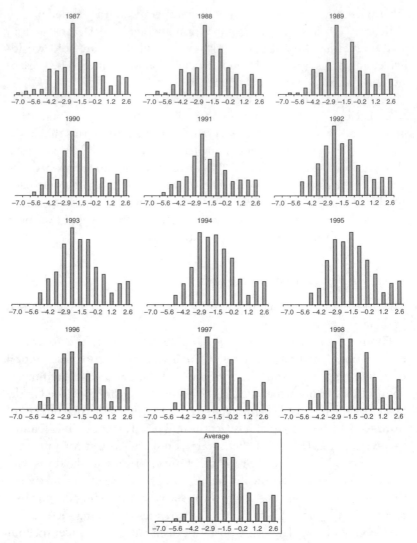

Figure 6.1. Histograms of observed data: the pharmaceutical industry

Several important facts can be discussed. First of all, notice that by forcing the units to have the same coefficients α_i and ρ_i (pool OLS) or just the same slope (fixed effect) we obtain the results generally

which represents the common part of ρ_i across firms, or the central value, to which ρ_i would collapse if there where no heterogeneity in the data.

Table 6.1 Estimation and testing: the pharmaceutical industry

	Pool OLS		Fixed effect		Mean group		Bayes	
	Posterior distribution		Posterior distribution of ρ					
1 16%	0.9924		0.9787		0.7612		0.8924	
2 Mean	0.9948		0.9827		0.8686		0.9142	
3 Median	0.9948		0.9827		0.8688		0.9141	
4 84%	0.9972		0.9868		0.9774		0.9361	
5 Posterior distribution	4455.59		4505.75		4704.13		4677.45	
			Testing $\rho = 1$					
6 $\rho_i = 1$	86.08		84.09		23.54		21.71	
7 ln(PO)	0.39		0.34		−0.99		−6.61	
8 ω^*	0.92		0.85		0.67		0.01	
	$\rho=1$	$\rho<1$	$\rho=1$	$\rho<1$	$\rho=1$	$\rho<1$	$\rho=1$	$\rho<1$
9 ln(PO)	1.92	−1.35	1.05	−0.48	0.67	−2.06	0.58	−1.67
10 ω^*	0.97	0.61	0.94	0.77	0.90	0.50	0.91	0.60
			Testing $\rho_i = \rho_j$					
11 ln(PO)	−107.12		−15.94		−131.19		−67.87	
12 ω^*	0.00		0.00		0.00		0.00	
			Testing $SS_i = SS_j$					
13 ln(PO)	−2.37		−2.84		na		−703.16	
14 ω^*	0.09		0.06		na		0.00	

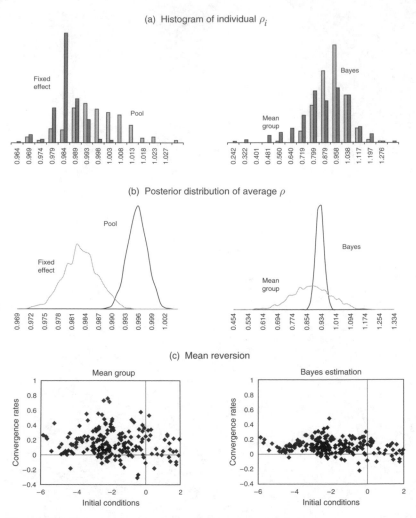

Figure 6.2. Convergence rates: the pharmaceutical industry

obtained in the literature. Therefore, under this set of restrictions the model is indeed able to reproduce the standard cross-sectional/pooling panel and the fixed-effect regression results (e.g. Goddard, Wilson, and Blandon, 2002, table 1, p. 417). In these two cases we cannot reject the null hypothesis that the average ρ is equal to one, because the log (PO) ratio is positive, meaning that the posterior odds favor the null, and the largest prior probability we should assign on the alternative in order for the data not to reject the null is very high ($\omega^* = 0.92$, and $\omega^* = 0.85$, respectively). The latter result means that we must have almost zero confidence in the null for the data not to overturn our prior beliefs. In other words, when we force coefficients to be the same across units, on average the null is a posteriori highly likely. Results are confirmed if we test for unit root firm by firm, under the same set of restrictions. Concretely, for 84 to 86 percent of the firms in the sample we cannot reject the null of $\rho = 1$ (line 6). The same information is contained in figures 6.2a and 6.2b, where the histograms of the posterior mean of the parameter ρ_i for each firm and the average posterior mean $\bar{\rho}$ are, respectively, plotted. The limiting distribution of the autoregressive parameter is not very dispersed around a mean *very close* to one.

When we allow for heterogeneous parameters across units (either $\Sigma \to \infty$ or $0 < \Sigma < \infty$), the average ρ ranges from 0.86 to 0.91 (line 2, *mean group* and *Bayes*). The average log (PO) is negative and favors the alternative versus the null of $\rho = 1$, while ω^* ranges from 0.01 (*Bayes*) to 0.61 (*mean group*), meaning that, especially for the Bayes estimator, a high prior probability should be attached on the null in order for the data not to reject it (lines 7–8). In other words, when the coefficients are estimated using the information contained both in the cross-section and in the time series dimension of the panel, the size of the firms does not follow a random walk on average. The same statistics by firm (lines 6, 9–10) confirm that only for 21 to 23 percent of firms can we not reject the null. Similar information is contained in figures 6.2a and 6.2b where, as for the *pool OLS* and the *fixed effect* cases, the histograms of the posterior mean of the parameter for each firm and the posterior distribution of $\bar{\rho}$ are plotted. The figures reveal that, when we use the information contained both in the cross-section and in the time series dimension, data show a considerable dispersion in the estimated distribution of ρ across firms, which renders the a posteriori probability of facing a random walk very unlikely on average. At the same time, the information contained in figure 6.2a is also a way of testing the null hypothesis, $\rho_i = \rho_j$. The substantial dispersion of ρ_i supports the view that the estimated autoregressive coefficients are far from collapsing towards the central value $\bar{\rho}$, and hence that the null hypothesis is likely to be rejected. The statistics shown in line 11–12 confirm these results under the four settings.

Line 5 shows that the posterior density is higher under these two settings than under the *pool* and *fixed effect* specifications. Therefore, the models that allow for heterogeneity in the parameters are to be preferred, according to what has been discussed in the previous section, at least in the sense that they have a better predictive power. Notice that when only the time series dimension is used (*mean group*), given that we have just twelve time observations, a small-sample downward bias in the estimation of the average ρ is present, as is well known. In this case the distribution of average ρ is centered on a lower value than the Bayes estimator, though being more dispersed.

Finally, note that the simple observation of $\rho < 1$ is not sufficient to indicate the existence of "catch-up" or "mean reversion." In contrast, there seems to be a weak relation between the initial conditions and the speeds of adjustment $(1 - \rho_i)$. As shown in figure 6.2c, there is a slight negative relationship between the two variables, but it does not seem enough to claim that firms that were initially larger grow more slowly than firms that were initially smaller. On the other hand, the chart and the above-mentioned results indicate that it is also not true on average that big firms follow a random walk while small firms don't, as has been argued in recent works (see Lotti, Santarelli, and Vivarelli, 2000, for instance). Therefore, it is not the leveling out in growth rates between large and small firms that bounds the overall rise in the variance of firm sizes but, rather, the absence of a unit root on average. In this respect, notice, however, that $\rho < 1$, and hence the failure of Gibrat's law is not incompatible with a growing variance. To show this, assume in our model specification that $\alpha_i = \lambda g_{i0}$. Then the model becomes

$$g_{it} = \lambda g_{i0} + \rho_i g_{it-1} + \eta_{it} \tag{6.9}$$

or, going backwards,

$$g_{it} = \lambda g_{i0} \sum_j \rho_i^j + \rho_i^t g_{i0} + \sum_j \rho_i^j \eta_{it-j}$$

Therefore, the variance of g_{it} is

$$Var(g_{it}) = \left(\lambda \sum_j \rho_i^j \right)^2 Var(g_{i0}) + \rho_i^{2t} Var(g_{i0})$$

$$+ \sum_j \rho_i^{2j} \sigma_{i\eta}^2 + \left(\lambda \sum_j \rho_i^j \right) \rho_i^t Var(g_{i0})$$

Notice that when $|\rho_i| < 1$ this expression converges to

$$Var(g_{it}) = \left(\frac{\lambda}{1 - \rho_i}\right)^2 Var(g_{i0}) + \frac{\sigma_{i\eta}^2}{1 - \rho_i^2}$$

as t becomes sufficiently large. Therefore, it may very well be the case that $Var(g_{i0}) < Var(g_{it})$, even in the case of a failure of Gibrat's law and without implying that predictions of g_{it+k} become increasingly uncertain as k gets larger.

Summarizing this first set of results, the above discussion suggests the view that only a small percentage of firm sizes drift unpredictably over time and clearly diverge within the industry, while the size of the vast majority of firms in the sample converges to a stable steady state. An interesting question now is to see whether firms converge to the same steady states or not. A negative answer to the question does not necessarily mean that differences in firm sizes are permanent and not transitory, because firms can converge to different steady states, and small firms may have steady states greater in size than those of large firms.

6.3.3 The steady state

Focusing on the cases with higher posterior predictive densities, which in our opinion are the most reasonable ones, the dispersion of steady states is substantial. Figure 6.3a plots a histogram of firms' estimated posterior steady states.[9] The histogram is constructed so that firms are grouped in ten classes of steady-state size: up to 5 percent, 6–10 percent, 11–35 percent, 36–60 percent, 61–85 percent, 86–100 percent, 101–115 percent, 116–125 percent, 126–135 percent, and above 136 percent, where 100 is the average size – i.e. the steady-state level of g_{it} that we would obtain if all the units converged to the same steady state (the bold line in the figure). Clearly, the estimated steady-state distribution is far from collapsing towards the central value. Table 6.1 (lines 13–14) reports, as has been said, the statistics for the hypothesis that the steady states are the same across units. Under the four settings $\log(PO)$ is negative and ω^* is zero, meaning that the null is highly unlikely or that, unless we assume that the alternative is impossible ($\omega^* = 0$), the null hypothesis is always

---

[9] Whenever $\rho > 1$, we compute steady states using the small-sample formula $S(i) = \alpha_i \frac{1 - \rho_i^{T+1}}{1 - \rho_i} + \rho_i^T g_{i0}$ at each draw of the Monte Carlo in the Gibbs sampler.

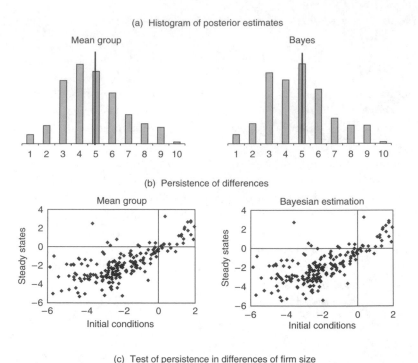

Figure 6.3. The steady state: the pharmaceutical industry

overturned by the data. Whenever the value of the statistics is not available, log (PO) must be regarded as minus infinity.

These results indicate that the estimated distribution of steady states is non-degenerate – i.e. firms converge to different steady states. The next question is to find the appropriate variables that may account for the cross-sectional dispersion in estimated steady states. This is not the purpose of this chapter, though we can at least propose a natural candidate to explain the limiting distribution. Figure 6.3b plots the estimated steady states against the initial (scaled) size levels. It is clear that there is a strong positive relationship – i.e. firms that were initially large also have the highest steady states and the initial ranking is largely maintained. Figure 6.3c measures the strength of this relation running a cross-sectional regression of estimated steady states on a constant and the initial condition (see equation 6.8 above). Clearly, the estimated b is positive – i.e. the distribution of the initial sizes of firms matters for the limiting distribution of the steady states. In other words, differences in firm size are persistent, and, given the estimated value of 0.7, one would argue that inequalities are *strongly* persistent. The $\bar{R}^2$ can be interpreted as a measure of long-run mobility (see Canova and Marcet, 1995, for instance). A small value would suggest that individual units may move up and down in the ranking, whereas a high value indicates that the ordering in the initial distribution is the same as in the steady states. The latter seems to adjust better to our estimation results. A slope of 0.7 in the cross-sectional regression suggests that, on average, the gap between the big and the small firms will be reduced in the limit only by 30 percent, while $\bar{R}^2 = 0.57$ indicates that the initial conditions alone explain almost 60 percent of the variation of the cross-sectional distribution of steady states. We regard these results as strong evidence in favor of the persistence of differences in firm sizes.

Summing up, the first set of results does not confirm the conventional wisdom that firms' size drifts erratically over time. Even more interestingly, the data show that firms converge to different steady states, that their speed of convergence does not depend on the initial size, and that the steady state is strongly correlated to initial size. In other words, there is only weak evidence of reversion to the mean, and firms' size differentials tend to persist over time.

6.4 Robust inference and sensitivity analysis

In this section we check the robustness of previous results by performing the analysis with a different data set and different prior assumptions.

6.4.1 A different data set: UK manufacturing firms

Compared to the pharmaceutical database, the new data set has a shorter time series dimension and includes different industries belonging to a single country. The analysis is performed on a sample of 267 UK manufacturing firms. The data constitute a balanced panel of five years, from 1988 to 1992. The histograms for each year are plotted in figure 6.4. The same remarks as for the previous data set can apply.

The aim of this subsection is twofold. On the one hand, we can verify if our main results on Gibrat's argument are robust to a different data

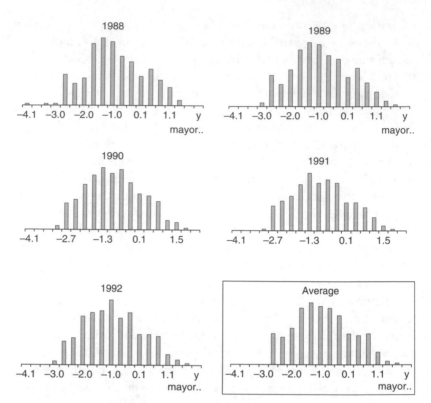

Figure 6.4. Histograms of observed data: UK manufacturing

set. On the other hand, given that the new data set contains different industries and hence another level of possible heterogeneity, one could cast light on some other features that explain firms' growth across industries.

The model specification is the same as before, and so are the prior assumptions. The estimation results, shown in table 6.2 and figures 6.5 and 6.7, confirm the previous findings. In particular, when both types of information (cross-sectional and time series) are controlled for, we reject the null hypothesis of unit root both on average (table 6.2, lines 7–8) and by firms (lines 6, 9–10). The average autoregressive parameter ρ ranges from 0.55 (*mean group*) to 0.81 (*Bayes*) in the settings where heterogeneity is controlled for, while it is 0.99 in the plain pool and in the fixed-effect cases (table 6.2, lines 1–4). The dispersion of the mean estimates of ρ_i by firms (figure 6.5a) is, again, a way of rejecting the null

Table 6.2 *Estimation and testing: UK manufacturing*

	Pool OLS		Fixed effect		Mean group		Bayes	
			Posterior distribution of ρ					
1 16%	0.9915		0.9874		0.4596		0.7726	
2 Mean	0.9949		0.9913		0.5516		0.8072	
3 Median	0.9952		0.9915		0.5508		0.8074	
4 84%	0.9983		0.9954		0.6456		0.8422	
5 Posterior distribution	421.48		445.14		1449.06		845.83	
			Testing $\rho = 1$					
6 %$\rho \geq = 1$	87		89.44		8.87		4.23	
7 ln(PO)	1.84		0.68		−11.49		−14.71	
8 ω^{*}	0.98		0.92		0.00		0.00	
	$\rho = 1$	$\rho < 1$	$\rho = 1$	$\rho < 1$	$\rho = 1$	$\rho < 1$	$\rho = 1$	$\rho < 1$
9 ln(PO)	1.55	−0.35	1.27	−0.32	0.44	−6.99	0.25	−2.26
10 ω^{*}	0.97	0.94	0.96	0.93	0.78	0.34	0.83	0.54
			Testing (i) = (j)					
11 ln(PO)	−19.39		−11.70		−115.51		−93.67	
12 ω^{*}	0.00		0.00		0.00		0.00	
			Testing ss(i) = ss(j)					
13 ln(PO)	−1.64		−1.79		na		na	
14 ω^{*}	0.16		0.14		na		na	

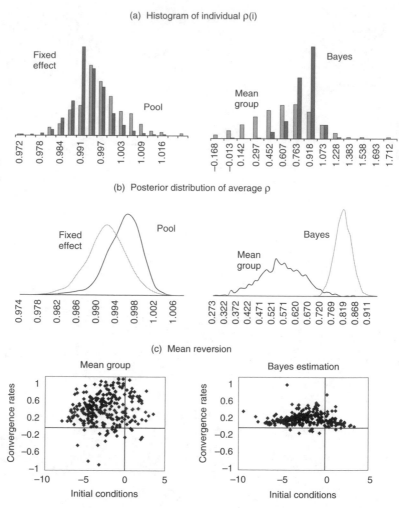

Figure 6.5. Convergence rates: UK manufacturing

hypothesis that $\rho_i = \rho_j$ in all settings (see also table 6.1, lines 11–12), although the mean group and the Bayes specification provide more dispersed estimators. There seems to be even less relation between the convergence rate and the initial conditions, with respect to the previous data set (figure 6.5c), confirming the view that no mean reversion is present in the sample data.

Taken together, these findings support again the *fact* that firm size *does not* drift unpredictably over time on average. Notice that, when we pool the data forcing the parameters to be the same across units, for most firms we do not reject the null that Gibrat's law holds (table 6.2, lines 6–8). Even for firms where the test rejects the unit root hypothesis, the largest prior probability to attach on the alternative in order for the data not to reject the null is very high (0.93–0.94), meaning that the specifications with little or no heterogeneity in practice would never reject the null of unit root. In contrast, when we do not impose this restriction, only 4 to 9 percent of the firms in the sample behave according to a random walk in size. Among these firms there seems not to be a clear pattern, at least across industries. In table 6.3 we report the number of firms and the percentage of firms with $\rho = 1$ for each industry, as well as the scaled size of the average firm in the industry. It can be noticed that only in one industrial group (motor vehicles and parts) do all firms have a ρ significantly less than one, while in the others the percentage of firms the size of which drifts unpredictably does not follow a clear pattern at first glance (see also figure 6.6). Thus, for instance, the simple argument that big firms may follow a random walk while small firms certainly do not is contradicted by the evidence that in the first industry (metal manufacturing), where the average size is relatively high, 50 percent of firms grow erratically, while in industry 14 (footwear and clothing), in which the firms are among the smallest, all firms follow Gibrat's law.

Conclusions on the limiting behavior of firm size can be appreciated from figure 6.7. As before, we reject the null hypothesis of equal steady states. The dispersion of estimated steady states is again substantial (figure 6.7a) and, unless we assume a priori that the alternative is impossible, the null hypothesis will always be overturned by the data (in table 6.2, lines 13–14, the log (PO) is negative and ω^* is approximately zero under the four settings). Moreover, differences in firm size are extremely persistent. The evidence contained in figures 6.7b and 6.7c is overwhelming. The position in the initial size distribution of a given unit strongly determines the position of the same unit in the steady-state distribution. On average, the gap between the big and small firms will be reduced in the limit only by 10 percent, while the initial conditions alone explain more than 80 percent of the variation of the cross-sectional distribution of estimated steady states.

Table 6.3 UK manufacturing firms by industry

Sectors (SIC 80-digit)	Industry Number	Times	Percentage of $\rho=1$	Average scaled size
Metal manufacturing (22)	1	8	0.50	0.520
Non-metallic manufacturing (24)	2	9	0.11	0.180
Chemical (25 and 26)	3	28	0.07	0.246
Other metal goods (31)	4	21	0.19	0.034
Mechanical engineering (32)	5	59	0.24	0.052
Office and data machinery (33)	6	12	0.08	0.086
Electrical and electronic machinery (34)	7	45	0.16	0.080
Motor vehicles and parts (35)	8	8	0.00	0.195
Other transport (36)	9	6	0.33	0.369
Instrument engineering (37)	10	18	0.17	0.036
Food, drink, and tobacco (41/42)	11	9	0.44	0.295
Textiles (43)	12	11	0.09	0.122
Leather goods (44)	13	1	1.00	2.718
Footwear and clothing (45)	14	3	1.00	0.002
Timber (46)	15	6	0.33	0.046
Paper and printing (47)	16	11	0.18	0.086
Rubber and plastics (48)	17	10	0.10	0.062
Other manufacturing (49)	18	2	0.50	0.059

Note: SIC = standard industrial classification.

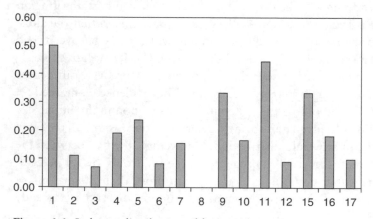

Figure 6.6. Industry distribution of firms with $\rho=1$

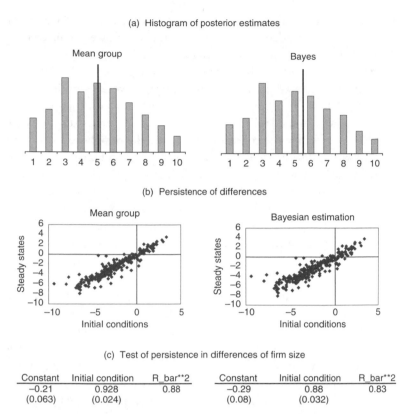

(a) Histogram of posterior estimates

Mean group

Bayes

(b) Persistence of differences

Mean group

Bayesian estimation

(c) Test of persistence in differences of firm size

Constant	Initial condition	R_bar**2		Constant	Initial condition	R_bar**2
−0.21	0.928	0.88		−0.29	0.88	0.83
(0.063)	(0.024)			(0.08)	(0.032)	

Figure 6.7. The steady state: UK manufacturing

Although results may suffer from a small-sample bias, they point out, once again, the importance of controlling for heterogeneity in the coefficients when a panel data set is used, as well as the importance of using efficiently all the information contained in the sample.

6.4.2 Different priors: model checking

In this section we assess the fit of the model to the data and the plausibility of the specification for the purposes for which the model has been used. The basic questions are these. How much does the posterior predictive distribution fit to the data? And how much does previous posterior inference change when other reasonable-probability models are used in place of the present model? It is important to distinguish between the two questions, because, even if the present model provides an adequate fit to the data, the posterior inferences can still differ under plausible alternative models.

The technique used for checking the fit of our model to the data is to draw simulated values from the posterior predictive distribution of the replicated data and to compare these samples to the observed data. This search is combined with a more general family of distributions for both the error term in the regression (6.3) and the population structure (6.4) to analyze the sensitivity of the results. Concretely, we assume the t_ν distribution in place of the normal. The t distribution has a longer tail than the normal and can be used for accommodating occasional unusual observations in the data distribution or occasional extreme parameters in the prior distributions or hierarchical model. Histograms of the observed data for the pharmaceutical industry and UK manufacturing, (figures 6.1 and 6.4, respectively) do not reveal extreme values in either data set. Therefore, we expect results based on the normal distribution for the data to fit the data better, and use the t_ν distribution just for sensitivity analysis. In other words, as far as the assumptions on data distribution are concerned, the t_ν is chosen simply as a robust alternative to the normal, provided the degrees of freedom ν are fixed at values no smaller than prior understanding dictates in our example. On the other hand, given the small time series dimension of the panel, we would expect a better predictive performance by assuming a t_ν distribution with ν for the parameters of the hierarchical model.

The procedure for carrying out both posterior predictive model checking and a sensitivity analysis is the following. Let y be the observed data and ψ the vector of parameters (including all hyperparameters). Define y^{rep} as the *replicated* data that *could have been* observed, or the data we *would* see tomorrow if the experiment that produced y today were replicated with the same model and the same value of ψ that produced the observed data. The distribution of y^{rep} given the current state of knowledge – i.e. the posterior predictive distribution – is

$$p(y^{rep}) = \int p(y^{rep} \,|\, \psi)p(\psi \,|\, y)d\psi$$

The discrepancy between the model and the data is measured, as suggested by Gelman et al. (1995), by defining a *discrepancy* measure $T(y, \psi)$, which is a scalar summary of parameters and data. Lack of fit of the data with respect to the posterior predictive distribution is then measured by the tail-area probability (*p-value*) of the quantity, and computed using posterior simulations of (ψ, y). This p-value is defined as the probability that the replicated data could be more extreme than the observed data, as measured by the test quantity,

$$\text{p-value} = \Pr(T(y^{rep}, \psi)) \geq \Pr(T(y, \psi))$$

where the probability is taken over the joint posterior distribution of (ψ, y^{rep}). Major failures of the model typically correspond to extreme tail-area probabilities (less than 0.01 or more than 0.99).

The discrepancy measure chosen is the χ^2 discrepancy quantity, an omnibus measure for routine checks of fit, and is defined as

$$T(y, \psi) = \sum_i (y_i - E(y_i \mid \psi))' \, [var(y_i \mid \psi)]^{-1} (y_i - E(y_i \mid \psi)) \quad (6.10)$$

We compute these statistics, the p-values, and the relevant results for inference of interest, first fitting a range of t_ν distributions with $\nu = 10,50,100$ for the errors in (6.3), maintaining the normality assumption on θ_i, and then fitting a range of t_ν distributions with $\nu = 5,10,15,50$ for the vector θ_i in (6.4), maintaining the normality assumption on ε_i. Infinite degrees of freedom have already been fitted (the normal–normal data model). Results are reported in figures 6.8 to 6.10 and table 6.4. Most of them refer only to the pharmaceutical industry, because those for UK manufacturing are very similar.

Some comments are in order. Figure 6.8 reports scatter plots showing prior and posterior simulation of the chosen test quantity (6.10) based on 2500 simulations from the posterior distribution of (ψ, y^{rep}) for different values of the degrees of freedom of the t. The p-value is computed as the proportion of points in the upper left half of the plot with respect to an imaginary $45°$ line. As expected, more extreme values are encountered when a t distribution is fitted for the error terms than when a t replaces the normal for the vector θ_i. The values of the posterior predictive density in table 6.4 confirm this finding. Among the specifications fitted, the one assuming normal errors $(\nu = \infty)$ has the highest predictive power (posterior distribution $= 4677.45$). Under the assumption of normality of the error term ε_i, the t_{15} distribution for θ_i also performs well (posterior distribution $= 4342.99$). Notice that the p-values when $\theta_i \sim t_\nu$ are always around 0.02, regardless of ν.

Table 6.4 Robust inference and sensitivity analysis: the pharmaceutical industry
(a) $y \sim t\text{-student } (\nu), \theta \sim normal$

	ν	10		50		100		inf	
Bayes	Median ρ	0.913		0.912		0.913		0.914	
	16% 84%	0.891	0.935	0.891	0.935	0.890	0.935	0.892	0.934
	Posterior distribution	2629.33		3460.83		3736.86		4677.45	
	In(PO)	−6.961		−6.782		−6.320		−6.610	
	ω_*	0.006		0.008		0.013		0.010	
	p-value	0.997		0.388		0.185		0.120	
Pool	Median ρ	0.994		0.994		0.994		0.995	
	16% 84%	0.991	0.997	0.991	0.998	0.991	0.997	0.992	0.997
	Posterior distribution	1942.25		3327.66		3514.02		4455.59	
	In(PO)	0.840		0.849		0.772		0.393	
	ω_*	0.937		0.937		0.945		0.921	
	p-value	0.994		0.411		0.180		0.130	

(b) $\theta \sim$ *t-student* (ν), $y \sim$ *normal*

	ν	5	10	15	30
Bayes	Median ρ	0.893	0.893	0.891	0.888
	16% 84%	0.834 0.950	0.857 0.931	0.860 0.924	0.835 0.918
	Posterior distribution	2340.19	4459.74	4532.62	3784.17
	ln(PO)	−2.501	−3.917	−5.066	−3.361
	ω^*	0.371	0.136	0.041	0.201
	p-value	0.017	0.020	0.018	0.022
Pool	Median ρ	0.980	0.987	0.990	0.987
	16% 84%	0.924 1.083	0.945 1.032	0.957 1.023	0.967 1.021
	Posterior distribution	1906.27	4262.26	4341.99	3529.69
	ln(PO)	0.424	0.825	1.584	0.298
	ω^*	0.905	0.934	0.947	0.841
	p-value	0.018	0.016	0.018	0.019

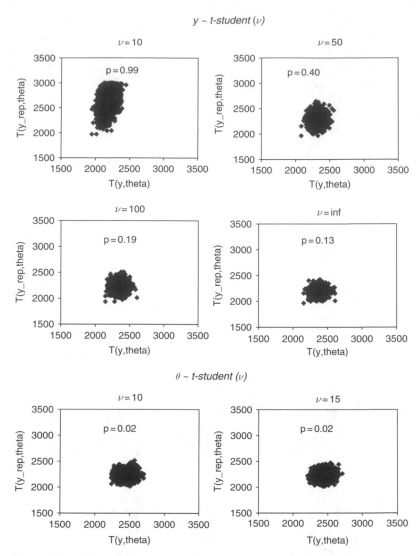

Figure 6.8. A sensitivity analysis: the pharmaceutical industry

Figures 6.9 and 6.10 report the histograms of replicated data for both data sets. They are computed as posterior averages over the simulation draws at each time across firms, with the benchmark specification: normal ε_i–normal θ_i. They look very similar to those already

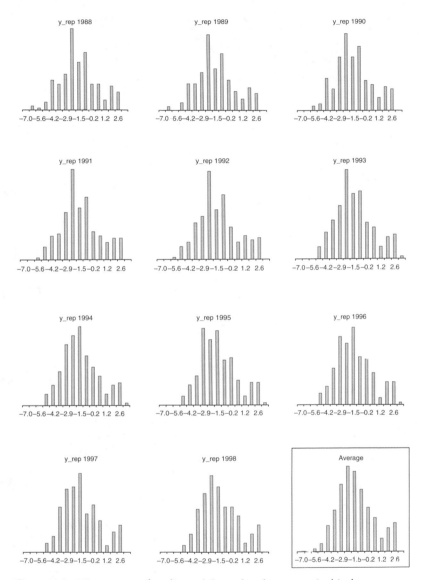

Figure 6.9. Histograms of replicated data: the pharmaceutical industry

shown in figures 6.1 and 6.4 for observed data, confirming that our basic assumptions are clearly able to capture the variation observed in the data. We regard these findings as evidence that no potential failings of our model are present, in the sense that it does not produce any

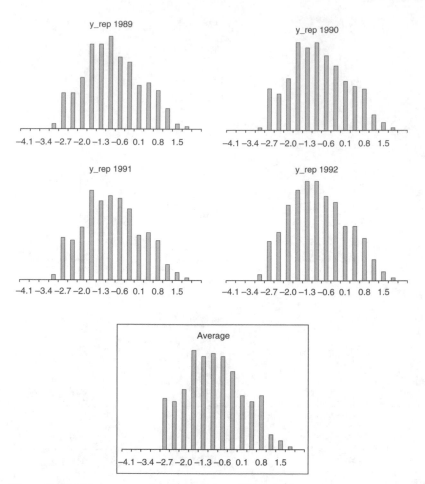

Figure 6.10. Histograms of replicated data: UK manufacturing

systematic differences between the simulations and the data, and hence that it fits well to the data.

Finally, for each value of ν, table 6.4 also reports the posterior distribution of $\bar{\rho}$ and the relevant statistics to test for unit root, for the cases *Bayes* and *pool*. It is easy to check that, under all values of ν tried, the results are essentially the same as those obtained under the normal–normal model as displayed in table 6.2, except for the fact that, under the assumption $\theta_i \sim t_\nu$, the 68 percent confidence interval is larger than under the normality assumption, as expected.

The *plain heterogeneous* model (Bayes) always provides estimates of ρ far from being equal to one a posteriori, while the pool model is not able to reject on average Gibrat's argument. The resulting inference, therefore, is unchanged, meaning that it is robust to a model expansion that uses the t_ν as a more general and robust alternative to the normal.

6.5 Summary and concluding remarks

The results discussed above can be summarized as follows.

(i) The estimated average speed of adjustment is far from being zero when the information contained both in the cross-sectional and in the time series dimension is used. This implies that the main assertion of Gibrat's law, that growth rates are erratic, is not true on average in the case of pharmaceuticals.

(ii) When we allow for heterogeneity, both in the intercepts and in the slope coefficients, the data show a considerable dispersion in the estimated distribution of ρ across firms, whereas when we force the parameters to be the same across units the distribution of ρ is centered around values very close to one. This confirms our initial suspicion that the previous results, based on cross-sectional or pooled panel data models, may be econometrically biased, because they do not exploit all the information contained in the data and hence they misspecify the econometric model without considering heterogeneity among firms. The null hypothesis $\rho_i = \rho_j$ is a posteriori very unlikely, meaning that the distribution across firms of the autoregressive parameter is far from collapsing to the central value $\bar{\rho}$, as a priori imposed in the cross-section or in the pool panel data models.

(iii) There is only weak evidence of mean reversion. Even if on average $\rho < 1$, this does not necessarily mean that firms that were initially larger grow more slowly than firms that were initially smaller. Therefore, the overall rise in the variance of firm growth turns out to be bounded, but for reasons different from the conventional one linked to the leveling out in growth rates between large and small firms. In any case, as shown in section 6.3.2, the variance may increase, as time goes by, even if $\rho < 1$.

(iv) Estimated steady states differ across units, and firm sizes do not converge within or across industries to a common limiting distribution. This fact does not imply per se that firm size drifts unpredictably over time, as argued by some authors (see Geroski et al., 2001, p. 6). It is true that a unit root in the process of firm size implies divergence, but the reverse causality does not necessarily hold, as shown in this chapter.

(v) Initial conditions are important determinants of the estimated distribution of steady states. Differences are likely to reduce at a very slow rate but they do not seem to disappear over time. A firm with an initial size below the average is going to narrow the gap with respect to bigger firms, but it does not seem to increase its relative size in the cross-sectional distribution. In other words, differences in firm size persist.

(vi) The model that we used to perform the analysis does not show failings in fitting to the data. Moreover, the results are unchanged with a robust alternative to the priors chosen.

Our conclusion is that the simple empirical fact regarding the growth of firms is that growth is not erratic and that firms' size does not drift unpredictably over time, as was claimed in many previous studies, and hence that Gibrat's argument does not hold on average. Moreover, the argument that small firms tend to grow more rapidly than large ones is also not confirmed. Rather, firms that were initially small tend to remain small in the steady state, and their convergence rate is not higher in comparison to larger firms. In other words, there are systematic differences in growth rates among firms. These differences are not size-specific and may depend on other firm-specific features that are not observable in our data.

These results open the way for further investigations into the determinants of firms' growth. With specific reference to pharmaceuticals, they may indicate that innovation is less erratic than usually thought and that more complex stochastic processes driving the arrival of innovations ought to be considered in analysis, indirectly confirming the results obtained by Bottazzi et al. (2002).

Moreover, it is highly likely that size is not the only correct variable that growth should be conditioned on. Other sources of heterogeneity may plausibly be responsible for the differential growth rates of firms over time. Age and marketing strategies are primary – but certainly not the only – candidates. Other firm-specific factors are likely to play an important role, related perhaps to the idiosyncratic capabilities, attitudes, and organizational practices of firms towards the adoption and use of the new scientific research that is associated with the "molecular biology revolution" and biotechnology (Henderson and Cockburn, 1996; Henderson, Orsenigo, and Pisano, 1999). In this respect, it would be particularly interesting to explore and, hopefully, identify some common features across clearly divergent/convergent firms as possible relevant explanatory factors of the cross-sectional dispersion in estimated steady states. Finally, the mechanisms through which selection operates in highly regulated markets such as pharmaceuticals to promote the growth and the decline of firms should also be explicitly modeled and tested.

At a more general level, the results of this chapter cast serious doubts on the validity and the robustness of all those conventional econometric exercises that do not treat heterogeneity appropriately. How pervasive these problems might be is a fundamental question for future research.

Appendix: The posterior distributions

Given the prior information previously specified, we look for the posterior density of the parameter vector $\psi = \left(\theta_i, \bar{\theta}, \Sigma_\theta^{-1}, \{\sigma_i^2\}_{i=1}^{N} \right)$, which is given by

$$p(\psi \mid y, y_{io}) \propto f\left(y \mid \theta_i, \bar{\theta}, \Sigma_\theta^{-1}, \{\sigma_i^2\}_{i=1}^{N}, y_{io}\right) p(\psi \mid y_{io})$$

Assuming a vague prior for σ_i^2 (i.e. taking $\nu_o = 0$), the joint density of all the parameters can be written as

$$
\begin{aligned}
p\left(\theta_i, \bar{\theta}, \Sigma_\theta^{-1}, \sigma_i^2 \mid y, y_{i0}\right) \propto & \prod_{i=1}^{N} \sigma_i^{-T} \exp\left[-\frac{1}{2}\sum_{i=1}^{N} \sigma_i^{-2}(y_i - X_i\theta_i)'(y_i - X_i\theta_i)\right] \\
& \times |\Sigma_\theta|^{-\frac{N}{2}} \exp\left[-\frac{1}{2}\sum_{i=1}^{N}(\theta_i - \bar{\theta})'\Sigma_\theta^{-1}(\theta_i - \bar{\theta})\right] \\
& \times |C|^{-\frac{1}{2}} \exp\left[-\frac{1}{2}(\bar{\theta} - \mu)'C^{-1}(\bar{\theta} - \mu)\right] \\
& \times |\Sigma_\theta|^{-\frac{1}{2}(s_o - k - 1)} \exp\left[-\frac{1}{2}tr\left((s_o S_o)\Sigma_\theta^{-1}\right)\right] \\
& \times \prod_{i=1}^{N} \sigma_i^{-2}
\end{aligned}
$$

where $y_i = (g_{i1}, \ldots g_{iT})'$ and $X_i = (X'_{i1}, \ldots, X'_{iT})'$. The first line of the formula represents the standard likelihood conditional on the initial conditions, and the others represent the prior information.

As mentioned in the text, in order to obtain the marginal posterior distributions of each component of ψ, a numerical integration is needed. We use the Gibbs sampler, a sampling-based approach, first introduced by Geman and Geman (1984) and subsequently popularized by Gelfand and Smith (1990), among others. If we dispose of the full conditional distributions of the parameters, the idea is to construct a Markov chain on a general state space such that the limiting distribution of the chain is the joint posterior of interest. The relevant conditional distributions are obtained from the above formula. For example, the conditional distribution for θ_i is obtained by combining line one with line two, completing the square for θ_i. The conditional distribution of $\bar{\theta}$ is obtained by combining line two with line three, completing the square, and so on. Concretely, the conditional distributions needed to implement the Gibbs sampler are the following:

$$p(\theta_i \mid y, \psi_{-\theta}) = N\left[A_i\left(\sigma_i^{-2}X'_iy_i + \Sigma_\theta^{-1}\bar{\theta}\right), A_i\right] \qquad i = 1, \ldots, n$$

$$p(\bar{\theta} \mid y, \psi_{-\bar{\theta}}) = N\left[B\left(n\Sigma_\theta^{-1}\tilde{\theta} + C^{-1}\mu\right), B\right]$$

$$p\left(\Sigma_\theta^{-1} \mid y, \psi_{-\Sigma_\theta^{-1}}\right) = W\left[\left(\sum_{i=1}^n \left(\theta_i - \bar{\theta}\right)\left(\theta_i - \bar{\theta}\right)' + s_oS_o\right)^{-1}, s_o + n\right]$$

$$p\left(\sigma_i^2 \mid y, \psi_{-\sigma_i^2}\right) = IG\left[T/2, \left((y_i - X_i\theta_i)'(y_i - X_i\theta_i)\right)/2\right] \qquad i = 1, \ldots, n$$

where $A_i = \left(\sigma_i^{-2}X'_iX_i + \Sigma_\theta^{-1}\right)^{-1}, B = \left(n\Sigma_\theta^{-1} + C^{-1}\right)^{-1}, \tilde{\theta} = (1/n)\sum_i \theta_i, W()$ denotes the Wishart, $IG()$ denotes the inverse gamma distribution, and $\psi_{-\gamma}$ denotes ψ without γ.

After iterating, say, M times, the sample $\psi^{(M)}$ can be regarded as a drawing from the true joint posterior density. Once this simulated sample has been obtained, any posterior moment of interest or any marginal density can be estimated using the ergodic theorem. Convergence to the desired distribution can be checked as suggested by Gelfand and Smith (1990).

In our exercise, the number of Monte Carlo iterations is set equal to 5000, and the first 1000 are discarded. Convergence is already achieved with the first 3000 iterations. Moreover, the third stage of the hierarchy is assumed vague, or non-informative – i.e. we set $C^{-1} = 0$. This means that the only hyperparameters to be assumed known are s_o and S_o – i.e. the degrees of freedom and the scale matrix of the Wishart prior for Σ_θ. These hyperparameters control the four settings under which we estimate the model. Concretely, the *pool OLS* ($\Sigma_\theta = 0$) is approximated by choosing $S_o = diag$ (0.0001 0.0001); for the *fixed effect* ($\sigma\rho = 0$) we choose $S_o = diag$ (0.1 0.0001). The setting *mean group* ($\Sigma_\theta \to \infty$) is approximated with $S_o = diag$ (100 100). Finally, for the *Bayes* setting we choose $S_o = diag$ (1. 1.). In the four cases, prior degrees of freedom are chosen, randomizing uniformly over the interval (3,10) – i.e. $s_o \sim$ *uniform* (3,10).

The conditional posterior distributions need to be modified when a t_ν distribution is used in place of the normal, either for the error term or for the population structure. Recalling that the t_ν can be interpreted as a mixture of normal distributions with variances distributed as scaled inverse-χ^2, the Gibbs sampler is easily extended to include one or more conditional distributions. For instance, in the case of $y_i \sim N(X_i\theta_i, \sigma_i^2 I_T)$ and $\theta_i \sim t_\nu(\tilde{\theta}, \Sigma_\theta)$, the latter is equivalent to

$$\theta_i \mid \nu_i \sim N(\tilde{\theta}, \nu_i\Sigma_\theta)$$
$$\nu_i \sim Inv\text{-}\chi^2(\nu, 1)$$

where $Inv\text{-}\chi^2(\nu, s^2)$ denotes a scaled inverse-χ^2 with scale factor s^2.

Therefore, the change in the distributive assumption on the population structure is equivalent to an expansion of the model. This, in turn, implies that the posterior distributions of the parameters must change accordingly and also that a new variable, ν_i, must enter the Gibbs sampler. There is a similar hierarchy y_i when $y_i \sim t_\nu (X_i\theta_i, \sigma_i^2 I_T)$.

The sensitivity analysis is performed by comparing different sets of results corresponding to different values for ν. To compute the test quantities $T(y, \theta)$ and tail-area probabilities we proceed as follows. If we already have, say, L simulations from the posterior density of ψ, we draw one y^{rep} from the predictive distribution for each simulated ψ. We now have L draws from the joint posterior distribution $p(y^{rep}, \psi \mid y)$. The posterior predictive check is the comparison between the realized test quantities, $T(y, \psi^l)$, and the predictive test quantities, $T(y^{rep,l}, \psi^l)$. The estimated p-value is just the proportion of these L simulations for which the test quantity equals or exceeds its realized value; that is, for which $T(y^{rep,l}, \psi^l) \geq T(y, \psi^l)$, $l = 1, \ldots, L$. For further technical details on model checking and sensitivity analysis, see also Gelman et al. (1995, chaps. 6 and 12).

References

Baily, M. N., and A. K. Chakrabarty (1985), "Innovation and productivity in US industry," *Brookings Papers on Economic Activity*, 2, 609–32.

Bottazzi, G., E. Cefis, and G. Dosi (2002), "Corporate growth and industrial structure: some evidence from the Italian manufacturing industry," *Industrial and Corporate Change*, 11 (4), 705–25.

Bottazzi G., G. Dosi, M. Lippi, F. Pammolli, and M. Riccaboni (2000), "Process of corporate growth in the evolution of an innovation-driven industry: the case of pharmaceuticals," *International Journal of Industrial Organization*, 19, 1161–87.

Breschi, S., F. Malerba, and L. Orsenigo (2000), "Technological regimes and Schumpeterian patterns of innovation," *Economic Journal*, 110, 388–410.

Canova, F., and A. Marcet (1995), *The Poor Stay Poor: Non-Convergence across Countries and Regions*, Discussion Paper no. 1265, Centre for Economic Policy Research, London.

Cefis, E., and L. Orsenigo (2001), "The persistence of innovative activities: a cross-countries and cross-sectors comparative analysis," *Research Policy*, 30 (7), 1139–58.

Chib, S., and E. Greenberg (1995), "Hierarchical analysis of SUR models with extensions to correlated serial errors and time-varying parameter models," *Journal of Econometrics*, 68, 339–60.

(1996), "Markov chain Monte Carlo simulation methods in econometrics," *Econometric Theory*, 12, 409–31.

Dosi, G., O. Marsili, L. Orsenigo, and R. Salvatore (1995), "Technological regimes, selection and market structure," *Small Business Economics*, 7, 1–26.

Dunne, T., M. J. Roberts, and L. Samuelson (1989), "The growth and failure of US manufacturing plants," *Quarterly Journal of Economics*, 104 (4), 671–98.

Ericson, R., and A. Pakes (1995), "Markov-perfect industry dynamics: a framework for empirical work," *Review of Economic Studies*, 62 (1), 53–82.

Evans, D. S. (1987a), "The relationship between firm growth, size, and age: estimates for 100 manufacturing industries," *Journal of Industrial Economics*, 35 (4), 567–81.

(1987b), "Tests of alternative theories of firm growth," *Journal of Political Economy*, 95 (4), 657–74.

Farinas, J. C., and L. Moreno (2000), "Firms' growth, size and age: a nonparametric approach," *Review of Industrial Organization*, 17 (3), 249–65.

Gelfand, A. E., S. E. Hills, A. Racine-Poon, and A. F. M. Smith (1990), "Illustration of Bayesian inference in normal data models using Gibbs sampling," *Journal of the American Statistical Association*, 85, 972–85.

Gelfand, A. E., and A. F. M. Smith (1990), "Sampling-based approaches to calculating marginal densities," *Journal of the American Statistical Association*, 85, 398–409.

Gelman, A., J. B. Carlin, H. S. Stern, and D. B. Rubin (1995), *Bayesian Data Analysis*, Chapman and Hall, London.

Geman, S., and D. Geman (1984), "Stochastic relaxation, Gibbs distributions and the Bayesian restoration of images," *IEEE Transactions on Pattern Analysis and Machine Intelligence*, 6 (6), 721–41.

Geroski, P. A. (1999), *The Growth of Firms in Theory and in Practice*, Working Paper no. 2092, Centre for Economic Policy Research, London.

Geroski, P. A., S. Lazarova, G. Urga, and C. F. Walters (2001), "Are differences in firm size transitory or permanent?," *Journal of Applied Econometrics*, 18 (1), 47–59.

Geroski, P. A., S. Machin, and J. Van Reenen (1993), "The profitability of innovating firms," *RAND Journal of Economics*, 24 (2), 198–211.

Gibrat, R. (1931), *Les inégalités économiques*, Librairie du Recueil Sirey, Paris.

Goddard, J., J. Wilson, and P. Blandon (2002), "Panel tests of Gibrat's law for Japanese manufacturing," *International Journal of Industrial Organization*, 20, 415–33.

Hall, B. H. (1987), "The relationship between firm size and firm growth in the U.S. manufacturing sector," *Journal of Industrial Economics*, 35 (4), 583–606.

Harhoff, D., K. Stahl, and M. Woywode (1998), "Legal form, growth and exit of West German firms: empirical results for manufacturing, construction, trade and service industries," *Journal of Industrial Economics*, 46 (4), 453–88.

Hart, P. E., and N. Oulton (1996), "Growth and size of firms," *Economic Journal*, 106, 1242–52.

Henderson, R. M., and I. Cockburn (1996), "Scale, scope and spillovers: the determinants of research productivity in drug discovery," *RAND Journal of Economics*, 27 (1), 32–59.

Henderson, R. M., L. Orsenigo, and G. Pisano (1999), "The pharmaceutical industry and the revolution in molecular biology: exploring the

interactions between scientific, institutional, and organizational change," in D. C. Mowery and R. R. Nelson (eds.), *Sources of Industrial Leadership: Studies of Seven Industries*, Cambridge University Press, Cambridge, 267–311.

Higson, C., S. Holly, and P. Kattuman (2002), "The cross-sectional dynamics of the US business cycle: 1950–1999," *Journal of Economic Dynamics and Control*, 26, 1539–55.

Hsiao, C., M. H. Pesaran, and A. K. Tahmiscioglu (1999), "Bayes estimation of short-run coefficients in dynamic panel data models," in C. K. Lahiri, L. -F. Lee, and M. H. Pesaran Hsiao, (eds.), *Analysis of Panels and Limited Dependent Variables: A Volume in Honour of G. S. Maddala*, Cambridge University Press, Cambridge, 268–96.

Jovanovic, B. (1982), "Selection and the evolution of industry," *Econometrica*, 50 (3), 649–70.

Klepper, S., and P. Thompson (2004), *Submarkets and the Evolution of Market Structure*, mimeo, Carnegie Mellon University, Pittsburgh.

Klette, T. J., and S. Kortum (2002), *Innovating Firms and Aggregate Innovation*, Working Paper no. 8819, National Bureau of Economic Research, Cambridge, MA.

Klevorick, A., R. Levin, R. R. Nelson, and S. G. Winter (1999), "On the sources and significance of interindustry differences in technological opportunity," in H. Hanusch (ed.), *The Legacy of Joseph A. Schumpeter*, Vol. I, *Intellectual Legacies in Modern Economics*, Edward Elgar, Cheltenham, 262–82.

Leamer, E. E. (1978), *Specification Searches: Ad Hoc Inference with Non-Experimental data*, Wiley, New York.

Lindley, D. V., and A. F. M. Smith (1972), "Bayes estimates for the linear model" (with discussion), *Journal of the Royal Statistical Society* B, 34, 1–41.

Lotti, F., E. Santarelli, and M. Vivarelli (2000), *Does Gibrat's Law Hold among Young, Small Firms?*, Working Paper no. 2000–5, Laboratory of Economics and Management, Sant' Anna School for Advanced Studies, Pisa.

Mata, J. (1994), "Firm growth during infancy," *Small Business Economics*, 6 (1), 27–39.

Matraves, C. (1999), "Market structure, R&D and advertising in the pharmaceutical industry," *Journal of Industrial Economics*, 47 (2), 169–94.

Mueller, D. C. (ed.) (1990), *The Dynamics of Company Profits: An International Comparison*, Cambridge University Press, Cambridge.

Pakes, A., and R. Ericson (1998), "Empirical implications of alternative models of firms' dynamics," *Journal of Economic Theory*, 79 (1), 1–45.

Pesaran, M.H., and R. Smith (1995), "Estimating long-run relationships from dynamic heterogeneous panels," *Journal of Econometrics*, 68, 79–113.

Sims, C. (1988), "Bayesian skepticism on unit root econometrics," *Journal of Economic Dynamics and Control*, 12, 463–74.

Sutton, J. (1997), "Gibrat's legacy," *Journal of Economic Literature*, 35, 40–59.

(1998), *Technology and Market Structure: Theory and History*, MIT Press, Cambridge, MA.

Winter, S. G., Y. M. Kaniovski, and G. Dosi (2000), "Modeling industrial dynamics with innovative entrants," *Structural Change and Economic Dynamics*, 11 (3), 255–93.

7 | Growth and diversification patterns of the worldwide pharmaceutical industry

GIULIO BOTTAZZI, FABIO PAMMOLLI,
AND ANGELO SECCHI

7.1 Introduction

This chapter investigates some statistical properties of the growth patterns of a large sample of pharmaceutical companies representing the top incumbents worldwide. We articulate our analysis along three complementary directions, which, with an evident lack of precision but, we hope, with a certain degree of suggestiveness, we denote as "temporal," "cross-sectional," and "disaggregated."

With the "temporal" direction of investigation we refer to statistical analyses of the size of the firm in its time evolution – i.e. the study of the time series of firm sizes and growth rates. The point of departure of this kind of analysis is usually a question *à la* Gibrat, addressing issues concerning the relationships between firm size and its dynamics. The Gibrat benchmark (Gibrat, 1931), also known as the "law of proportionate effect," postulates that the growth of business firms is a random walk that ultimately yields an asymptotic log-normal size distribution. The evidence concerning these properties, as explored by a rich and growing literature (see, for instance, Dunne, Roberts, and Samuelson, 1988; Evans, 1987; Hall, 1987; Hart and Prais, 1956; and Mansfield, 1962; and, more recently, Amaral et al., 1997; Lotti, Santarelli, and Vivarelli, 2001; and De Fabritiis, Pammolli, and Riccaboni, 2003), roughly supports two main findings: Gibrat's hypothesis is confirmed by the lack of any relationship between the (log) size of firms and their average growth rates, but is violated by a clear negative dependence of the growth rates' variance on size (see Sutton, 1997, for a review).

Support from the Merck Foundation (as part of the EPRIS program), from the Italian Ministry of University and Research (grant A.AMCE.E4002GD), and from the Sant'Anna School for Advanced Studies (grant E6003GB) is gratefully acknowledged.

The "cross-sectional" analyses are performed on the universe of firms considered at a certain moment in time. In literature, one finds as possible objects of these analyses a variegated set of variables intended as measures of market concentration or dispersion, together with the study of particular scaling relations among the sizes of the top firms of the sector, such as the Pareto (Pareto, 1896) or Zipf (Zipf, 1949) laws. The ultimate object of these investigations is, however, the empirical size distribution of firms. All the other variables can, in general, be obtained from this one via algebraic manipulation. Thus, in the present work, the "cross-sectional" analysis will focus on the size distribution of firms, and its evolution over time.

Finally, by "disaggregated" analysis we mean, essentially, the simultaneous analysis of the firm size and growth dynamics performed over "disaggregated" data – i.e. performed on all the different sectors (or markets) in which the firm operates. This analysis can be run on one single sector at a time or, more interestingly, can be used to study the firm's diversification structure and the relation existing between the performance of the firm in the different sectors in which it operates.

Both the "temporal" and "cross-sectional" approaches to the analysis of industrial data have a long tradition, and their relationships with economic theory can, at least in first approximation, be tracked down quite easily. Concerning the "temporal" approach, the presence of autocorrelation structure in the growth rates time series, for instance, can denote the presence of persistency in relative performances, and, hence, the presence of a high degree of heterogeneity in competitive advantages among firms, with some companies consistently performing better than others. The relationship between the average growth rate of a firm and its size can be thought of as an indicator of the presence/absence of an increasing return to scale in production, while the relationship between the variance of the growth rates distribution and the size of the firm can be explained as the reaction of the internal organization of the firm to the firm's expansion (Stanley et al., 1996), or as a simple scale effect due to independence between the productive activities taking place in the different plants, roughly of the same size, that constitute the firm.

The link between the "cross-sectional" kind of analysis and economic theory often originates from the idea that the degree of concentration in a market can be intended as an indirect measure of the level of competition existing in that market. In this respect, the "upper-tail laws" have been the subject of a plethora of models, from the venerable Champernowne analysis (Champernowne, 1953) to recent developments inside the Simon-inspired literature (among many contributions, see Krugman, 1996, and Gabaix, 1999, and the review in Kleiber and Kotz, 2003).

The third direction of analysis, in contrast, has seldom been discussed in the literature, the reason being, we believe, the lack of available data (see the recent Matia et al., 2004, and Teece et al., 1994, among the exceptions). Typically, companies' activities, even if spread across different markets, are reported at the aggregate level in their consolidated balance sheets. These consolidated data are those usually available to researchers. Moreover, even when data on firm's activities in different sectors are available, they are reported with respect to some industrial classification scheme (such as SIC or ISIC) the classes of which hardly correspond to the different markets on which the activity of the firm is organized, either at the level of production or of marketing.

In this chapter we try to merge the different approaches described above in order to provide a wider and richer statistical description of the industry. Remarkably, this kind of effort is seldom undertaken in the literature, where the different analytical approaches are often presented separately. The narrowing of the scope of the analysis is perfectly understandable when the empirical investigations are used for the purposes of statistically proving, or falsifying, a theory or model, checking for the appearance in the data of a specific property implied by that theory or model. The econometric analyses, however, do not always address a specific, theory-oriented, question, and, even when they do so, the testable implications of the theory are often so general (or vague) that a great deal of descriptive effort is required from the researcher. A perfect example is provided by the literature on the empirical tests of Gibrat's law. If it is true that the point of departure for these tests is the verification of a precise statement, namely the lack of relation between the firm's growth rate and its size, the violations of this statement can take so many forms that, without further specifications, no statistical analysis is possible in practice. It is precisely in this descriptive dimension that the "wider scope" analysis we present here is relevant.

The main aim of the work is to present a list of statistical features and discuss their degree of generality, comparing with previous analyses on different databases.

Even if the present study is of a purely empirical nature and we do not propose or discuss any theoretical explanation for the various findings, we believe that this kind of wide-scope exercise is not devoid of theoretical interest. Indeed, if one is able to specify a relatively small number of stylized facts that robustly characterize the behaviors of firms with respect to the most relevant dimensions of their activities (and the dynamics of performances, in time and with respect to competitors, and the diversification structure are definitely among them), then the design of a phenomenological model of the firms' behavior becomes possible. It is quite unlikely that one would be able to produce such a model analyzing a single aspect of the firm at a time.

Following this idea, we organize our analysis along different but complementary directions, focusing, when possible, on the study of precise "stylized facts" and exploring them with a mix of old, reliable techniques and a few new ones that, we believe, can bring new insights in the matter.

The structure of the chapter is as follows. In section 7.2 we briefly present the database under study. The statistical analyses are organized in two sections. In section 7.3 we present the analysis of the firms' size distributions and of the firms' dynamics across time. In this section we mix the "cross-sectional" and "temporal" approaches, since we present the analysis of one variable at a time. The "disaggregated" approach is presented in section 7.4. Finally, in section 7.5 we summarize the results of our analyses and offer some concluding remarks.

7.2 Data description

The statistical analyses presented in this chapter are based on data from the Pharmaceutical Industry Database produced by CERM/EPRIS, which covers sales figures for the top incumbent firms in seven Western markets (the United States, the United Kingdom, France, Germany, Spain, Italy, and Canada) over a time window of eleven years, from 1987 to 1997. This original database contains information about sales of 7654 different drugs. These single-product sales are aggregated in different micro-classes according to the Anatomical Classification System (ACS), developed and maintained by the European Pharmaceutical Marketing Research Association (EPhMRA) together with the Pharmaceutical Business Intelligence and Research Group (PBIRG) (for details, see EPhMRA, 2004). For the purposes of the

ACS, a product represents a discrete pharmacological pack or unit that can be dispensed, prescribed, or purchased. In the ACS, products are classified in different groups at four different levels. The EPhMRA/PBIRG classification has as its primary objective the satisfaction of the marketing needs of the pharmaceutical companies. This observation, together with the fact that the substitutability of products belonging to different classes (at least at the fourth level) is practically zero, suggests that this classification is able to identify clearly the different sub-markets.

In what follows we consider a balanced panel comprising the top 198 firms that are present worldwide, for the whole period 1987–97, in all the countries covered by the database. In this way we restrict our analysis to firms that can be considered multinational companies operating on a worldwide scale. This choice is dictated by the obvious problems that would arise from the pooling of firms based in different countries and operating in different national markets.

Moreover, since this work is focused on the processes of internal growth and diversification, in order to take into account M&As during the period of observation we have constructed "super-firms" that correspond to the end-of-period actual entity (so, for example, if any two firms merged during the observed history, we consider them merged from the start). This procedure clearly biases intertemporal comparisons on actual size distributions, but it helps to highlight those changes in the distributions that are due to processes of intra-market competition and inter-market diversification.

Given the overwhelming problems of deflation in this industry, all sales in the database are in dollars at current prices and exchange rates; therefore, the analyses entail a normalization procedure (for example, in terms of market shares or equivalent measures) to get rid of any nominal effect.

7.3 Firm size distributions and the dynamics of growth

We structure our investigation of the size distribution and growth dynamics of firms in three successive steps, analyzing in turn three aspects of the firm growth process that have been widely discussed in the literature and are commonly suggested as typical characterizations of both firm behavior and market structure.

First, we start with a non-parametric analysis of the size distribution of firms, focusing in particular on the degree of stationarity and on the analysis of its peculiar shape. Second, we take a parametric approach and analyze the autoregressive structure of the firm growth process. Third, we move to a multivariate analysis and, with the help of both parametric and non-parametric methods, explore the joint probability density of the size and of the growth rate of firms, and the possible relations between different moments of the firm growth rates' marginal density and firm size.

Since we believe that the use of non-parametric methods is still establishing itself inside the empirical literature on industrial dynamics and cannot be considered a standard component of the industrial economist's "tool bag," when we introduce these methods we accompany them with a short description and a few references.

7.3.1 Size distribution

In this study, as mentioned in section 7.2, we proxy firms' size by their total sales. Let $S_i(t)$ be the size (total sales) of firm i at time t. Due to the possible presence of a nominal trend, we introduce a normalized version of the (log) size obtained by subtracting from each observation at time t the average (log) size of all the firms at the same time,

$$s_i(t) = \log(S_i(t)) - \frac{1}{N} \sum_{i=1}^{N} \log(S_i(t)) \tag{7.1}$$

where I stands for the total number of firms in the panel.

We start our analyses with the study of the basic properties of the firms' size distribution. To this purpose, we build a kernel estimation of the probability density of the firm size at a given time t. The kernel density estimate based on the observations $s_1(t), \ldots, s_N(t)$ is defined (Silverman, 1986) as

$$\hat{f}(x, t; h) = \frac{1}{Nh} \sum_{i=1}^{n} K\left(\frac{x - s_i(t)}{h}\right) \tag{7.2}$$

where h is a bandwidth parameter controlling the degree of smoothness of the density estimate and K is a kernel density – i.e. $K(x) \geq 0$, $\forall x \in (-\infty, +\infty)$ and $\int dx K(x) = 1$. The kernel density estimate can be considered a smoothed version of the histogram obtained by counting the observations in different bins. The density estimation in (7.2) relies, however, on the provision of two objects: the kernel K and

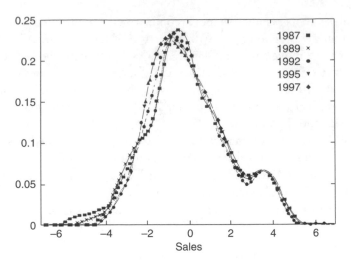

Figure 7.1. Kernel density estimates of normalized (log) sales for five different years

the bandwidth h. It turns out that the obtained result is quite insensitive to the choice of the first[1] but depends strongly on the second.

Many methods have been proposed for the choice of the optimal value for h, but none can be considered infallible. For qualitative purposes, however, it is sufficient to follow the heuristic method suggested in Silverman (1986) and to take a bandwidth value

$$h = \frac{0.9}{n}A \tag{7.3}$$

where n is the number of observations and A is the minimum between the sample standard deviation and the sample interquartile range divided by 1.34. In what follows, if not stated otherwise, we apply this rule.

Following the above procedure, we compute a kernel density estimate of the firm-normalized log size, $s_i(t)$. In figure 7.1 we report the estimated density for five different years: 1987, 1989, 1992, 1995, and

---

[1] Throughout this chapter the kernel function will always be the most efficient Epanenchnikov density. The use of different kernels, such as the Gaussian or the triangular, does not change our results noticeably.

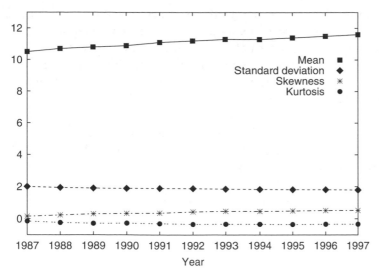

Figure 7.2. The time evolution over the range 1987–1997 of the mean, standard deviation, skewness, and kurtosis of the firm size distribution

1997. Visual inspection of this plot reveals two notable features. First, the density seems to display a remarkable degree of stationarity. This finding is confirmed by figure 7.2, where the time evolution of the first four moments of $\log(S)$ is plotted over the whole range 1987–97. Except for an expected upward trend in the mean, due possibly to a nominal effect and removed with the procedure in (7.1), all the moments reported appear very stable over time.

The second relevant feature of the probability densities reported in figure 7.1 is their apparent bimodal shape. Even if the presence of bimodality seems indubitable from the picture, one can suspect that its origin depends on an under-smoothing of the density due to the choice of too small a value for the parameter h in (7.2). To dispel this suspicion we need some quantitative method to assess the degree of significance of the observed bimodality. To this end, we continue our non-parametric approach and follow a technique, proposed in Silverman (1981) and based on the kernel density estimate, that consists essentially of two steps.

Suppose that we wish to test the null hypothesis that a given estimated density f possesses at most k modes against the alternative that the same f possesses more than k modes. First, we need to compute the "critical value" h^* for the bandwidth parameter, defined as the larger value of the parameter h that guarantees a kernel density estimate (7.2) with at least k modes. This definition is meaningful, since the number of modes is a decreasing function of the bandwidth parameter:[2] for $h > h^*$ the formula in (7.2) would give an estimated density with fewer than k modes, while for $h \leq h^*$ the estimated density would have at least k modes.

Second, once the value h^* has been found we need to assess its significance. Suppose that the true density under the null hypothesis f possesses exactly k modes, and let the critical bandwidth obtained with n observations be h_0. Then the probability of getting a critical bandwidth larger than h_0 under the true density f, $Pr\,(h_{crit} > h_0)$, is defined according to

$$Pr\left\{\hat{f}(s; h_0) \text{ has more than k modes}/\{s_1, \ldots, s_N\} \text{ is drawn from f}\right\}$$

$$(7.4)$$

To compute this probability one can repeatedly draw n observations from the true density f and count the modes of the kernel density estimate $\hat{f}(s;\, h_0)$ obtained from these observations. The fraction of times in which these modes are greater than k is an estimate of the probability in (7.4) and of the p-value of the statistics $p(h^*)$. The problem with this method is that, in general, the underlying true density is unknown. Silverman (1981) suggests, as a natural candidate density from which to simulate, a rescaled version of $\hat{f}(s;\, h_0)$ derived from data equating the variance of $\hat{f}$ with the sample variance. Hall and York (2001) show that this choice is biased towards conservatism, and propose an improved procedure to achieve asymptotic accuracy. Following their suggestion we compute, pooling the observations of all the years together, the critical bandwidth h^* and the p-values of the test where the null is "the (log) size distribution is unimodal" and the alternative is "the (log) size distribution presents more than one mode." Our results confirm the first impression obtained from figure 7.1, since we reject the null hypothesis at the 1 percent significance level.

---

[2] To be precise, this has been proven to be true only for a relatively small family of kernels. In any event, the kernels that we use belong to this family. For further details, see Silverman (1981).

7.3.2 Gibrat violation and the growth rates density

A large part of the empirical literature on firm dynamics has focused, since the early contributions, on the relations between the size of companies and their growth rates. The point of departure for these studies is often the test of Gibrat's law of proportionate effect.

Consider the (log) growth rates, defined as the first difference of (log) size according to

$$g_i(t) = s_i(t + 1) - s_i(t) \qquad (7.5)$$

Notice that, from (7.1), the distribution of the g's is by construction centered around zero for all t.

A detailed analysis of the autoregressive structure of the time series defined by firms' sales and growth rates is presented in Bottazzi and Secchi (2005). We do not repeat this analysis here. The results can be summarized by saying that the analysis of the size times series $\{s_i(t)\}$ reveals a clear unit root nature. Moving to the first-difference process $\{g_i(t)\}$, a weak autoregressive structure has been found, which, however, accounts for less than 10 percent of the time series variability. In the following section we take a different approach and consider the linear regression of conditional central moments of the firm growth rates on the firm size.

We partition the firms into twenty bins (quantiles) according to their (log) sizes $s_i(t)$. We then compute the mean, standard deviation, skewness, and kurtosis of the growth rates $g_i(t)$ of the firms in each bin and we fit the linear relation

$$\Psi(g|s) = \alpha + \beta s \qquad (7.6)$$

where $\Psi(g|s)$ is in turn the mean, the log of the standard deviation, the skewness, and the kurtosis of the growth rates of the firms belonging to the bin with an average size of s. Results of these fits are reported in table 7.1.

First, no clear relation emerges between the average growth rate and the size of the firm. Analogously, we do not find any evidence that skewness or kurtosis scale with the size of the firm. In contrast, we observe a strongly significant negative relation between the size and the standard deviation of growth rates: the estimated slope is $\beta = -0.21$ with a standard error of 0.02. This coefficient is strikingly similar to the one found in US manufacturing industry by Amaral et al. (1997) and

Table 7.1 *Slopes and standard errors of the linear fit in*
equation (7.6) of the relations between firm size and the
first four moments of the growth rate density

Relation	β	Standard error
Size–mean	−0.11	.08
Size–standard deviation	**−0.21**	.02
Size–skewness	−0.12	.09
Size–kurtosis	−0.03	.11

Note: The only significant slope is in bold.

Bottazzi and Secchi (2003). In figure 7.3 we report, on a log–log scale, the growth rates' standard deviation for each bin versus the average bin size together with the linear fit in equation (7.6). The significance of the observed slope is very high (more than .99), but we are also interested in investigating if the fitted line provides "locally" a good description of the data behavior. To this end we exploit the tool of non-parametric local kernel regression. Consider a set of coupled observations (x_i, y_i) with $i = \{1, \ldots, n\}$. One is interested in knowing what relation, if any, exists between the two variables and to obtain a model-free description of this relation. From a statistical point of view the relation between y and x is well described by the conditional expectation $E(y\,|\,x)$ of the former variable once the latter is specified. A kernel method can be used to estimate this quantity. With K denoting the kernel and h the band-width, one has[3]

$$\hat{E}(y|x;h) = \frac{1}{\hat{f}(x;h)} \frac{1}{n\,h} \sum_{i=1}^{n} y_i\, K\!\left(\frac{x - x_i}{h}\right) \tag{7.7}$$

where $\hat{f}$ is the estimation of the independent variable density obtained with the same kernel K and the bandwidth h according to (7.2). The previous method can easily be extended to higher central moments of the conditional distribution $E[y^m\,|\,x]$:

$$\hat{E}(y^m|x;h) = \frac{1}{\hat{f}(x;h)} \frac{1}{n\,h} \sum_{i=1}^{n} y_i^m\, K\!\left(\frac{x - x_i}{h}\right) \tag{7.8}$$

[3] See Pagan and Ullah (1999, sec. 3.2) for a detailed description of the method.

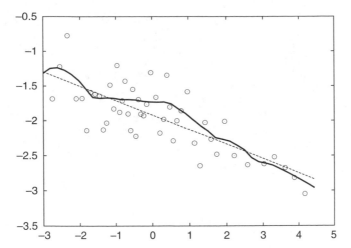

Figure 7.3. The relation between firm size and the standard deviation of growth rates (on a log scale) together with the fit $\log(\sigma(g|s)) = \alpha + \beta s$ and a non-parametric estimate of the relation between $\log(\sigma(g|s))$ and s

In the present case we are interested in the relation between firm size and the standard deviation of its growth rates. From the above equations we can estimate the behavior of this relation using

$$\hat{E}(\sigma_y|x;h) = \sqrt{\hat{E}(y^2|x;h) - \hat{E}(y|x;h)^2} \qquad (7.9)$$

The non-parametric local regression obtained using (7.9) and the coupled observations $(s_i(t), g_i(t))$ are reported in figure 7.3 with a solid line. As can be seen, the non-parametric local regression remains near the linear fit for the whole support of s, suggesting that the latter provides a reliable description of the variance–size relation for any firm size.

In conclusion, the dependence of the growth rate on the size of the firm is completely described by the variance–size relation. We can exploit this statistical property to define a "scale-invariant" growth variable. Using equation (7.6), we define the "rescaled" growth rates according to

$$\hat{g}_i(t) = g_i(t)/\exp(\alpha + \beta s_i(t)) \qquad (7.10)$$

with α and β as estimated in (7.6). These rescaled growth shocks have, by construction, unit variance and zero mean, and possess statistical properties that are independent from the size of the firm. It is then possible to perform a univariate analysis on these variables, pooling

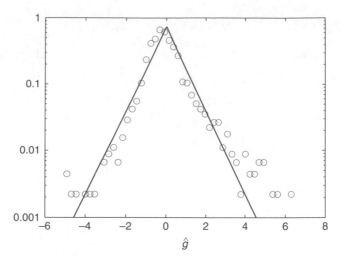

Figure 7.4. Log of the empirical densities of rescaled growth rates $\hat{g}$ together with a Laplace fit; growth rates of different years are pooled together

together rescaled growth rates coming from firms in different size bins, without the risk of introducing "artificial" statistical properties.

In figure 7.4 we plot the empirical probability densities of the rescaled growth rates $\hat{g}$ on a log scale. This density displays a characteristic tent shape very similar to a Laplace (symmetric exponential) density.[4] In Bottazzi and Secchi (2005) a large family of densities, known as a Subbotin family (Subbotin, 1923), is used via a maximum likelihood procedure on the same database to provide statistical support for the hypothesis of a Laplace behavior.

As can be seen in figure 7.4, the agreement between the empirical density and the Laplace fit is quite good, especially in the central part. The right tail, however, appears fatter in the data than in the Laplace fit. In fact, the empirical density seems asymmetric, with the right tail fatter than the left one. The original Subbotin family, however, contains only symmetric density and is, consequently, unable to provide any insight into the possible asymmetric nature of the growth rates. As a result, we extend our previous analysis to consider the family of asymmetric Subbotin densities. This family is defined by five

---

[4] Remember that an exponential density, when plotted on a log scale, appears as a straight line.

Table 7.2 Coefficients of the asymmetric Subbotin densities estimated on the rescaled growth rates ĝ

Parameter	Estimate	Standard error
b_l	.99	.10
b_r	.82	.05
a_l	.49	.04
a_r	.66	.03
m	−0.23	.08

Note: Standard errors have been obtained with bootstrap analysis based on 500 independent resamplings.

parameters: a positioning parameter μ, two scale parameters a_l and a_r, and two shape parameters b_l and b_r characterizing the width and the shape of the left and the right part of the density, respectively. Its functional form reads

$$
f_S(x) = \begin{cases} \dfrac{1}{2ab_l^{1/b_l}\Gamma(1/b_l+1)}\, e^{-\frac{1}{b_l}\left|\frac{x-\mu}{a_l}\right|^{b_l}} & x \leq \mu \\[3ex] \dfrac{1}{2ab_r^{1/b_r}\Gamma(1/b_r+1)}\, e^{-\frac{1}{b_r}\left|\frac{x-\mu}{a_r}\right|^{b_r}} & x > \mu \end{cases}
\tag{7.11}
$$

where $\Gamma(x)$ is the gamma function. The smaller the shape parameter b, the fatter the relative tail. For $b = 2$ the relative tail displays a Gaussian behavior, which becomes exponential (Laplace) for $b = 1$.

Using a maximum-likelihood procedure, as detailed in Bottazzi (2003), we estimate the five parameters on the observation $\hat{g}$, obtaining for b_l and b_r values of .99 and .82, respectively.[5] The estimated coefficients are reported in table 7.2, together with their standard errors. Due to the non-smooth nature of the asymmetric Subbotin distribution, the estimation of these errors using the Fisher information matrix can be troublesome, especially for the parameters in the region we are interested in. The values provided in table 7.2 have been obtained with a bootstrap technique using repeated sampling with replacement from the original data. As can be seen, the values of the two tail coefficients

---

[5] The maximization is performed after removing three extreme observations from the total of 1980.

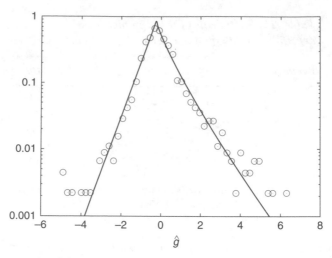

Figure 7.5. Log of the empirical densities of rescaled growth rates $\hat{g}$ together with an asymmetric Subbotin fit; growth rates of different years are pooled together

b_l and b_r are quite a way apart, but less than two standard errors. A deeper analysis of the problem, however, reveals that the maximum-likelihood estimates of the two tail coefficients, b_l and b_r, are not orthogonal. Consequently, a better quantity by which to judge the real presence of asymmetry in the growth rates density is provided by the distribution of the difference of the two coefficients, $b_l - b_r$. This difference has a negative value in only 2 percent of the bootstrapped estimations, so the hypothesis of a positive difference is confirmed with a relatively high significance.

These results suggest that, while the left part of the density is almost exactly described by a Laplace, the right part presents a fatter tail and decreases less than exponentially. In figure 7.5 we report the empirical probability densities of the rescaled growth rates $\hat{g}$ together with the asymmetric Subbotin fit.

7.4 Firm diversification

As mentioned in section 7.2, one of the peculiar features of the PHID database is the possibility of disaggregating the sales figures of each individual firm down to the fourth level of the ACS. This level of disaggregation permits the identification of sub-markets that are "specific" enough to be considered, roughly speaking, the loci of competition between firms: the products belonging to a given sub-market possess similar therapeutic characteristics and can then be considered substitutable, while products belonging to different sub-markets are usually targeted at different pathologies. Moreover, Orsenigo, Pammolli and Riccaboni (2000) suggest that this disaggregation level is the one at which single research projects develop. This evidence supports the hypothesis that the different therapeutic sub-markets provide the natural "scale" at which firm diversification should be studied.

In the present section we use data disaggregated at the fourth level of ACS to study the diversification structure of firms, looking at the scope of their activities, at the relative size with which a firm is present in different sub-markets, and at the possible relation between the performance of a firm in the different sub-markets in which it is present.

To maintain a notation similar to that of section 7.3, let $S_{i,j}(t)$ be the size (sales) of firm i in sub-market j at time t and $N_i(t)$ the number of sub-markets in which it operates ("active" sub-markets). We denote with lower-case letters the log size $s_{i,j}(t) = \log(S_{i,j}(t))$ and the (log) growth rates[6] $g_{i,j}(t) = s_{i,j}(t + 1) - s_{i,j}(t)$.

7.4.1 Scope–scale relation

We start our "disaggregated" investigation with the analysis of the relation between the number of sub-markets in which a firm is active $N(t)$ and its size $s(t)$. In figure 7.6 we plot on a log–log scale the average number of active sub-markets of the firms belonging to different size bins against the average bin size. A clear positive relation between the two variables emerges.

Then we fit a linear relation between the firm (log) aggregated size and the log of the number of its active markets

$$\log(N_i(t)) \sim A \log(S_i(t)) + B \qquad (7.12)$$

We obtain a strongly significant slope $A = 0.35$ with a standard error of 0.01. In figure 7.6 we report the linear fit in equation (7.12) and a

---

[6] Note that the disaggregated variables used in this section, unlike the aggregated log variables defined in section 7.3, are not detrended.

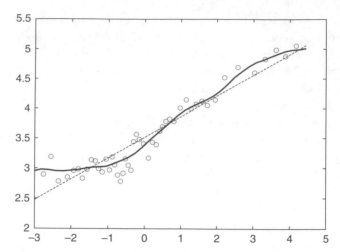

Figure 7.6. Log of the number of active sub-markets versus firm size together with a linear fit (dotted line) and a non-parametric estimate of the relation

non-parametric estimate (kernel local regression) of the relation between the size and number of sub-markets obtained with (7.7).

Next, we investigate if size has any role in shaping the relation between firm growth dynamics and the evolution of the number of active sub-markets. In order to do this we split the whole sample of firms into two size groups: for each year we consider the average firm size, and put firms with a size lower than the average into the first class and firms larger than the average into the second. For each class separately we analyze the relation between firm growth rates $g_i(t)$ and the variation in the number of its active sub-markets $\delta N_i(t) = N_i(t+1) - N_i(t)$. Figures 7.7 and 7.8 report the scatter plots of $\delta N_i(t)$ versus $g_i(t)$ for the two groups. They clearly suggest that small and large firms are characterized by different ways in which growth shocks are linked with modification in the number of sub-markets they are active in. First of all, note that the scales in the two plots are different: relatively small firms present several events in which more than ten new sub-markets are entered during one year, whereas, inevitably, larger firms – already being active in many sectors – cannot display such large increases in the scope of their activities. A second difference concerns the way in which growth is linked with entry in new sub-markets. For small firms (figure 7.7) a large growth event involves, in general, a large

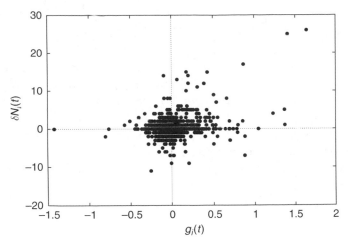

Figure 7.7. Scatter plot of the variation in active sub-markets δN_i versus firm growth rates g_i for smaller than average firms

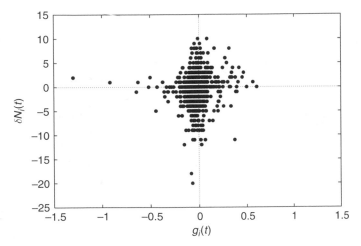

Figure 7.8. Scatter plot of the variation in active sub-markets δN_i versus firm growth rates g_i for larger than average firms

dynamics in the number of submarkets. The opposite seems true for big firms, where even small growth shocks are possibly associated with an intense dynamics of opening or closing new sub-markets, hinting at a more relevant role for diversification strategies.

7.4.2 Sub-market cross-correlations

Let $\bar{g}_{i,j}$ be the sample average of the growth rate of firm i in sub-market j in the ten years covered by our database and $\sigma_{i,j}$ the sample standard deviation. If the firm is not active in the sub-market j for the whole period, we count only the couple of consecutive years in which it is present, so that a growth rate can be defined. Let $T_{i,j,j'}$ denote the set of years in which it is possible to define the rate of growth of firm i in both sub-markets j and j' (for our database, this set contains at most ten elements).

We consider the covariance of the growth rates of firm i in sub-markets j and j', with $j \neq j'$ defined as

$$c_{i,j,j'} = \frac{1}{T_{i,j,j'}} \sum_{t \in T_{i,j,j'}} (g_{i,j}(t) - \bar{g}_{i,j}) \, (g_{i,j'}(t) - \bar{g}_{i,j'}) \qquad (7.13)$$

where $T_{i,j,j'}$ denotes the number of elements of $T_{i,j,j'}$. The associated cross-correlation coefficient is defined according to

$$r_{i,j,j'} = \frac{c_{i,j,j'}}{\sigma_{i,j} \, \sigma_{i,j'}} \qquad (7.14)$$

We ignore sub-markets in which the firm never operates for three consecutive years, in order to have a meaningful definition of the growth rate sample variance. For the nature of the data, the variables in (7.13) and (7.14) are based on estimates of the central moments obtained with very small samples (at most ten observations) and are, consequently, extremely "noisy." Nevertheless, we are not interested in the cross-correlation existing between the performance of a firm in two specific sub-markets but in an average value, describing the degree of cross-correlation among all the active sub-markets of a firm. Let M_i denote the set of sub-markets in which firm i operates for more than three consecutive years. We consider the average value of the quantity defined in (7.14) over all the possible couples (j, j') with $j, j' \in M_i$ and $j \neq j'$. The average cross-correlation coefficient for firm i reads

$$r_i = \frac{1}{M_i(M_i - 1)} \sum_{j \in M_i} \sum_{j' \in M_i, j' \neq j} r_{i,j,j'} \qquad (7.15)$$

where M_i is the number of elements of M_i.

The distribution of the coefficients r on the population of firms has a mean value of .0063 and a standard deviation of .057. Very surprisingly, the coefficients are sharply centered around zero and the average cross-correlation is absent not only "on average" but for all the firms in the database. We can conclude that the growth processes of a firm in different active sub-markets are independent processes.

7.4.3 Diversification patterns

As the final step in our analysis of firms' diversification, we focus on the relation between the size of the firm and its "diversification pattern" – i.e. the way in which the total sales of the firm are distributed in its active markets. One can indeed think of different ways in which the activities of a company can potentially be distributed in the markets in which it operates. For instance, it may be the case that a given company realizes sales of a comparable level in its different active markets, so that the aggregate company size is the sum of several components of comparable size. On the other hand, it is also possible that the diversification pattern of a company consists of a very strong market, in which the company realizes almost all its sales, and a group of smaller markets, in which the company is present but its sales are relatively low. Relatedly, it is interesting to check if the diversification pattern of a firm bears any relation with the firm's size.

For the present analysis, we need an index that captures the degree of heterogeneity in the diversification pattern of a firm. Let us define the "diversification heterogeneity" index as

$$\tilde{\Delta}_i(t) = \frac{\sqrt{\frac{1}{N_i(t)} \sum_{j=1}^{N_i(t)} (\Delta_{i,j}(t) - 1)^2}}{\sqrt{N_i(t)}} \qquad (7.16)$$

where $\Delta_{i,j}(t) = N_i(t) \, S_{i,j}(t)/S_i(t)$.

It is easy to see that $\Delta_i(t)$ ranges from zero to one. In fact, if a firm evenly distributes its sales among its active markets, it is

$$\frac{S_{i,j}(t)}{S_i(t)} = \frac{1}{N_i(t)} \qquad \forall j$$

and then $\tilde{\Delta}_i(t) = 0$. On the other hand, if the sales of the firm are completely concentrated in a single market j' while it retains negligible positions in the others, we have

$$\frac{S_{i,j'}(t)}{S_i(t)} \sim 1 \quad \text{and} \quad \frac{S_{i,j}(t)}{S_i(t)} \sim 0 \qquad \forall j \neq j'$$

so that the index $\tilde{\Delta}$ takes its maximum value of one.

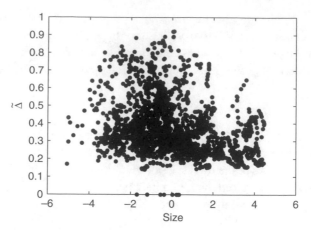

Figure 7.9. Scatter plot of firm size s_i versus our measure of diversification heterogeneity $\tilde{\Delta}_i$, for all firms and all years

In figure 7.9 we report a scatter plot of the size $s_i(t)$ versus the diversification index $\tilde{\Delta}_i(t)$ for all the firms i and all the years t in our database. As can be seen, the support of the $\tilde{\Delta}$'s spans almost the whole range zero to one, but the great majority of firms have values around .3. The scatter plot appears like an amorphous blob, suggesting the lack of any relation between the size of the firm and its diversification pattern. To verify this impression we fit a linear relation between s_i and $\tilde{\Delta}$, finding a slope of -0.025 (with a standard error of .0035). The negative slope is significant, and is essentially due to the presence of a few large values for $\tilde{\Delta}$ associated with small and average-sized firms. Nevertheless, the value of the slope is so small that we do not dare to give it any economic relevance.

Hence, we conclude that, in the pharmaceutical industry, we do not observe any particular difference between large and small firms in the way in which they distribute their sales among different sub-markets.

7.5 Generic and particular properties of the worldwide pharmaceutical industry

In this chapter we have presented a wide range of analysis on the worldwide pharmaceutical industry, in which we have taken into consideration both the short-run dynamics of the sectoral structure and the behavior of individual firms.

One of the aims of the investigation was to check if well-known "stylized facts" about the growth dynamics of firms also characterize the worldwide pharmaceutical industry, or if this industry presents some distinctive features. Our results are mixed, even if, at least as far as the general properties analyzed in this study are concerned, they suggest the emergence in the pharamaceutical industry of almost the same regularities observed when other sectors are studied.

Indeed, we found a unit root structure for the growth process and a lack of relation between firms' average growth rates and their sizes, which also characterizes many sectors of US (Bottazzi and Secchi, 2003) and Italian (Bottazzi et al., 2003) manufacturing industry and which can be considered common to the majority of the analyses found in the literature, at least when balanced panels are considered and the effect of the entry and exit of firms is ignored (for a review of the early contributions, see Sutton, 1997). In principle, it would be interesting to check the effect that the inclusion of entry/exit processes could have on the statistical properties we found. Notice, however, that the database we used covers only the top pharmaceutical firms. In our case, though, an extension of the analysis to an unbalanced panel in order to take into account the entry and exit dynamics would be, from an economic point of view, meaningless, since the "entry" and "exit" processes would refer to the scope of the database, not to the actual industry. We can safely conclude, therefore, that our findings are valid, at least as far as the core of the industry is concerned.

We also found a scale relation between the size of the firm and the variance of its growth rates. It is difficult to say if this is an industry-specific property or has a more general validity. While there are several studies, mainly focused on large US companies, that report the presence of this property (Hymer and Pashigian, 1962; Mansfield, 1962; Amaral et al., 1997; Bottazzi and Secchi, 2003), we have shown recently that it is absent in all the sectors of Italian manufacturing industry (Bottazzi et al., 2003). Undoubtedly, more studies, and on more heterogeneous data, are needed before deciding about the degree of generality of this "second-order" violation of Gibrat's law.

In contrast, one characteristic that can be considered peculiar to the pharmaceutical industry is the observed bimodal shape of its firms' size distribution. In our analysis we found that this bimodality derives from the existence of an industry "core," composed of a relatively small group of firms[7] that persistently hold a predominant position in the market. In this respect, an interesting exercise would be to study the emergence process of this group of firms, analyzing the similarities and diversities of their growth dynamics with respect to the rest of the industry. Ideally, one would hope to be able to explain the evolution of this dichotomous structure as an effect of the particular competition processes that have characterized the pharmaceutical industry since World War II. Nevertheless, as is clear from section 7.2, the time horizon of our database did not allow us to follow this direction of research. In fact, the bimodal structure of the pharmaceutical industry was already present at the beginning of the period of observation considered in our analysis, and, as discussed in section 7.3, the length of the time window we considered was not enough to observe any relevant change in the firm size distribution. Moreover, the formation of an industrial "core" of relatively large firms can, plausibly, be the outcome of the intense mergers and acquisitions activity that we know has characterized the pharmaceutical industry in the last forty years, but that has been deliberately ignored in the present analysis, which has focused on the properties of "internal" growth.

In a different perspective, a few interesting results also come from the analysis of the firms' diversification structure. Essentially, we found a substantial lack of structure in the way in which firms diversify. We were unable to identify any robust relation between firms' size and their diversification "pattern." At the same time, we did not find any evidence for the existence of cross-correlation between the growth dynamics of firms in the different sub-markets in which they operate. Do not forget, however, that we considered only multinational firms – that is, very large companies operating in all the major Western markets. This picture could change if the analysis were extended to pharmaceutical firms of a smaller size and operating in only a few national markets.

Concerning the extent of our database, however, we can conclude that the way in which firms distribute their activities in different markets seems independent from their sizes and performances, and the degree of "diversification" of a firm is simply described by the

---

[7] There are more or less twenty of them, including firms such as Abbot, Roche, Bayern, Merck, Novartis, Pfizer, Schering, GlaxoSmithKline, and Bristol–Myers.

number of sub-markets in which it operates. Moreover, we found that the relation between the size of the firm and the number of its active sub-markets is governed by a power–law scaling. These properties are suggestive, and hint at the possibility of developing stochastic models *à la* Gibrat for the diversification dynamics of firms. A first effort in this direction has been made in Bottazzi (2001), where a model is proposed that describes the diversification dynamics as a simple stochastic process characterized by a self-reinforcing mechanism that cumulatively drives the "proliferation" of active sub-markets. This model can easily account for the observed scaling relation between the size of the firm and the number of active sub-markets.

A better understanding of the nature and of the degree of generality of the various "stylized facts" reported above – and, ultimately, the potential for acquiring a satisfactory phenomenological understanding of firm growth processes – would require extending the analysis to different industries, sectors, and countries, in order to identify both the general regularities and the sectoral specificities that govern these processes. We believe, however, that – for this extension to be effective – it is essential to move on from the "temporal" models *à la* Gibrat, and from the Pareto–Zipf "cross-sectional" tradition, encompassing them, to develop a wider, more descriptive, approach to the empirical study of firm dynamics.

References

Amaral, L. A. N., S. V. Buldyrev, S. Havlin, M. A. Salinger, H. E. Stanley, and M. H. R. Stanley (1997), "Scaling behavior in economics: the problem of quantifying company growth," *Physica A*, 244, 1–24.

Bottazzi, G. (2001), *Firm Diversification and the Law of Proportionate Effect*, Working Paper no. 2001–1, Laboratory of Economics and Management, Sant'Anna School for Advanced Studies, Pisa.

  (2003), *Subbotools: A Users Manual*, available at http://www.sssup.it/~bottazzi.

Bottazzi, G., E. Cefis, G. Dosi, and A. Secchi (2003), *Invariances and Diversities in the Evolution of Manufacturing Industry*, Working Paper no. 2003–21, Laboratory of Economics and Management, Sant'Anna School for Advanced Studies, Pisa.

Bottazzi, G., and A. Secchi (2003), "Common properties and sectoral specificities in the dynamics of U.S. manufacturing companies," *Review of Industrial Organization*, 23, 217–32.

(2005), "Growth and diversification patterns of the worldwide pharmaceutical industry," *Review of Industrial Organization*, 26, 195–216.

Champernowne, D. G. (1953), "A model of income distribution," *Economic Journal*, 163, 318–51.

De Fabritiis, G., F. Pammolli, and M. Riccaboni (2003), "On size and growth of business firms," *Physica A*, 324, 38–44.

Dunne, T., M. J. Roberts, and L. Samuelson (1988), "The growth and failure of U.S. manufacturing plants," *Quarterly Journal of Economics*, 104, 671–98.

EPhMRA (2004), *Anatomical Classification Guidelines*, available at http://www.ephmra.org/pdfs/ATCguidelines.pdf.

Evans, D. S. (1987), "The relationship between firm growth, size and age: estimates for 100 manufacturing industries," *Journal of Industrial Economics*, 35, 567–81.

Gabaix, X. (1999), "Zipf's law for cities: an explanation," *Quarterly Journal of Economics*, 114, 739–67.

Gibrat, R. (1931), *Les inégalités économiques*, Librairie du Recueil Sirey, Paris.

Hall, B. H. (1987), "The relationship between firm size and firm growth in the U.S. manufacturing sector," *Journal of Industrial Economics*, 35, 583–606.

Hall, P., and M. York (2001), "On the calibration of Silverman's test for multimodality," *Statistica Sinica*, 11, 515–36.

Hart, P. E., and S. J. Prais (1956), "The analysis of business concentration," *Journal of the Royal Statistical Society*, 119, 150–91.

Hymer, S., and P. Pashigian (1962), "Firm size and rate of growth," *Journal of Political Economy*, 70, 556–69.

Krugman, P. (1996), *The Self-Organizing Economy*, Basil Blackwell, Cambridge, MA.

Kleiber, C., and S. Kotz (2003), *Statistical Size Distributions in Economics and Actuarial Sciences*, Wiley, New York.

Lotti, F., E. Santarelli, and M. Vivarelli (2001), "The relationship between size and growth: the case of Italian newborn firms," *Applied Economics Letters*, 8, 451–4.

Mansfield, E. (1962), "Entry, Gibrat's law, innovation and the growth of firms," *American Economic Review*, 52, 1023–51.

Matia, K., D. Fu, A. O. Schweiger, S. V. Buldyrev, F. Pammolli, M. Riccaboni, and H. E. Stanley (2004), "Statistical properties of structure and growth of business firms," *Europhysics Letters*, 67, 493–503.

Orsenigo, L., F. Pammolli, and M. Riccaboni (2000), "Technological change and network dynamics: lessons from the pharmaceutical industry," *Research Policy*, 30, 485–508.

Pagan, A., and A. Ullah (1999), *Nonparametric Econometrics*, Cambridge University Press, Cambridge.

Pareto, V. (1896), *Cours d'Economie Politique*, Droz, Geneva.

Silverman, B. W. (1981), "Using kernel density estimates to investigate multi-modality," *Journal of the Royal Statistical Society B*, 43, 97–9.

(1986), *Density Estimation for Statistics and Data Analysis*, Chapman and Hall, London.

Stanley, M. H. R., L. A. N. Amaral, S. V. Buldyrev, S. Havlin, H. Leschhorn, P. Maass, M. A. Salinger, and H. E. Stanley (1996), "Scaling behavior in the growth of companies," *Nature*, 379, 804–6.

Subbotin, M. T. (1923), "On the law of frequency of errors," *Matematicheskii Sbornik*, 31, 296–301.

Sutton, J. (1997), "Gibrat's legacy," *Journal of Economic Literature*, 35, 40–59.

Teece, D., R. Rumelt, G. Dosi, and S. G. Winter (1994), "Understanding corporate coherence: theory and evidence," *Journal of Economic Behavior and Organization*, 23, 1–30.

Zipf, G. (1949), *Human Behavior and the Principle of Least Effort*, Addison-Wesley, Cambridge, MA.

8 | Entry, market structure, and innovation in a "history-friendly" model of the evolution of the pharmaceutical industry

CHRISTIAN GARAVAGLIA, FRANCO
MALERBA, AND LUIGI ORSENIGO

8.1 Introduction

In this chapter we explore the relationships between entry, market structure, and innovation in the pharmaceutical industry, on the basis of a previous "history-friendly" model (HFM) developed by Malerba and Orsenigo (2002). The motivations underlying this modeling style have been discussed extensively in previous papers (Malerba et al., 1999, 2001), and we will not come back to this issue here. For the purposes of the present chapter, suffice it to say that HFMs are an approach to the construction of formal evolutionary economic models aiming at capturing – in stylized form – qualitative theories about mechanisms and factors affecting industry evolution and technological and institutional change suggested by empirical research.[1]

In this respect, the pharmaceutical industry constitutes an ideal subject for history-friendly analysis. Pharmaceuticals are traditionally a highly R&D-intensive sector, which has undergone a series of radical technological and institutional "shocks." However, the core of leading innovative firms and countries has remained quite small and stable for a very long period of time, but the degree of concentration has been consistently low, whatever the level of aggregation being considered.

In a previous study (Malerba and Orsenigo, 2002), we claimed that the patterns of the evolutionary dynamics that emerged were related to the following main factors.

(a) The nature of the processes of drug discovery – i.e. to the properties of the space of technological opportunities and of the search procedures through which firms explore it. Specifically, innovation processes have been characterized for a very long time by low

[1] For discussion about the history-friendly modeling style, see also Garavaglia (2004), Werker and Brenner (2004), and Windrum (2004).

234

degree of cumulativeness and by "quasi-random" procedures of search (random screening). Thus, innovation in one market (a therapeutic category) does not entail higher probabilities of success in another one.

(b) The fragmented nature of the relevant markets. The pharmaceutical market is actually composed of a large number of independent sub-markets (therapeutic categories); for example, cardiovascular products do not compete with antidepressants. And, given the "quasi-random" nature of the innovative processes, innovation in one therapeutic category has few consequences for the ability to innovate in another market.

(c) The type of competition and the role of patents. Pharmaceuticals represent a case where competition is similar to the model of patent races, even despite the relevance of processes of imitation and "inventing around."

The model was encouragingly successful – in our view – in reproducing the main stylized facts of the evolution of the pharmaceutical industry. Moreover, the model responded, according to our appreciative understanding, to changes in key parameters concerning costs and economies of scale, the structure of demand, the features of opportunity conditions. The main result – in a nutshell – was that the model is quite robust to these changes in its essential features: it is quite difficult to raise concentration substantially, with the exception of substantial cost increases in drug discovery and development.

A key issue that was only partially addressed in the previous study, however, concerned entry. Despite the observed low levels of concentration, the entry of new firms has not been a relevant phenomenon in pharmaceuticals, especially with regard to the innovative core of the industry, until the advent of the molecular biology revolution. And, even in that case, new entrants were not able to displace incumbents. Why, given such low levels of concentration, has the industry not been characterized by any significant *de novo* entry of new firms? And why have biotechnology firms not been able – on the whole – really to challenge older industry leaders?

In the previous study, entry was treated as an exogenous phenomenon, and it was essentially linked to the advent of major technological discontinuities. A given number of firms entered the industry at the beginning of the simulation and no further entry occurred until the technological discontinuity represented by molecular biology. New

biotech firms, however, could not compete with incumbents successfully, because the emergence of a new technological paradigm was not
unequivocally competence destroying. New entrants lacked crucial
capabilities in product development and marketing and they did not
compete in "protected niches." Rather, they had to face the competition of older firms and products, which might have been worse in terms
of quality, but were protected by their past marketing expenditures.
Moreover, new biotech firms did not compete in the whole market, but
only in those sub-markets where they were actually able to discover
and perhaps develop their new products. Hence, they could not "win
the whole market." Incumbents, by contrast, continued to earn revenues on older products and gradually learned the new technology.
Bearing in mind also the fact that the discovery and development of
new drugs is a costly process and takes time, new firms had little chance
of displacing the leaders, even in the long run.

In this chapter we address this question by examining alternative
"counterfactual" patterns of entry, and analyzing the outcomes in
terms of innovative performance and market dynamics generated by
different structures of the entry process. The main question concerns
the small relevance of entry in a quite competitive industry, in which
the lack of cumulativeness of technological advances removes a fundamental source of barriers to entry and first-mover advantages. Thus, in
order to sharpen the focus of our exercise, in what follows we restrict
the analysis of the dynamics of the market structure to the so-called era
of "random screening."

In the next section we sketch a brief historical account of the evolution of the pharmaceutical industry. The third section presents the
model, and section 8.4 discusses the results. Section 8.5 concludes.

8.2 Relevant features of the evolution of the pharmaceutical industry

Here we briefly present the significant characteristics of the historical
evolution of pharmaceuticals.[2] Different eras can be recognized in
which the technological environment was exposed to relevant changes.

---

[2] For a detailed discussion of the evolution of the pharmaceutical sector, see, among
others, Gambardella (1995), Henderson, Orsenigo, and Pisano (1999), Sutton
(1998), Pammolli (1996), and Grabowski and Vernon (1977, 1994).

Up to the middle of the twentieth century the industry was not characterized by intensive R&D projects: few new drugs[3] were introduced into the market, their effectiveness was limited, and the research methods were primordial, associated with a relatively stationary technology. The pioneering firms were mainly chemical firms that exploited the knowledge spillovers and links between the chemical experience and the pharmaceutical field. Essentially, it was Swiss and German companies, such as Bayer, Hoechst, Ciba, and Sandoz, that first entered the industry, exploiting their competencies and knowledge accumulated in the related area of organic chemicals and dyestuffs. Later on, American and British producers, such as Eli Lilly, Pfizer, Warner-Lambert, Burroughs-Wellcome, and Wyeth (i.e. American Home Products), also entered the pharmaceutical industry.

After the 1940s the industry experienced deep changes: the role of science increased continuously during this period, and firms started to formalize in-house R&D programs. The R&D/sales ratio increased rapidly, and firms also started to invest heavily in sales efforts, aimed at contacting the physicians directly instead of the patients, as in the previous period. Massive public investment in basic research sustained the advancement of research capabilities, and the creation and development of the welfare state provided producers with large organized markets for drugs. New research opportunities and the existence of a number of unmet needs represented a rich environment for pharmaceutical firms. This translated into a period of rapid introduction of new drugs: more than 54 percent of the drugs available in 1947 had been unknown just ten years before. On the other hand, the detailed knowledge of the biological bases of specific diseases was still limited. Pharmaceutical companies still followed during this period an approach to research now referred to as "random screening." Under this approach, natural and chemically derived compounds were randomly screened in test tube experiments and laboratory animals for potential therapeutic activity. Pharmaceutical companies maintained enormous "libraries" of chemical compounds, and added to their collections by searching for new compounds in places such as swamps, streams, and soil samples. Thousands of compounds might be analyzed

---

[3] Basically, the most important innovations during this period were the introduction of alkaloids, of coal tar derivatives, and of sulfa drugs, between 1938 and 1943 (Sutton, 1998).

before researchers could identify a promising compound (Henderson, Orsenigo, and Pisano, 1999).

Despite the massive investments in R&D and major marketing efforts, the industry did not experience a significant rise in the degree of concentration.[4] However, the industrial structure has been characterized by a stable core of leading firms. On the one hand, patent protection granted temporary monopoly power to companies, but, on the other hand, no dominant positions in the global market emerged. Competition during this era relied extensively on the introduction of new products, but also on the incremental refinements of existing drugs, and on the imitation of drugs when their patent protection expired. Many firms did not specialize in R&D and innovation but, rather, in imitation/inventing around, as well as in the production and marketing of products often invented elsewhere. This group of firms included companies such as Bristol-Myers, Warner-Lambert, Plough, and American Home Products, as well as almost all the firms in such countries as France, Italy, Spain, and Japan. The "oligopolistic core" of the industry has continued to be composed of the early innovative Swiss and German firms, joined after World War II by a few American and British entrants, all of which have maintained over time an innovation-oriented strategy. Many of the leading firms during this period – companies such as Roche, Ciba, Hoechst, Merck, Pfizer, and Lilly – had their origins in the "pre-R&D" era of the industry. Price competition was also intense: usually markups were lower than 5 percent.

The 1970s represented another turning point in the evolution of the pharmaceutical industry. Progress in pharmacology, physiology, enzymology, and biology led to a deeper understanding not only of the mechanisms through which the drugs worked but also of the diseases themselves. In turn, this advance opened up the way for new techniques of searching, named "guided search" and "rational drug design," that made it possible for researchers to design compounds with specific therapeutic effects. In addition, the way in which the compounds were screened changed substantially. The transition from random to guided search was still under way when advances in DNA technologies and molecular genetics ushered in a radical transformation to the

---

[4] For example, the Herfindahl index in 1985 was equal to 0.0164, in 1990 to 0.0163, and in 1995 to 0.0195 (Pammolli, 1996).

knowledge base of the industry. The "biotechnological revolution" had a deep influence on the industrial structure. The required competencies for drug discovery and the development process were profoundly affected by the advent of biotechnology, and the existing firms had to face up to this transformation.

In accordance with this revolution, the industry experienced a significant entry of new firms – the first in all the time since World War II.[5] Despite the high rates of entry, it took several years for the biotechnology industry to start to have an impact on the pharmaceutical market.[6] Moreover, the great majority of these new companies never managed to become fully integrated drug producers. The growth of new biotechnology firms (NBFs) as pharmaceutical companies was constrained by the need to develop competencies in a number of crucial areas.

First, it was necessary to understand better the biological processes in which proteins were involved and to identify the specific therapeutic effects of such proteins. Companies, in fact, turned immediately to produce those proteins (e.g. insulin and the growth hormone) that were sufficiently well known. The subsequent progress of individual firms and of the industry as a whole was, however, predicated on the hope of being able to develop much deeper knowledge of the working of other proteins in relation to specific diseases. In the event, progress along these lines proved more difficult than expected. Second, these companies lacked competencies in other crucial aspects of the innovation process: in particular, knowledge and experience of clinical testing and other procedures related to product approval, on the one hand, and marketing, on the other. As a consequence, they fell back on their core area of expertise and acted primarily as research companies and specialized suppliers of high-technology intermediate products, performing contract research for and in collaboration with established pharmaceutical corporations.

---

[5] We refer explicitly to *de novo* entry. We are abstracting here from entry by M&As.

[6] The first biotechnology product, human insulin, was approved in 1982, and between 1982 and 1992 sixteen biotechnology drugs were approved for the US market. Sales of biotechnology-derived therapeutic drugs and vaccines had reached $2 billion by the early 1990s, and two new biotech firms (Genentech and Amgen) entered the club of the top eight major pharmaceutical innovators (Grabowski and Vernon, 1994).

Collaboration allowed NBFs to survive and – in some cases – to pave the way for subsequent growth in other areas. First, clearly, it provided the financial resources necessary to fund R&D. Second, it provided much-needed access to organizational capabilities in product development and marketing. Established companies faced the opposite problem. While they needed to explore, acquire, and develop the new knowledge, they had the experience and the structures necessary to control testing, production, and marketing.

Indeed, large established firms approached the new scientific developments mainly from a different perspective – i.e. as tools to enhance the productivity of the discovery of conventional "small molecule" synthetic chemical drugs. There was enormous variation across firms in the speed with which the new techniques were adopted. The adoption of biotechnology was much less difficult for those firms that had not made the transition from "random" to "guided" drug discovery. For them, the tools of genetic engineering were initially employed as another source of "screens" with which to search for new drugs. Their use in this manner required a very substantial extension of the range of scientific skills employed by the firm: a scientific workforce that was tightly connected to the larger scientific community and an organizational structure that supported a rich and rapid exchange of scientific knowledge across the firm (Gambardella, 1995; Henderson and Cockburn, 1994). The new techniques also significantly increased returns to the scope of the research effort (Henderson and Cockburn, 1996).

The embodiment of the new knowledge was in any case a slow and difficult process, because it implied a radical change in research procedures and a redefinition of the disciplinary boundaries within laboratories, and, in some cases, in the divisional structure of the company as well. Collaborative research with the NBFs and with universities allowed these companies to get access to the new technology and to experiment with alternative directions. The advantages stemming from these interactions could be fully exploited, however, only through the contextual development of in-house capabilities, which made it possible to absorb and complement the knowledge supplied by external sources (Arora and Gambardella, 1990). Internal research and collaboration with universities and NBFs were indeed strongly complementary.

Thus, a dense network of collaborative relations emerged, with the startup firms positioned as upstream suppliers of technology and

R&D services, and established firms positioned as downstream buyers that could provide capital as well as access to complementary assets. Networking was facilitated by the partly "scientific" – i.e. abstract and codified – nature of the knowledge generated by NBFs (Arora and Gambardella, 1998; Gambardella, 1995), which made it possible, in principle, to separate the innovative process in different vertical stages: the production of new scientific knowledge, the development of this knowledge in applied knowledge, and the use of the latter for the production and marketing of new products. In this context, different types of institutions specialized in that stage of the innovative process in which they were more efficient: university in the first stage, the NBFs in the second stage, and large firms in the third. A network of collaboration between these actors provided them with the necessary coordination to manage the innovation process. The new firms acted as "middlemen" in the transfer of technology between the universities – which lacked the capability to develop or market the new technology – and the established pharmaceutical firms, which lacked technical expertise in the new realm of genetic engineering but which had the downstream capabilities needed for commercialization (Orsenigo, 1989).

However, substantial costs remained in transferring knowledge across different organizations, especially for the tacit and specific component of knowledge. Moreover, the innovation process still involves the effective integration of a wide range of elements of knowledge and activities, which are not ordered in a linear way and which may not be easily separated (Orsenigo, 1989). Thus, the processes of drug discovery and – *a fortiori* – drug development still require the integration of different disciplines, techniques, search, and experimental procedures and routines, which are not generally separable and codified.

Moreover, since knowledge is still fragmented and dispersed, and since the rate of technological change is still very high, no single institution is able to develop internally in the short run all the necessary ingredients for bringing new products to the market place. Each NBF, in effect, represents a possible alternative approach to drug discovery and a particular instantiation of the opportunities offered by the progress of science (Orsenigo, Pammolli, and Riccaboni, 2001). New generations of NBFs have been created that adopt different approaches to the use of biotechnology in the pharmaceutical industry. Large established corporations continue, therefore, to explore these new developments through collaborative agreements.[7]

If we look at the worldwide market shares of the leading firms, we observe that none of them is above 4.5 percent: the market shares of the top twenty firms[8] ranged in 1996 from 1.3 to 4.4 percent. The four-firm concentration ratio (CR4) index[9] in 1994 was equal to 16.1 percent – i.e. a decidedly low value when compared to other industries characterized by the existence of an oligopolistic core of firms. Indeed, this is a peculiar feature of the pharmaceutical industry: the existence of an oligopolistic core of firms with an individual worldwide market share that is, nonetheless, not as significant as in other industries. Part of the explanation for this puzzle is provided by the observation that these companies hold a relevant dominant position in individual therapeutic categories (TCs). In some TCs the CR4 index was above 80 percent in 1995,[10] and in many others just two or three drugs account for more than 50 percent of market sales (Chong, Crowell, and Kend, 2003). These firms also

---

[7] The proliferation of NBFs was essentially an American (and partly British) phenomenon. The development of the biotechnology segment in Europe and Japan lagged considerably behind the United States and rested on the activities of large established companies. The British and the Swiss companies moved earlier and more decisively in the direction pioneered by the large US firms in collaborating or acquiring American startups. But those firms that had smaller research organizations, that were more local in scope, or that were more orientated towards the exploitation of well-established research trajectories – in short, those firms that had not adopted the techniques of "rational" or "guided" drug discovery – have found the transition more difficult (Henderson and Cockburn, 1994; Gambardella, 1995): almost all the established French, Italian, and Japanese companies – as well as the German giants – have been slow to adopt the tools of biotechnology as an integral part of their drug research efforts. More generally, ever since the mid-1970s the American, British, and Swiss companies appear to have gained significant competitive advantages vis-à-vis the other firms, including the German ones (Gambardella, Orsenigo, and Pammolli, 2001). And, traditionally, the Continental European (apart from the German and Swiss) and Japanese industries have been much less oriented towards innovation than towards strategies based on imitation, production, and marketing mainly for the domestic market.

[8] The top twenty pharmaceutical companies in 1996 were: Novartis, Glaxo Wellcome, Merck, Hoechst Marion Roussel, Bristol-Myers Squibb, Johnson and Johnson, American Home Products, Pfizer, SmithKline Beecham, Roche Holdings, Bayer, Astra, Eli Lilly, Rhone-Poulenc Rorer, Abbot Laboratories, Schering-Plough, Pharmacia and Upjohn, Boehringer Ingelheim, Takeda, Warner-Lambert (Hayes and Fagan, 1998).

[9] Source: author's computations based on Hayes and Fagan (1998) data.

[10] In antiviral products, for example, the CR4 was 86 percent in 1995.

represent the most active firms in terms of innovative output (measured by the introduction of NCEs in the market). Yet, even in sub-markets, dominant positions are often temporary and contestable.

As specified above, in this chapter we concentrate our analysis on the "random screening" period, focusing our attention only on a few relevant variables – basically, the entry process and the degree of concentration. We do not explore, in this work, the role of two other essential dimensions: the evolution of scientific knowledge and, subsequently, that of technological opportunities; and the institutional environment.

8.3 The model

The model we base our simulations on was developed, in its fundamental aspects, in Malerba and Orsenigo (2002). The model aims to capture the fundamental features of the dynamics that characterized market evolution and the effects of different patterns of entry of new firms into the industry during the era of "random screening."

8.3.1 Qualitative description of the model

A number of firms compete to discover, develop, and market new drugs for a large variety of diseases. Firms face a large space of unexplored opportunities. The process of searching for promising new compounds is essentially random, because the knowledge as to why a certain molecule can "cure" a particular disease, and where that particular molecule can be found, is limited. This reflects a modest role for "scientific knowledge" in our model. Thus, firms randomly explore the "molecule space" until they find one that might become a useful drug, and then they patent it. The patent provides protection from imitation for a certain amount of time and over a given range of "similar" molecules. After the discovery, firms engage in the development of the drug, without knowing either how difficult, time-consuming and costly the process is or what the quality of the new drug will be. Then, if a required minimum level of quality is reached, the drug is sold on the market. Marketing expenditures allow firms to increase the number of patients they can access. At the beginning the new drug is the only product available in its own particular therapeutic class, but later on other firms can either discover competing drugs or imitate the original one.

It has been argued by several scholars (for a detailed discussion, see Sutton, 1998) that the innovation process in pharmaceuticals has been characterized by a low degree of scope economies in R&D and search activity: knowledge and experience acquired in the search for and development of a new drug do not usually entail advantages in the discovery and introduction of further drugs belonging to the same TC. It follows that innovation in one market (a TC) does not entail higher probabilities of success in another one (Malerba and Orsenigo, 2002).

An appropriate representation of the search process for promising new compounds (which we call molecules) is given by the "lottery model" (Sutton, 1998). With the aim of treating a given medical condition, firms design a series of laboratory tests in which they randomly screen a considerable number of candidate agents in order to try out their efficacy. Accordingly, in our representation we set a space of discrete points in each therapeutic category, which firms can screen at some cost.

Given that the discovery of a drug in a particular TC does not result in any advantage in the discovery of another drug in a different category, firms will start searching randomly again for a new product everywhere in the space of molecules. The only advantage that firms get from the discovery of new drugs is related to the volume of profits they can reinvest in search and development for new molecules. There seems to be no serial correlation in the introduction of successful new drugs by a company. It follows that "an innovating firm has little tendency to develop a competence that cannot quickly be matched by several rivals. The simple model of a sequence of independent lotteries would seem a good first approximation in this case" (Sutton, 1998).

Besides trying to discover entirely new drugs, firms can also adopt imitative strategies, developing and commercializing products when their patent has expired. Each firm is characterized by a different propensity towards innovation, on the one hand, and imitation and marketing, on the other.[11]

Market structure is also heavily influenced by institutional variables, which are represented in this model by the structure of the patenting regime and by the procedures for product approval by the regulatory

---

[11] The propensity to research determines whether a firm is an innovator or an imitator. In this version of the model we set half of the firms as innovators and half as imitators. In a more complex version, we let the firms behave randomly, either as innovators or imitators, in each period.

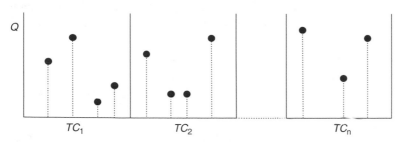

Figure 8.1. The topography of the industry

agency.[12] On the one hand, the degree of patent protection affects the appropriability of the revenues related to the introduction of an innovation. A tight patent regime, then, *ceteris paribus*, is expected to increase the success of innovative activities vis-à-vis imitative strategies, and to lead to higher levels of concentration. On the other hand, the less stringent the procedures for product approval the higher the number of new drugs, but the lower the average quality of the products in the market. A stringent approval system should also lead to more favorable conditions for the growth of more innovative companies.

8.3.2 The topography of the model

The industry environment is constructed as a series of therapeutic categories (*TC*), each of which is composed of several molecules (*M*) that firms can screen in order to develop a drug. The topography of the industry is represented in figure 8.1.

Each therapeutic category has a different economic size, expressed by the total potential sales (V_{TC}), which is given exogenously in the model: V_{TC} is set at the beginning of each simulation and is a random number drawn from a normal distribution [$V_{TC} \sim N(\mu_V, \sigma_V)$]. At the beginning of the simulation runs we set the number n of therapeutic categories in the model. The value of each *TC* grows in every period at a certain rate (the growth rate ranges randomly for each *TC* between 0 and 2 percent). Firms in a certain *TC* get a share of V_{TC} equal to their market share.

---

[12] We do not consider the forms of price regulation in this version of the model.

Each TC is composed of a certain number M of molecules, which firms aim to discover and which represent the base of pharmaceutical products that firms will launch in the market. Each molecule is characterized by a certain quality Q, which can be imagined as the "height" of the molecule in figure 8.1. The uncertainty of the market environment is represented by setting the value of Q greater than zero only with a given probability λ (with $\lambda < 1$):[13]

$$\text{Prob}\{Q > 0\} = \lambda \qquad (8.1)$$

otherwise the molecule is zero-valued.

8.3.3 Firms

The industry is populated by F potential entrants. Each of them is endowed with a given initial budget (B) – equal for all firms in the parameterization of the model used here – that is spent trying to discover a "good" molecule and then develop and commercialize a drug.

Firms are boundedly rational and there is imperfect information. Firms do not know the value Q of the molecules. Once firms engage in a search process in a specific TC, they may "discover" a molecule, say i, with $Q_i > 0$ or not. If they do, they start a development process (see below) and a patent for that molecule is then obtained. In this case, the molecule gives origin to its correlated product, which, in case it passes the regulatory agency's quality check, is commercialized by the discovering firm. This quality threshold (FDA) is set at a fixed value at the beginning of the simulations.

Let us analyze more precisely firms' processes of discovery, development, and commercialization.

8.3.3.1 Firms' routines
Firms engage in three activities: research, which is divided into search and development, and marketing. Each firm has a given individual propensity to these activities. Some of them may want to spend somewhat more on research and less on marketing, while other firms do the opposite. The firms' budget B is divided into sub-budgets according to their propensity: B_M gives the resources for marketing expenditures, B_S

---

[13] In these cases the value is drawn from a normal distribution $N(\mu_Q; \sigma_Q)$.

the resources for search activities, and B_D the resources for the development of the product.[14] According to their own propensity, firms are either innovators or imitators.

Search activity
Firms that follow innovation strategies search for new molecules in every period. The amount of money invested in search activities, B_S, gives the number X of therapeutic categories that are explored by a firm during its current project according to the following linear function:

$$X = B_S/Drawcost \qquad (8.2)$$

where *Drawcost* is a fixed parameter. If X is lower than one then the firm is assumed to be able to explore only one *TC*. In each *TC* that has been picked up, firms randomly draw a molecule. Consistently with our assumption of bounded rationality, firms do not know the "height" Q_i of the molecule; they know only whether Q_i is greater than zero – that is to say, if the molecule is potentially "interesting" or not.

On the other hand, if a firm is an imitator, after having drawn a certain number of *TCs* (defined by equation (8.2)) it looks for an existing molecule for which the patent has been expired. These "free molecules" are rated according to their quality Q_i, which is, however, only imperfectly known.[15]

If the discovered molecule is potentially "interesting" (in the case of innovating firms[16]), then the firm obtains a patent, which has a given duration (*pd*) and width (*w*). When the patent expires, the molecule becomes free for all the firms. A patent gives the firm the right to extend

---

[14] In the model, the marketing propensity of the firms, ϕ, gives the resources for marketing expenditures, B_M. The parameter ϕ is defined as a share of the budget randomly set in the interval [0.2–0.8]. Consequently, the firms' research propensity is characterized by a share of the budget that complements the share of the propensity to marketing, which is $(1 - \phi)$. This amount is then divided into B_S and B_D according to a firm-specific share $\omega \in U[0.05-0.15]$, which is also invariant.

[15] Imitators select the molecule with the highest "perceived" quality, R_i, that also gives the measure of the probability of choosing that molecule, defined as $R_i = (1 + \beta) Q_i$, where $\beta \in U[-0.25; +0.25]$. As a consequence, the high-quality molecules will be more frequently picked up by imitators, generating a sort of congestion effect.

[16] In the case of imitating firms: if a "free molecule" has been selected.

protection also on the molecules situated in the "neighborhood," width w, of the molecule associated with the product that has been patented. This form of protection has significant effects on the search processes of competing firms, which are in fact blocked as to the development of potential molecules situated near the patented one.

Once firms obtain a patent, the development of the product based on that molecule begins.

Development activity

If the molecule that has been drawn is potentially interesting (i.e. $Q_i > 0$) then the firm starts a development project. Given the amount of resources destined to the development process B_D, the firm progresses towards the full development of the drug (attaining the value Q_i of the drug). We imagine firms progressively "climbing" Q_i steps in order to develop a drug having a quality Q_i.

The total cost of developing a drug having a quality Q_i is equal to $CS \cdot Q_i$, where CS represents the unit cost of development. Firms are heterogeneous and differ in their efficiency: specifically, they differ in the speed of their development process. The progress SP that a firm makes each period in "climbing" Q_i is randomly assigned, and is firm-specific.[17] Higher speed implies higher costs; in fact, firms that move ahead more rapidly per period pay more for each unit step.[18] According to its development resources, a firm may be able to reach Q_i. In this case, then, the launch of the product in the market begins. Otherwise, if Q_i is too "high" for the resources B_D of the firm, the project fails.

Moreover, a product is required to have a minimum quality, defined by the regulatory agency, fixed at FDA, in order to be allowed to be commercialized in the market. Below this value the drug cannot be sold and the project fails. When these development activities are over (i.e. after a product has been created), the firm starts another process of search.

---

[17] In the model SP ranges from one to five.

[18] The unitary cost CS of each step increases as SP increases according to the following relationship, $CS = (C_{ur} \sum_{i=1}^{SP} i)/SP$, where C_{ur} is the cost of a single step for a firm that has a SP equal to one (i.e. it progresses with one step each period: only in this case does $CS = C_{ur}$). CS for imitative firms is set at one-quarter of the CS of innovating firms. In our simulations, C_{ur} is equal to fifteen and is fixed for every firm.

If product development is successful, the product gets an economic value PQ. The value of the i-th product PQ_i is a function of the value Q_i of the i-th molecule.[19] PQ_i enters in the utility function of the consumers, influencing in this way the consumers' demand for such a drug.

For simplicity, in the model we assume that the manufacturing cost of the drug is equal to zero, so that the drug can be sold in the market place immediately after the development process has been completed.

Marketing activities

The available resources for marketing, B_M, are divided by the firm into two parts, with shares h and $(1 - h)$, where h is equal for all firms: hB_M and $(1 - h)B_M$. The former represents the marketing investment A_{iT_L} for the launch of the product i (T_L being the launch period of the product). The latter gives the total yearly marketing expenditures YA_t that will be spent over T_A periods after the launch.[20] Intuitively, the first term is the investment spent only once, for the launch of the product with an attempt to create a certain level of awareness as to its properties (let us call it generally "product image"), while the second term captures the firm's expenditures for maintaining the level of the "image" over time. We also consider that the level of the "product image" A_{it} is eroded with time at a rate equal to eA in each period, and that firms benefit from a marketing spillover (θ) from their own previous k products ($k \neq i$).[21]

8.3.4 Market demand

Market demand is implicitly defined as the potential customers in each therapeutic category that represent the total potential sales (V_{TC}) of that TC.

---

[19] According to the following relationship, $PQ_i = (1 + \alpha)Q_i$, where α is a noise term drawn from a normal distribution, $U[-0.25; +0.25]$.

[20] YA_t is given by the total amount $(1 - h)B_M$ divided by the number of periods T_A.

[21] Formally, the level of the "image" A_{it} of product i in period t is given by

$$A_{it} = A_{iT_L} + \theta \sum_{k \neq i} A_k \qquad \text{for } t = T_L$$

$$A_{it} = A_{it-1} \cdot (1 - eA) + \theta \sum_{k \neq i} A_k \qquad \text{for } t = T_L + 1, \ldots, TT$$

where TT is the end of the simulation period. In our simulations, h is equal to 0.5, erosion (eA) = 0.01, and $T_A = 20$.

Each drug i is characterized by a specific "merit," U_i, that defines the relative "appeal" of the drug vis-à-vis the other products belonging to the same therapeutic category. Formally, we have

$$U_{it} = PQ_i^a \cdot (1/mup)^b \cdot A_{it}^c \cdot YA_{it}^d \tag{8.3}$$

where PQ_i is the economic value of the product, previously defined; mup is the desired rate of return that each firm wants to obtain from its drug[22] (we assume that mup is double for innovative products, mup_{inn}, compared to the rate of return applied to imitative products, mup_{imi}), A_{it} is the product "image" derived by the marketing investment for that product, and YA_{it} is the yearly marketing expenditure (see footnote 18). Exponents a, b, c are parameters, specific to each therapeutic category,[23] while d is equal in all TC.

8.3.5 Firms' market share and revenues

For each product i of each firm f in all periods t of the simulation, we compute the product market share PMS_{fit} in each TC. In accordance with the definition of U_{it} given above, the product market share, PMS_{fit}, is a function of its relative merit as compared to other competing drugs (j) in the same TC. Formally, we have

$$PMS_{fit} = \frac{U_{fit}}{\sum\limits_{j \in TC} U_j} \tag{8.4}$$

The first sold product in a given TC gives a temporary monopoly power to the firm, which lasts until other competing products are developed by other firms in the same TC.

Given that firms may have several products, the aggregate firms' market share, FMS_{ft}, is then computed as the sum of the market shares of all the marketed products:

$$FMS_{ft} = \sum\limits_{i \in TC} PMS_{fit} \tag{8.5}$$

---

[22] In order to keep consistent with the evidence presented above, we set the markup value at a low level; in the standard run for innovating firms it is equal to 0.2.

[23] a, b, and c are drawn from uniform distributions.

Firm's revenues in a TC are given by the share of the TC's value corresponding to the firm's market share FMS_{ft} in that TC. Accordingly, firm's total revenues (Π_{ft}^{TOT}) are the sum of revenues obtained from all the products Π_{ftr} in each r-th explored TC:

$$\Pi_{ft}^{TOT} = \sum_{r=1}^{numofTC} \Pi_{ftr} = \sum_{r=1}^{numofTC} \left(FMS_{ft} \cdot V_{TCr} \right) \qquad (8.6)$$

In each period, the difference between total revenues and the current costs of search, development, and launch of the new products (when relevant) and the yearly expenditures on marketing sums to the budget account B, which is used to finance a new project.

8.3.6 Entry and exit dynamics

As has already been mentioned, the simulation is characterized by F potential entrants. Each of them undertakes research activities with the aim of developing a drug to be sold in the market place. When a firm succeeds in marketing its first product, it is then recorded as a new entrant in the industry. The timing according to which the potential entrants are allowed to try to enter the market (i.e. to search and develop a product) is defined below (see Section 8.4.2), and it is the focus of our subsequent analysis.

Firms' exit may occur for different reasons. First of all, given the initial amount of the budget B, if a potential entrant fails in the search activity for ρ periods in a row (i.e. if it does not start a drug development project, no matter if successful or not), it exits from the set of potential entrants. Secondly, if during the development project the firm's budget B falls to zero, the firm then exits the market.

8.4 Simulation runs

The aim of this section is to examine the market dynamics generated by different patterns of entry. In order to avoid outlier results, we present the outcomes as the average values over 100 runs.

8.4.1 Market fragmentation

We start our simulations by exploring a simplified version of the model where the industry is composed by a single market (TC). Then we investigate what the role of market fragmentation is in generating low levels of aggregate concentration. In order to be able to compare the simulation outcomes with the standard setting, in which we have n TCs, each composed by M molecules, we set the number of molecules of the unique TC equal to nM. Again, for comparability, the mean value of the single market $V_{UniqueTC}$ is set equal to n times the mean value of V_{TC} (μ_V), as expressed in section 8.3.2.

The result that emerges from our simulation runs is that there is no significant difference between the case of a unique TC and the case of several TCs. We believe that the explanation of this result is to the lack of cumulativeness inside each TC that characterizes the model. Having one market or many separated markets (keeping equal the number of molecules and the total value of the markets in the two cases), then, does not make any significant difference for the processes of search and development for the existing firms. A slight difference emerges when we consider the sequence of entry of potential entrants – more on this later.

The market structure is characterized by a low level of concentration. Both in one TC and n TCs, the global Herfindahl index H is approximately 0.036. When we consider the case of n TCs, the H index computed in each TC on average reaches values above 50 percent. This denotes a high level of concentration in the single therapeutic categories but a low level in the global market, as has been shown in the historical analysis of the industry. The rate of potential entrants that succeed in entering the market is around 78 percent. At the beginning of the evolution of the industry there are more innovative products in the market, while, as time goes on, imitative products overtake the innovative ones. The "innovative index," which we define as the ratio innovative firms/total number of firms, peaks at the very first period of the simulation run and then steadily declines towards a value of approximately 0.21 to 0.23. These basic features of the evolution of market structure have been extensively analyzed in Malerba and Orsenigo (2002).

Before turning to our main discussion about entry, for our purposes it suffices to analyze a few different setups of the model in order to understand how the results are a function of the initial setting. In particular, we want to investigate if and how the initial number of potential entrants and the level of the initial budget are relevant for the evolution of the market structure.

The initial number of potential entrants, F, does not seem to be relevant in determining the entry rate for low values of F, but it is quite obviously crucial in affecting the number of firms and concentration. We let F range from ten to 100; *ceteris paribus*, when the initial budget is equal to 10,000, the entry rate does not show particular variations, varying between 75 and 80 percent. As a consequence, the H index (being a function of the number of existing firms) declines as the number of initial potential entrants rises, but in all cases it settles around low levels. Also, the innovative index does not undergo much variation. This result holds both for the case of a unique TC and the situation with n TCs. On the other hand, for higher values of F, the entry rate seems to be influenced by the initial number of firms: for example, we let F vary up to 1500. The simulation result shows a substantial decrease in the number of entering firms as the parameter F increases significantly.

This result is mainly an effect of the "richness" of the search space. In the current parameterization of the model, it is quite easy to find a "good" molecule. In fact, results change when the initial budget/ Drawcost ratio is allowed to vary. As the ratio increases, the entry rate rises, the H index falls, and the innovative index increases as well. Consequently, the number of total products, both innovative and imitative, grows. It follows that the initial budget/Drawcost setup is crucial in determining the market structure, even if the level of concentration remains low in all cases. This is reasonable if we consider that the initial budget determines the number of screened molecules at the beginning of firms' lives, and consequently their probability of success and survival.

8.4.2 Patterns of entry

During the discussion about the evolution of the pharmaceutical industry, we emphasized that no significant *de novo* entry occurred along the years until the biotechnological revolution. In this chapter, we do not model entry endogenously. Rather, we limit ourselves to studying the effects of different patterns of entry on market dynamics. This is clearly a major limitation of this study. However, as a partial justification, it might be noted that most models of entry basically assume an infinite queue of would-be entrants that might decide actually to enter the industry when expected profits are positive. Nonetheless, the assumption that the pool of potential entrants is infinite is quite strong, and nothing is usually said about the determinants of the size of such pool. In some cases, it may well be that low rates of entry are observed not (only) because of insufficient incentives (profits below the norm), but because the pool of potential entrants is too small. For example, basic scientific and technological capabilities in a given country may be too low to generate a sufficiently large number of potential entrants, just for lack of adequate capabilities.[24] Thus, our following exercise could be interpreted as a preliminary analysis of (simplified) contexts whereby the pool of potential entrants is determined exogenously by a different set of variables having to do with, for example, technological competencies, rather than with direct economic incentives.

Having said this, we focus the attention on the outcomes of different *timings* of entry. Specifically, we set the initial budget equal to 10,000 and the total number F of potential entrants to fifty. Then we let five generations of equal size (i.e. ten firms in each generation) of potential entrants try to enter the market at the pre-set timing: t_1, t_2, t_3, t_4, t_5, as indicated below. More precisely, we explore four cases in order to understand if the timing is relevant in determining the entering rate:

case (a) timing: 0, 5, 10, 15, 20;
case (b) timing: 0, 10, 20, 30, 40;
case (c) timing: 0, 20, 40, 60, 80; and
case (d) timing: 0, 30, 60, 90, 120.

First, we compare the results in the two different setups of the market: the case of a single *TC* and *n TCs*. For all the four cases above, in the situation of several *TCs*, the simulation runs show a slightly higher probability of entering the market and gaining a higher market share for the late potential entrants. The reason for this result is that direct competition among firms is lower when the market is composed of several sub-markets: in this case, late entrants have more possibilities to differentiate (i.e. to search and develop a new drug in an unexplored *TC*) without facing the direct competition of incumbent firms in these sub-markets, and consequently to survive. As a result, we claim that the more differentiated the industry the higher the chances for late potential entrants to enter and survive.

---

[24] Incidentally, we believe that this type of story is quite important in explaining the dearth of entries in biotechnology in countries such as Italy.

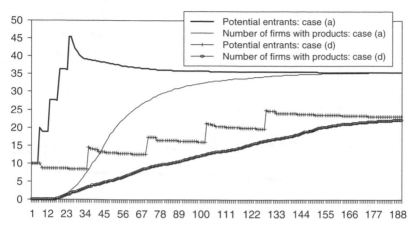

Figure 8.2. The number of potential entrants and firms with products in case (a) and case (d)

Under both setups, however, the timing of entry produces similar results and turns out to be quite important. In case (a), all but the very last generation have the same chance of entering and surviving: between 73 and 78 percent of potential entrants enters the market, as is shown in figure 8.2 (the figure reports the total number of potential entrants and the number of firms with a product in the market[25]), and all gain approximately the same share of the total market, around 20 to 24 percent (figure 8.3). In this case, the "innovative index" is around 0.23 to 0.24. When we expand the time sequence, examining cases (b), (c), and (d), we observe significant changes. In particular, in the extreme case, (d), the scenario is completely altered (figure 8.2): more than 75 percent of the

---

[25] The difference between the number of potential entrants and the number of the firms with a product represents the number of potential entrants still trying to search and develop a product. It is evident in the graph (see figure 8.2, and also the standard simulation in figure 8.7) that the two series converge to the same level, given that the potential entrants either market a drug or, sooner or later, "die," according to the exit rules. The timing of convergence is in some cases, like the standard simulation run, quite long; this reflects the fact that some potential entrants are trying to search, in this length of time, but always find a molecule the quality of which does not pass the "quality check" control (see "Development activity" in subsection 8.3.3.1), or that some potential entrants are working on a good molecule (i.e. one with high quality) at a "low speed" (as defined in subsection 8.3.3.1). This may require a long period of time before the firm either enters, or fails and exits because of the budget constraint.

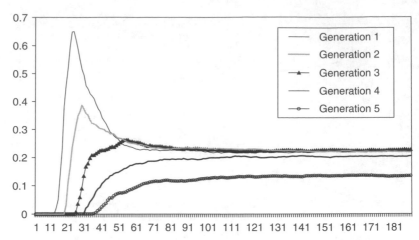

Figure 8.3. Market shares of the five generations of firms in case (a)

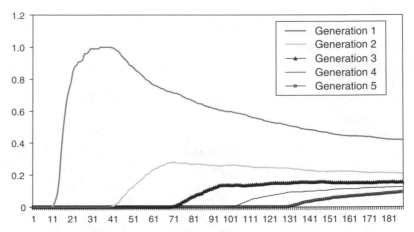

Figure 8.4. Market shares of the five generations of firms in case (d)

potential entrants of the first generation still reach the market and account for approximately 50 percent of total sales, while only one-third of the other generations' firms is able to enter. Their market share, in any event, ranges between 20 percent for all the firms of the second generation and 5 percent for the last generation (figure 8.4). Moreover, it emerges that late entrants rely rather more on innovative products than on imitative ones in order to enter and survive: the ratio Innovative/

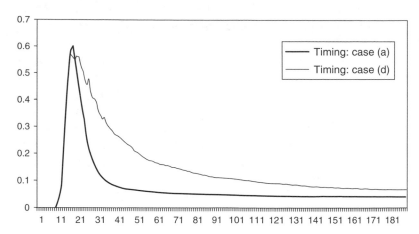

Figure 8.5. Overall concentration: the Herfindahl index in the overall market in case (a) and case (d)

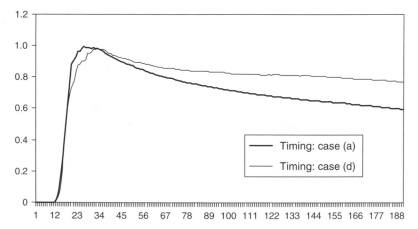

Figure 8.6. Concentration in each therapeutic category (mean values): the Herfindahl index in case (a) and case (d)

Imitative products for the first generation of firms is, in fact, lower in comparison with the ratio of the last generation. Consistent with this result, the "innovative index" increases as we proceed from case (a) to case (d). The total entry rate (i.e. considering all the generations together) then decreases from case (a) to case (d), from around 70 percent to 44 percent (figure 8.2), and consequently the H index increases both in the total market and in the individual TCs (figures 8.5 and 8.6).

Another interesting result is related to the simulation run in which the initial number of potential entrants is set equal to $F = 1000$. In case (a) now, only the first three generations of firms have the same chance of entering and surviving, while the last two generations exhibit lower entering rates and market shares, differently from the previous case ($F = 50$) in which only the last generation was significantly behind the others. In the other extreme case, (d), the same effect is shown in a higher entering rate and market share for the first generation's firms and in a lower entering rate and market share for all the other generations, compared to the case in which $F = 50$. This result is attributed to the fact that, the higher the number of firms that have entered previously, the lower the probability for later entrants to succeed and find a market niche.

We also analyze the case in which the process of entry is "continuous" – in other words, we let the fifty potential entrants try to enter continuously as time goes by. More precisely, we analyze the case in which one firm tries to enter the market in each period over fifty periods (the last potential entrant, then, is allowed to enter at time 50).[26]

The shape of the entry path (figure 8.7) shows how the entry rate declines significantly compared to the standard case, reaching a value equal to 55 percent. Figure 8.7 also shows the difficulties the latest entrants face: after the initial periods, late potential entrants fail to reach the market. This reflects the difficulties faced by later entrants in "finding room" in an already explored market. Market opportunities are depleted and the advantage accumulated by early entrants is significant.

8.4.3 What stimulates or obstructs entry in pharmaceuticals?

In this subsection, we investigate the role that some key variables of the model may have in stimulating or obstructing entry.[27] Taking the timing

---

[26] In the "continuous case" we analyzed these different setups: (I) ten firms try to enter in each period over five periods; (II) five firms try to enter in each period over ten periods; (III) two firms try to enter in each period over twenty-five periods; and (IV) one firm tries to enter the market in each period over fifty periods (the last potential entrant in this case is allowed to enter at time 50). The results show that, in the first two cases, there is no significant difference from the standard simulation run, in which all firms are allowed to enter at time 0. On the other hand, in the last two cases the entering rate declines significantly, reaching the value of 55 percent in case (IV).

[27] The results of the following experiments presented here refer to the various cases of "discrete" entry. However, we also conducted simulations in the case of

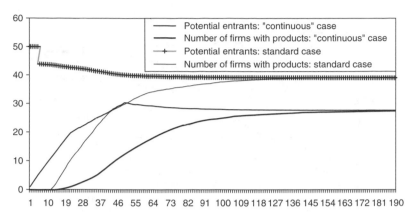

Figure 8.7. The number of potential entrants and firms with products in the standard simulation and in the "continuous" case

of entry (c) and the continuous entry (IV) (as defined in footnote 26) as the benchmark cases, we examine different settings, characterized respectively by: (1) richer opportunities for discovery; (2) longer patent protection; (3) higher cumulativeness in the search process; (4) an increasing cost of searching for new compounds; and (5) a smaller size of the market.

8.4.3.1 Richer opportunities for discovery

In the situation with higher market opportunities we set the value of λ in equation (8.1) equal to 0.9, as compared to 0.3 in the standard setting. Results show a larger number of firms, a lower degree of concentration (both in the global and single markets), a larger number of innovative and imitative products in absolute terms, and a higher value for the innovative index. The number of discovered TCs is larger too. With regard to the process of entry, results show (figure 8.8) that the entry rate of the late generations is higher and that these firms are able to gain a larger market share, as compared to the standard situation. This means that the level of technological opportunities is crucial in facilitating successful late entry.[28]

"continuous" entry. The results were not qualitatively different, and for reasons of space are not reported in this chapter.

[28] This is consistent with the interpretation of the considerable entry of new companies after the biotech revolution, as the possibility for new firms to explore and exploit newly available technological opportunities. As the biotech discoveries opened up a rich series of new opportunities, then, a flow of new entrants

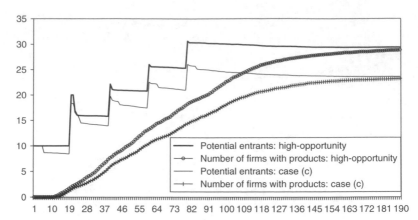

Figure 8.8. High-opportunity simulation run

Established firms, however, still hold a first-mover advantage in exploring compounds and markets. Thus, the gap between the first generation and the others remains significant. If it is true that high opportunities facilitate late entry, it is also true that the first entrants have higher probabilities of being successful from the beginning and consequently to grow, creating a gap that is still difficult for later entrants to catch up with.

8.4.3.2 Longer patent protection

Next we explore the case in which patent duration has been extended to sixty time periods along the simulation run (instead of twenty periods as in the standard setup). Surprisingly, no significant differences in the entry rate of the various generations of potential entrants emerged with respect to the standard case. The only relevant result is that the late entrants have to rely more on innovative products in order to penetrate the market. This is reasonable, given that imitation is more difficult, and it might indicate that, in pharmaceuticals, longer patent protection may stimulate more innovative entrants than imitative ones. This leads the innovative index to be much higher than in the standard case.

immediately followed. Note, however, that the analysis of the biotech revolution in our previous model (Malerba and Orsenigo, 2002) also implied – differently from the case discussed here – an advantage for new entrants in the ability to "use" biotechnology as compared to incumbents.

8.4.3.3 Higher cumulativeness in the search process

Nor did any relevant differences result from a setup characterized by stronger "cumulativeness" in search activities. We generated a cumulative effect in the search process by letting the number X of therapeutic categories that are explored by a firm, as expressed in equation (8.2), be a function of the number of already developed products of that firm. This means that, the higher the number of products a firm has, the higher its capabilities to screen markets and molecules. To be more rigorous, this effect represents the existence of economies of scope in the search process, which we interpret here as cumulativeness. At first sight, the result of no significant differences between this setup and the standard run might seem surprising. However, we believe that this outcome is explained in our model as follows: after firms have developed and sold a new product, the amount of resources they gain (i.e. the TC value) is huge, and this allows firms to invest more and to screen more markets and molecules. In other words, the added screening opportunities given by the expertise developed in the development of the "old" products are irrelevant in comparison with the first effect.[29]

8.4.3.4 Increasing cost of searching for new compounds

As many scholars have argued (see, for instance, Pammolli, 1996, for evidence and a survey of the literature), the costs of drug discovery and development have been substantially increasing over the last three or four decades. In order to examine if increasing costs of search played a key role in impeding new entry, we allow the search expenditures of firms required to screen the markets (i.e. the parameter *Drawcost* in equation (8.2)) to increase over time. While the presence of increasing costs does not obviously affect the entry rate of the first generation of firms, the following generations are exposed to a fiercer selection; lower entry rates of "late" generations and significantly lower market shares result from this setup (figures 8.9 and 8.10). As a consequence, the industry is populated by a smaller number of firms and, in addition,

---

[29] A different way of modeling more correctly the presence of cumulativeness in the model would be the introduction of the ability of firms to understand the value Q of the molecules they are researching on, after they have already developed a product in the same TC. But one of the basic assumption of our model, based on the pharmaceutical industry's history, is that during the "random screening" era firms were not able to discern in advance the "goodness" of the compound they were investigating. Accordingly, we do not consider this case in our experiments.

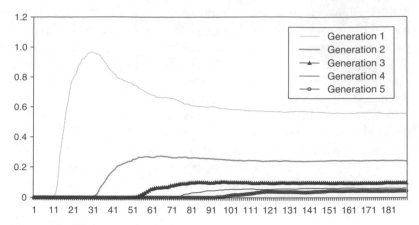

Figure 8.9. High-costs simulation run

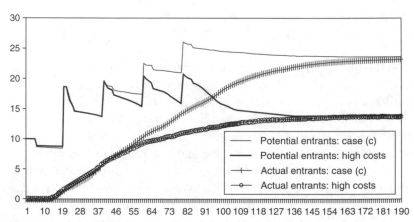

Figure 8.10. Market shares of generations of firms in the high-costs simulation run

the number of products is significantly lower, in particular with regard to innovative ones. Also, the innovative index decreases.

8.4.3.5 Smaller size of the market
A further set of variables that we claim might influence entry relates to the size of the market (in a static sense) and to the resulting "success breeds success" effects that might arise (in a dynamic interpretation).

The first effect would predict that the incentives for new firms to enter the market are higher with larger markets. However, in our model the number of entrants is set exogenously, and therefore we cannot capture this effect. Nonetheless, market size may also influence the survival ability of potential entrants, by allowing them to capture higher sales and profits, if successful. On the other hand, larger markets might entail a strengthening of the "success breeds success" effect: early successful entrants earn larger profits and therefore higher probabilities of finding new products in the future, at the expense of later entrants. Thus, in order to investigate if and how the amount of resources that a product yields to a producer affects the patterns of entry, we examine a setup of the model where the mean value of each *TC* is set equal to 10 percent of the standard value. The results show that the gap between the first generation of entering firms and the later generations reduces. In fact, in our simulations the second generation's firms overcome the first entrants in the production of innovative products, and they are also caught up by the third and fourth generation of potential entrants.

8.5 Conclusions

In this chapter we have explored a simulation model of the evolution of the pharmaceutical industry, focusing on the factors that influence entry and survival. Indeed, one major puzzle in the interpretation of the history of pharmaceuticals is why entry has not been a relevant feature of this industry – especially with regard to the innovative core and prior to the advent of the biotechnology revolution – despite low cumulativeness in the process of drug discovery and low levels of concentration.

While we did not directly model entry as an endogenous process, we did explore the implications for market structure and innovation of alternative patterns of entry, in terms of the number of entrants and the timing of entry. The main results can be summarized as follows.

(a) Market fragmentation per se does not significantly matter for market concentration and entry, in the absence of strong cumulativeness in the process of discovery.

(b) The timing of entry is crucial. First entrants face an unexplored scenario, characterized by unexploited opportunities and high possibilities for profits. This leads to an increasing gap over time between the first generation of entrants and the subsequent ones. The gap becomes more significant the later the subsequent potential entrants try to enter the industry.

(c) Two main factors influence the fate of entrants. First, high discovery opportunities make it easier for later potential entrants to prosper. Second, increasing search costs over time significantly obstruct survival, especially for later generations of entrants.

(d) The size of the market has an important effect on survival. But, in a dynamic setting, larger market size might imply first-mover advantages linked to "success breeds success" processes, making life harder for later entrants.

Clearly, these are preliminary results. Future developments of the model may permit a more systematic and thorough analysis. In particular, modeling the entry process explicitly constitutes a necessary step in that direction.

References

Arora, A., and A. Gambardella (1990), "Complementary and external linkages: the strategies of the large firms in biotechnology," *Journal of Industrial Economics*, 38, 361–79.

(1998), "Evolution of industry structure in the chemical industry," in A. Arora, R. Landau, and N. Rosenberg (eds.), *Dynamics of the Long-Run Growth in the Chemical Industry*, Wiley, New York, 379–413.

Chong, J., H. Crowell, and S. Kend (2003), *Merck: Refocusing Research and Development*, Blaisdell Consulting, Pomona College, Claremont, CA.

Gambardella, A. (1995), *Science and Innovation: The US Pharmaceutical Industry During the 1980s*, Cambridge University Press, Cambridge.

Gambardella, A., L. Orsenigo, and F. Pammolli (2001), *Global Competitiveness in Pharmaceuticals: A European Perspective*, Working Paper no. 1, Directorate-General Enterprise, EU Commission, Brussels (available at http://pharmacos.eudra.org/).

Garavaglia, C. (2004), *History-Friendly Simulations for Modelling Industrial Dynamics*, Working Paper no. 04.19, Eindhoven Centre for Innovation Studies, Eindhoven University of Technology.

Grabowski, H., and J. Vernon (1977), "Consumer protection regulation in ethical drugs," *American Economic Review*, 67, 359–64.

(1994), "Innovation and structural change in pharmaceuticals and biotechnology," *Industrial and Corporate Change*, 3 (2), 435–49.

Hayes, R. H., and P. L. Fagan (1998), *The Pharma Giants: Ready for the 21st Century?*, Harvard Business School Publishing, Cambridge, MA.

Henderson, R. M., and I. Cockburn (1994), "Measuring competence? Exploring firm effects in pharmaceutical research," *Strategic Management Journal*, 15, 63–84.

(1996), "Scale, scope and spillovers: the determinants of research productivity in drug discovery," *RAND Journal of Economics*, 27 (1), 32–59.

Henderson, R. M., L. Orsenigo, and G. Pisano (1999), "The pharmaceutical industry and the revolution in molecular biology: exploring the interactions between scientific, institutional, and organizational change," in D. C. Mowery and R. R. Nelson (eds.), *Sources of Industrial Leadership: Studies of Seven Industries*, Cambridge University Press, Cambridge, 267–311.

Malerba, F., R. R. Nelson, L. Orsenigo, and S. G. Winter (1999), "History-friendly models of industry evolution: the computer industry," *Industrial and Corporate Change*, 8 (1), 3–41.

(2001), "Competition and industrial policies in a history-friendly model of the evolution of the computer industry," *International Journal of Industrial Organization*, 19, 613–34.

Malerba, F., and L. Orsenigo (2002), "Innovation and market structure in the dynamics of the pharmaceutical industry and biotechnology: towards a history-friendly model," *Industrial and Corporate Change*, 11 (4), 667–703.

Orsenigo, L. (1989), *The Emergence of Biotechnology*, Francis Pinter, London.

Orsenigo, L., F. Pammolli, and M. Riccaboni (2001), "Technological change and network dynamics," *Research Policy*, 35, 485–508.

Pammolli, F. (1996), *Innovazione, Concorrenza a Strategie di Sviluppo nell'Industria Farmaceutica*, Guerini Scientifica, Milan.

Sutton, J. (1998), *Technology and Market Structure*, MIT Press, Cambridge, MA.

Werker, C., and T. Brenner (2004), *Empirical Calibration of Simulation Models*, Working Paper no. 04.13, Eindhoven Centre for Innovation Studies, Eindhoven University of Technology.

Windrum, P. (2004), "Neo-Schumpeterian simulation models," forthcoming in H. Hanusch and A. Pyka (eds.), *The Elgar Companion to Neo-Schumpeterian Economics*.

9 | The growth of pharmaceutical firms: a comment

BOYAN JOVANOVIC

9.1 Introduction

Of the three preceding chapters that are the subject of my review, two concentrate on the stochastic growth equation for firm size and the other on the process of invention and how it may influence the degree of concentration. My remarks here are chiefly concerned with relating the results of the three studies to two well-known neoclassical models.

Section 9.2 describes (parts of) the two models. It starts by interpreting equation (6.2) of Cefis, Ciccarelli, and Orsenigo's chapter and, indirectly, equation (7.6) of Bottazzi, Pammolli, and Secchi's chapter in terms of the technology of the firm and in terms of shocks. The model will also explain a particular finding of those two studies concerning the autocorrelation of growth. The interpretation follows Lucas and Prescott (1971). Section 9.3 outlines another model, based on Chari and Hopenhayn (1991), that relates to a cornerstone of Garavaglia, Malerba, and Orsenigo's chapter. Finally, section 9.4 contains some brief conclusions.

9.2 Gibrat's law and the Q-theory of growth

This section refers to chapter 6, by Cefis, Ciccarelli, and Orsenigo, and chapter 7, by Bottazzi, Pammolli, and Secchi. Equation (6.2) models a firm's growth rate as

$$g_t = \alpha + \rho g_{t-1} + \eta_t \qquad (9.1)$$

where the firm subscript has been dropped to ease notation. I shall now provide a structural interpretation of this equation that links it to the Q-theory of investment. Define a firm's state of technology as z and its capital as K. Its output therefore is

$$y = zK \qquad (9.2)$$

The firm is competitive and we normalize its output price to unity. The firm-specific shock z follows the first-order Markov process

$$Pr\ \{z_{t+1} \leq z' \mid z_t = z\} = F(z', z) \tag{9.3}$$

Let X be the firm's investment. Then the firm's capital stock evolves as

$$K' = (1 - \delta)K + X \tag{9.4}$$

The firm also faces a cost of adjustment

$$C(x)K \tag{9.5}$$

where

$$x = \frac{X}{K}$$

is its investment *rate*.

The feature we are seeking to explain is the *firm's growth rate*. The growth rate of the firm's capital stock, $g \equiv \frac{K'}{K} - 1$, is

$$g = x - \delta \tag{9.6}$$

and it may be either positive or negative.[1] Since δ is exogenous, explaining the growth rate means explaining x. Let p be the price of capital. Total profit is

$$zK - pX - C(x)K$$

and is homogeneous of degree 1 in X and K. Then profit per unit of capital does not depend on K^2 and it is just

$$z - px - C(x)$$

Since the firm's capital stock is given at the start of the period, maximizing the total value of the firm is the same as maximizing its value per unit of capital. Let $Q(z)$ denote the unit value of the firm's

---

[1] Both chapters look at the growth of sales and not of assets (which would be the counterpart of k). The growth of sales equals g plus the growth of z, so the equations for assets and for sales are closely related.

[2] This is an approximation to truth at best, and figure 7.3 of Bottazzi, Pammolli, and Secchi's chapter falsifies it. The figure shows that standard deviations of growth rates fall off as the size of firm rises. In contrast, the version of constant returns that I am using here implies that the standard deviation of growth rates should not depend on firm size.

capital. If the firm is traded on a stock exchange and if it has no debt, then $Q(z)$ is the price of a claim to the dividends that flow from a unit of capital. Then[3]

$$Q(z) = \max_{x \geq 0}\{z - px - C(x) + (1 - \delta + x)Q^*(z)\} \qquad (9.7)$$

where $Q^*(z)$ is the firm's discounted expected present value of capital tomorrow given today's realization of z,

$$Q^*(z) = \frac{1}{1+r}\int Q(z')\, dF(z', z) \qquad (9.8)$$

where r is the rate of interest that the firm faces. The first-order condition is

$$p + C'(x) = Q^*(z) \qquad (9.9)$$

The firm's investment rate is

$$x = (C')^{-1}[Q^*(z) - p] \qquad (9.10)$$

Combining (9.10) with (9.6), the firm's growth rate is

$$g_t = -\delta + (C')^{-1}(Q_t^* - p) \qquad (9.11)$$

where

$$Q_t^* = Q^*(z_t)$$

Equation (9.11) is similar in form to (9.1) except for the absence of the lagged growth term. But the difference is deceptive. To clarify things further, consider a concrete example, the adjustment-cost function

$$C(x) = \frac{\gamma}{2}x^2$$

Then (9.9) reads $p + \gamma x = Q^*$, so that $x = \frac{1}{\gamma}(Q^* - p)$, and (9.11) becomes

$$g_t = -\left(\delta + \frac{p}{\gamma}\right) + \frac{1}{\gamma}Q_t^* \qquad (9.12)$$

---

[3] Note that the total value of the firm is $Q(z)\,K$. It has to be linear in K because the production function is linear, and the cost function is homogeneous of degree 1 in X and K.

Lagging the equation by a period, we obtain

$$g_{t-1} = -\left(\delta + \frac{p}{\gamma}\right) + \frac{1}{\gamma}Q^*_{t-1} \qquad (9.13)$$

Substituting back into (9.12) we find that

$$g_t = g_{t-1} + \frac{1}{\gamma}\left(Q^*_t - Q^*_{t-1}\right) \qquad (9.14)$$

If we set $\rho = 1$ and

$$\alpha + \eta_t = \frac{1}{\gamma}\left(Q^*_t - Q^*_{t-1}\right)$$

this equation is quite similar in form to (9.1). The difference, however is that g_{t-1} is negatively correlated with the "error term" $\frac{1}{\gamma}\left(Q^*_t - Q^*_{t-1}\right)$, in view of (9.13). Therefore, a regression of g_t on g_{t-1} would for that reason deliver a biased coefficient on g_{t-1} of less than unity. And that is precisely what Cefis, Ciccarelli, and Orsenigo (tables 6.1 and 6.2) and Botazzi, Pammolli, and Secchi find.

The Cefis et al. parameter α is firm-specific. Here γ is a natural candidate for a firm-specific parameter. A higher γ implies a faster response to both positive and negative shocks and, hence, a higher variability of growth. If we say that smaller firms have smaller γ's, then we can explain the observation in figure 7.3 of Bottazzi et al. that higher firms have smaller growth variability.[4]

This is one way to put the *Q-theory of growth*. Firms with a stock price Q^*_t that exceeds their past-period stock price, Q^*_{t-1}, will grow more rapidly than they did the year before. Because z_t is stationary, however, Gibrat's law holds in that all firms grow at the same long-run rate. A firm's deviation from this long-run average may be autocorrelated, but is not permanent.

But, if z is positively autocorrelated, high-z firms have the highest Q^*. Therefore, we explain dispersion of growth by dispersion of z,

---

[4] However, a technical problem arises with assuming that γ falls with firm size. Suppose we were to write

$$\gamma = \Gamma(K)$$

where $\Gamma' < 0$. This would imply increasing returns to scale, and would invalidate the above analysis. I therefore offer this only as a suggestive remark.

whatever that may stand for – technology shocks, demand shocks, management shocks.

9.3 Entrants versus incumbents and the adoption of technology

Chapter 8, by Garavaglia, Malerba, and Orsenigo, simulates some effects of the biotechnology revolution using behavioral rules. Here I shall isolate just one issue that their study raises. In spite of lacking adequate finance, the new biotech firms are assumed to have an edge over incumbents. Indeed, there is plenty of evidence that disruptive and major technological change favors the new firm, the high-tech startup. But it is also true that established companies can and do acquire promising startups, and this is a possibility that the authors recognize and allow for via agreements struck between NBFs and incumbents.

As Garavaglia et al.'s model emphasizes, research and invention play a central role here – the pharmaceutical industry is one of the most innovative, at least as measured by patent counts and R&D spending. New drugs arrive all the time. Market leadership then depends on being able to invent the best products, or on the ability to imitate the best inventions of others. To stay on top, or at least to stay in the game for a long time, a firm must be able to move up the technological ladder. Either it must invent good new products, or it must be able to work around the patents of others and somehow imitate the best products of its rivals.

Whether a firm can adapt to technological change depends on how much of its knowledge transfers from the old technology that it has mastered to the new technology that it has never used before. How much of the expertise acquired in the old technology can be transferred to the new? One study that estimates the degree of such transfer in the semiconductor industry is the paper by Irwin and Klenow (1994). They estimate that intergenerational spillovers are fairly weak, which may be why semiconductor incumbents have not held on to their market shares as firmly as the pharma incumbents have been able to hold on to theirs.

An information-theoretic analysis of whether experience with one technology promotes or deters a switch to the next technological generation is in Jovanovic and Nyarko (1996). In that model, when the nature of knowledge is sufficiently general that it transfers from one technology to the next, a firm that has produced with a given technology will be the first to switch to a new one: experience promotes technological upgrading. The converse is also true: when technological knowledge is specific, then experience with a given technology creates a greater attachment to that technology, and makes a firm less disposed to moving to the next level. In that case, a firm falls into a technology trap and is eventually forced out of the market. This relates to the discussion in Garavaglia, Malerba, and Orsenigo's section 9.2, which states that the adoption of biotechnology was much easier for those incumbents that had *not* made the transition from "random" to "guided" drug discovery. For them, the tools of genetic engineering were helpful in their random screening, and, in that sense, their experience with the random screening technologies was helpful in the adoption of biotechnology itself.

The model I shall isolate here, however, is that of Chari and Hopenhayn (1991), partly because it can be described with just a few simple lines of algebra. What kind of a technological ladder is it that allows incumbents to maintain their positions for a long time? Chari and Hopenhayn (1991) build a model that states that the answer is: when the role of experienced workers is sufficiently essential. I shall outline this model. A vintage t technology has the constant returns to scale (CRS) production function

$$\gamma^t f(z, n)$$

where n is the number of workers who have no prior experience with the technology at hand, and z is the number of workers who have experience *with that technological vintage*. We may call z the number of managers of firms, and, if there is one manager per firm, then z is also the number of firms operating the technology.

Chari and Hopenhayn analyze only constant-growth paths in which the age distribution of technologies is fixed, and so it is simpler to talk about a technology's age than its vintage. Divide the output of all technologies by γ^t. Then the output of a technology of age τ is

$$\gamma^{-\tau} f(z, n)$$

where n is inexperienced labor input in technology of age τ, and z is labor that has some experience with technology τ – as stated earlier, I think of z as management. There is no capital.

Let w_τ be the wage that technology τ pays to inexperienced workers, and υ_τ the wage that it pays to experienced workers. A firm that operates an age-τ technology maximizes its profits as follows:

$$\max_{n,\, z}\{\gamma^{-\tau}f\,(z,n) - \upsilon_\tau z - w_\tau n\} \tag{9.15}$$

Since returns to scale are constant, profits must be zero.

Chari and Hopenhayn show that a range of technologies $\tau = 0, 1, \ldots,$ T is used in equilibrium. The elasticity of substitution between z and n is the main determinant of how long old technologies survive. A low elasticity of substitution means that T will be large. It also means that z – i.e. experience – is relatively essential in producing output. In that case, starting a new technology in which no one is experienced leads to low output, and so the adoption of new technologies is costly on those grounds. Therefore, older technologies, or generations of workers whose initial experience derives from older technologies, will survive for a longer period of time, as has apparently happened in the pharmaceutical industry.

Because it is a steady-state model in which the age distribution of technologies used is fixed, the Chari–Hopenhayn model is not suitable for analyzing a technological revolution, which is the aim of chapter 8. But, as its cornerstone, the model has a force – technology-specific human capital – that makes incumbents more attached or less attached to old methods: the role of experience, as manifested in the elasticity of substitution in production between experienced and inexperienced workers.

9.4 Concluding remarks

These three studies will no doubt stimulate further work on the topics they cover. I would hope that some of that work will interpret the evidence in terms of the two neoclassical models that I have outlined. These two models are almost certainly inadequate for the data at hand – I have indicated in each case one or two reasons why – but they may provide a useful place to start.

References

Chari, V. V., and H. Hopenhayn (1991), "Vintage human capital, growth, and the diffusion of new technology," *Journal of Political Economy*, 99 (6), 1142–65.

Irwin, D., and P. Klenow (1994), "Learning-by-doing spillovers in the semi-conductor industry," *Journal of Political Economy*, 102 (6), 1200–27.

Jovanovic, B., and Y. Nyarko (1996), "Learning by doing and the choice of technology," *Econometrica*, 64 (6), 1299–310.

Lucas, R. E., Jr., and E. C. Prescott (1971), "Investment under uncertainty," *Econometrica*, 39 (5), 659–81.

PART III

Policy implications

10 | The effects of research tool patents and licensing on biomedical innovation

JOHN P. WALSH, ASHISH ARORA,
AND WESLEY M. COHEN

10.1 Introduction

There is widespread consensus that patents have long benefited biomedical innovation. A forty-year empirical legacy suggests that patents are more effective, for example, in protecting the commercialization and licensing of innovation in the drug industry than in any other.[1] Patents are also widely acknowledged as providing the basis for the surge in biotechnology startup activity witnessed over the past two

This is an abridged version of an article by the same name that appeared in Wesley M. Cohen and Stephen A. Merrill (eds.), *Patents in the Knowledge-Based Economy*, National Academies Press, Washington, DC, 2003. We publish this abridged version with the kind permission of the National Academies Press.

The authors would like to thank the Science, Technology and Economic Policy (STEP) Board of the National Academy of Sciences, and the National Science Foundation (award no. SES-9976384), for financial support. We thank Jhoanna Conde, Wei Hong, JoAnn Lee, Nancy Maloney, and Mayumi Saegusa for research assistance. We would like to thank the following for their helpful comments on earlier drafts of this chapter: John Barton, Bill Bridges, Mildred Cho, Robert Cook-Deegan, Paul David, Rebecca Eisenberg, Akira Goto, Lewis Gruber, Janet Joy, Robert Kneller, Eric Larson, Richard Levin, Stephen Merrill, Ichiro Nakayama, Pamela Popielarz, Arti Rai, and participants in the STEP Board conference on New Research on the Operation and Effects of the Patent System, 22 October 2001, Washington, DC, and the OECD Workshop on Genetic Inventions, Intellectual Property Rights and Licensing Practices, 24–25 January 2002, Berlin, Germany, as well as the School of Information Seminar at the University of Michigan.

[1] See Scherer et al. (1959), Levin et al. (1987), Mansfield (1986), and Cohen, Nelson, and Walsh (2000). For pharmaceuticals, there is near-universal agreement among our respondents that patent rights are critical to providing the incentive to conduct R&D. Indeed, data from the Carnegie Mellon Survey of Industrial R&D (see Cohen, Nelson, and Walsh, 2000) show that the average imitation lag for the drug industry is nearly five years for patented products, whereas, for the rest of the manufacturing sector, the average is just over 3.5 years ($p < 0.01$). Moreover, recent evidence shows that the profits protected by patents constitute an important incentive for drug firms to invest in R&D (Arora, Ceccagnoli, and Cohen, 2002).

decades.[2] Heller and Eisenberg (1998) and the National Research
Council (NRC) (1997) have suggested, however, that recent policies
and practices associated with the granting, assertion, and licensing of
patents on research tools may now be undercutting the stimulative
effect of patents on drugs and related biomedical discoveries. In this
chapter we report the results of seventy interviews with personnel at
biotechnology and pharmaceutical firms and universities in considering the effects of research tool patents on industrial or academic biomedical research.[3] We conceive of research tools broadly to include any
tangible or informational input into the process of discovering a drug
or any other medical therapy or method of diagnosing disease.[4]

Heller and Eisenberg (1998) argue that biomedical innovation has
become susceptible to what they call a "tragedy of the anticommons,"
which can emerge when there are numerous property right claims to
separate building blocks for some product or line of research. When
these property rights are held by numerous claimants (especially if they
are from different kinds of institutions), the negotiations necessary for
their combination may fail, quashing the pursuit of otherwise promising lines of research or product development. Heller and Eisenberg
suggest that the essential precondition for an anticommons – the need
to combine a large number of separately patentable elements to form
one product – now applies to drug development because of the patenting of gene fragments or mutations (e.g. expressed sequence tags and
single-nucleotide polymorphisms) and a proliferation of patents on
research tools that have become essential inputs into the discovery of
drugs, other therapies, and diagnostic methods. Heller and Eisenberg

---

[2] For example, in one of our interviews, a licensing director for a large pharmaceutical firm said: "Patents are critical for startup firms. Without patents, we won't
even talk to a startup about licensing."

[3] The National Research Council also considers the challenges for biomedical
innovation posed by the patenting of research tools and upstream discoveries
more generally. In a series of case studies, the NRC (1997, chap. 5) documents
pervasive concern over limitations on access due to the price of intellectual
property rights and concern over the potential blocking of worthwhile innovations due to IPR negotiations, but no actual instances of worthwhile projects that
were blocked.

[4] Examples include recombinant DNA (Cohen–Boyer), polymerase chain reaction,
genomics databases, micro-arrays, assays, transgenic mice, embryonic stem cells,
and knowledge of a target – that is, any cell receptor, enzyme, or other protein
that is implicated in a disease and consequently represents a promising locus for
drug intervention.

argue that the combining of multiple rights is susceptible to a break-down in negotiations, or, similarly, a stacking of license fees, to the point of overwhelming the value of the ultimate product. Shapiro (2000) has raised similar concerns, using the image of the "patent thicket." He notes that technologies that depend on the agreement of multiple parties are vulnerable to holdup by any one of them, making commercialization potentially difficult.

The argument that an anticommons may emerge to undercut inno-vation emphasizes factors that could frustrate private incentives to realize what would otherwise be mutually beneficial trades. Merges and Nelson (1990) and Scotchmer (1991) have argued, however, that the self-interested use of even just one patent – although lacking the encumbrances of multiple claimants characterizing an "anticommons" – may also impede innovation where a technology is cumulative (i.e. where invention proceeds largely by building on prior invention). An example of such an upstream innovation in biomedicine is the discovery that a particular receptor is important for a disease, which may make that receptor a "target" for a drug development program. A key concern regarding the impact of patents in such cumulative technologies is that, "unless licensed easily and widely," patents – espe-cially broad patents – on early, foundational discoveries may limit the use of these discoveries in subsequent discovery and consequently limit the pace of innovation (Merges and Nelson, 1990).[5] The revolution in molecular biology and related fields over the past two decades and coincident shifts in the policy environment have now increased the salience of this concern for biomedical research and drug innovation in particular (NRC, 1997). Drug discovery is now more guided by prior scientific findings than previously (Gambardella, 1995; Cockburn and Henderson, 2000; Drews, 2000), and in the United States those findings are now more likely to be patented after the 1980 passage of the Bayh–Dole Act and related legislation that simplified the patenting of

---

[5] Scotchmer (1991) focuses on the related issue of the allocation of rents between the holder of a pioneer patent and those who wish to build on that prior discovery, suggesting that there is no reason to believe that markets left to themselves will set that allocation in such a way that the pace of innovation in cumulative technol-ogies is maximized. Barton (2000), in fact, suggests that the current balance "is weighted too much in favor of the initial innovator." Scotchmer (1991) has suggested that *ex ante* deals between pioneers and follow-on innovators can, however, be structured to mitigate the problem.

federally supported research outputs that are often upstream to the development of drugs and other biomedical products.

In this chapter we consider whether biomedical innovation has suffered because of either an anticommons or restrictions on the use of upstream discoveries in subsequent research. Notwithstanding the possibility of such impediments to biomedical innovation, there is still ample reason – and recent scholarship (Arora, Ceccagnoli, and Cohen, 2002) – to suggest that patenting benefits biomedical innovation, especially via its considerable impact on R&D incentives or via its role in supporting an active market for technology (Arora, Fosfuri, and Gambardella, 2001). Although any ultimate policy judgment requires a consideration of the benefits and costs of patent policy, an examination of the benefit side of this calculus is outside the scope of our current study.

In the second section of this chapter we provide background to the anticommons and restricted access problems. The third section describes our data and methods. In the fourth section we provide an overview of the results from our interviews and assess the extent to which we witness either "anticommons" or restricted access to intellectual property (IP) on upstream discoveries and research tools. To prefigure the key result, we find little evidence of routine breakdowns in negotiations over rights, although research tool patents are observed to impose a range of social costs and there is some restriction of access. In the fifth section of the chapter we describe the mechanisms and strategies employed by firms and other institutions that have limited the negative effects of research tool patents on innovation. The final section discusses our findings and our conclusions.

10.2 Background

10.2.1 Science and policy

Changes in the science underlying biomedical innovation, and in policies affecting what can be patented and who can patent, have combined to raise concerns over the impact of the patenting and licensing of upstream discoveries and research tools on biomedical research. Over the past twenty years fundamental changes have revolutionized the science and technology underlying product and process innovation in drugs and the development of medical therapies and diagnostics.

Advances in molecular biology have increased our understanding of the genetic bases and molecular pathways of diseases. Automated sequencing techniques and bioinformatics have greatly increased our ability to transform this understanding into patentable discoveries that can be used as targets for drug development. In addition, combinatorial chemistry and high-throughput screening techniques have dramatically increased the number of potential drugs for further development. Reflecting this increase in technological opportunity, the number of drug candidates in phase I clinical trials grew from 386 in 1990 to 1512 in 2000.[6] The consequence of these changes is that progress in biomedical research is now more cumulative; it depends more heavily than heretofore on prior scientific discoveries and previously developed research tools (Drews, 2000; Henderson, Orsenigo, and Pisano, 1999).

As the underlying science and technology have advanced, policy changes and court decisions since 1980 have expanded the range of patented subject matter and the nature of patenting institutions. In addition to the 1980 *Diamond v. Chakrabarty* decision that permitted the patenting of life forms, and the 1988 Harvard OncoMouse patent that extended this to higher life forms (and to a research tool), in the 1980s gene fragments, markers, and a range of intermediate techniques and other inputs key to drug discovery and commercialization also became patentable. Moreover, Bayh–Dole and related legislation have encouraged universities and national labs, responsible for many such upstream developments and tools, to patent their inventions. Thus, coincident changes in the science underpinning biomedicine and the policy environment surrounding IPR have increased both the generation and patenting of upstream developments in biomedicine.

10.2.2 Conceptual

When is either an "anticommons" problem or restricted access to upstream discovery likely to emerge and why, and what are the welfare implications of their emergence?

Consider the anticommons. The central question here, as posed by both Heller and Eisenberg (1998) and Eisenberg (2001), is: if there is a cooperative surplus to be realized in combining property rights

[6] We thank Margaret Kyle for making these data available to us.

to commercialize some profitable biomedical innovation, why might it not be realized? They argue that biomedical research and innovation may be especially susceptible to breakdowns and delays in negotiations over rights for three reasons. First, the existence of numerous rights holders with claims on the inputs into the discovery process or on elements of a given product increases the likelihood that the licensing and transaction costs of bundling those rights may be greater than the ultimate value of the deal. Second, when there are different kinds of institutions holding those rights, heterogeneity in goals, norms, and managerial practice and experience can increase the difficulty and cost of reaching agreement. Such heterogeneity is manifest in biomedicine given the participation of large pharmaceutical firms, small biotechnology research firms, large chemical firms that have entered the industry (e.g. DuPont and Monsanto), and universities. Third, uncertainty over the value of rights, which is acute for upstream discoveries and research tools, can spawn asymmetric valuations that contribute to bargaining breakdowns and provide opportunities for other biases in judgment. This uncertainty is heightened because the courts have yet to interpret the validity and scope of particular patent claims.

Regarding the restriction of access to upstream discoveries highlighted by Merges and Nelson (1990; 1994), one can ask why that should be a policy concern. From a social welfare perspective, nothing is wrong with restricted access to IP for the purpose of subsequent discovery as long as the patentholder is as able as potential downstream users to exploit fully the potential contribution of that tool or input to subsequent innovation and commercialization.[7] This, however, is unlikely for several reasons. First, firms and, especially, universities are limited in their capabilities. Second, there is often a good deal of uncertainty about how best to build on a prior discovery, and patentholders will be limited in their views about what that prior discovery may best be used for and how to go about exploiting it. Consequently, a single patentholder is not able to exploit fully the research and commercial potential of a given upstream discovery, and society is better off to the extent that such upstream discoveries are

---

[7] That patents imply some type of output restriction due to monopoly is taken as given. The question here is whether there is any social harm if the patentholder chooses to exploit the innovation himself exclusively.

made broadly available.[8] For example, if there is a target receptor, it is likely that there are a variety of lines of attack, and that no single firm is capable of mounting or even conceiving of all of them. The notion that prior discoveries should be made broadly available rests, however, on an important assumption: that broad availability will not compromise the incentive to invest the effort required to come up with that discovery to begin with (see Scotchmer, 1991).

In this chapter we are, therefore, concerned with whether the access to upstream discoveries essential to subsequent innovation is restricted. Restriction is, however, a matter of degree. If a discovery is patented at all, then it is to be expected that access will be restricted – reflecting the function of a patent. Indeed, any positive price for a license implies some degree of restriction. Therefore, we are concerned with more extreme forms of restricted access, which may come in the form of the exclusive licensing of broadly useful research tools, high license fees that may block classes of potential users, or decisions on the part of a patentholder itself to exploit some upstream tool or research finding that it developed.

10.2.3 Historical

The possibility that access to a key pioneering patent may be blocked, or that negotiations over patent rights might break down (even when a successful resolution would be in the collective interests of the parties concerned), is not a matter of conjecture. There is historical precedent. Merges and Nelson (1990) and Merges (1994), for example, consider the case of radio technology, where the Marconi Company, De Forest, and De Forest's main licensee, AT&T, arrived at an impasse over rights that lasted about ten years, and was resolved in 1919 only when the Radio Corporation of America (RCA) was formed at the urging of the US Navy. In aviation, Merges and Nelson argue that the refusal of the Wright brothers to license their patent significantly retarded progress in the industry. The problems caused by the initial pioneer patent (owned by the Wright brothers) were compounded as

---

[8] The premise of this argument, well recognized in the economics of innovation (Jewkes, Sawers, and Stillerman, 1958; Evenson and Kislev, 1973; Nelson, 1982), is that, given a technological objective (e.g. curing a disease) and uncertainty about the best way to attain it, that objective will be most effectively achieved to the extent that a greater number of approaches to it are pursued.

improvements and complementary patents, owned by different companies, came into existence. Ultimately, World War I forced the Secretary of the Navy to intervene to work out an automatic cross-licensing arrangement. "By the end of World War I there were so many patents on different aircraft features that a company had to negotiate a large number of licenses to produce a state-of-the-art plane" (Merges and Nelson, 1990, p. 891).

Although breakdowns in negotiations over rights may therefore occur, rights over essential inputs to innovation are routinely transferred and cross-licensed in industries, such as the semiconductor industry, where there are numerous patents associated with a product and multiple claimants (Levin, 1982; Hall and Ziedonis, 2001; Cohen, Nelson, and Walsh, 2000). In Japan, where there are many more patents per product across the entire manufacturing sector than in the United States, licensing and cross-licensing are commonplace (Cohen et al., 2002).

Thus the historical record provides instances both of when the existence of numerous rights holders and the assertion of patents on foundational discoveries have retarded commercialization and subsequent innovation and of when no such retardation emerged. The history suggests several questions. Have anticommons failures occurred in biomedicine? Are they pervasive? To what degree do we observe restricted access to foundational discoveries that are essential to the subsequent advance of biomedicine? What factors might affect biomedicine's susceptibility (or lack thereof) to either anticommons or restrictions on the use of upstream discoveries in subsequent research?

10.3 Data and methods

To address these issues, we conducted seventy interviews with IPR attorneys, business managers, and scientists from ten pharmaceutical firms and fifteen biotech firms, as well as university researchers and technology transfer officers from six universities, patent lawyers, and government and trade association personnel. Table 10.1 gives the breakdown of the interview respondents by organization and occupation.

This purposive sampling was designed to solicit information from respondents representing various aspects of biomedical research and drug development (Whyte, 1984). We used the interviews to probe

Table 10.1 The distribution of interview respondents, by organization and occupation

	Pharmaceutical	Biotech	University	Other
IPR lawyer	12	7	–	12 (7)
Scientist	3	4	10	3
Business manager	9	7	3	–

Note: "Other" includes outside lawyers (7) and government and trade association personnel. University technology transfer office personnel are classified as "business managers," although some are also lawyers. In addition, many of the lawyers and business managers had also worked as R&D scientists before their current position.

whether there has been a proliferation and fragmentation of patent rights and whether this has resulted in the failure to realize mutually beneficial trades, as predicted by the theory of anticommons. We also looked for instances in which restricted access to important upstream discoveries has impeded subsequent research. In addition, we asked our respondents how these conditions may have changed over time, including whether the character of negotiations over IPR has changed. Finally, we asked about strategies and other factors that may have permitted firms to overcome challenges associated with IP.

10.4 Findings

10.4.1 Preconditions for an anticommons

Do conditions that might foster an "anticommons" exist in biomedicine? The essential precondition for an anticommons is the existence of multiple patents covering different components of some product, its method of manufacture, or inputs into the process through which it was discovered.

We have no direct measure of the number of patents covering a new product. There has, however, been a rapid growth in biotechnology patents over the past fifteen years, from some 2000 issued in 1985 to over 13000 in 2000 (http://www.bio.org/er/statistics.asp). Such rapid growth is consistent with a sizable number of patents being granted for research tools and other patents related to drug development. Our interview respondents also suggest that there are indeed more patents now related to a given drug development project. One biotechnology executive responsible for IP stated:

The patent landscape has gotten much more complex in the eleven years I've been here. I tell the story that, when I started and we were interested in assessing the third-party patent situation, back then, it consisted of looking at [four or five named firms]. If none were working on it, that was the extent of due diligence. Now, it is a routine matter that, when I ask for some search for third-party patents, it is not unusual to get an inch or two thick printout filled with patent applications and granted patents ... In addition to dealing with patents over the end product, there are a multitude of patents, potentially, related to intermediate research tools that you may be concerned with as well.

Almost half of our respondents (representing all three sectors of our sample: big pharmaceutical firms, small biotech firms, and universities) addressed this issue, and all of them agreed that the patent landscape has indeed become more complex. How complex is, however, an important issue. Although there are often a large number of patents potentially relevant to a given project, the actual number needed to conduct a drug development project is often substantially smaller. We asked about ten of our industry respondents to tell us how many pieces of IP had to be in-licensed for a typical project. They said that there may be a large number of patents to consider initially, sometimes in the hundreds, and that this number is surely larger than in the past. However, respondents then went on to say that in practice there may be, in a complicated case, about six to twelve that they have to address seriously, but that more typically the number was zero. Thus, although most R&D executives report that the number of licenses they must obtain in the course of any given project has increased over the past decade, that number is considered to be manageable.

In addition to a larger number of patents typically bearing on a given project, the numbers and types of institutions involved have also grown. Preceding the recent growth in biotechnology patenting, the number of biotechnology firms grew rapidly in the 1980s (Cockburn et al., 2000). More recently, we observe NBFs acquiring significant patent positions. Hicks et al. (2001), for example, report that the number of US NBFs receiving more than fifty patents in the prior six years grew from zero in 1990 to thirteen by 1999.

Universities have also become major players in biotechnology, as sources of both patented biomedical inventions and startup firms that are often founded on the strength of university-origin patents. Many respondents (fourteen from industry and six from universities) noted that this new role of universities is one of the significant changes over the last two decades in the drug and related industries. Universities have increased their patenting dramatically over the last two decades, and, although still small, their share of all patents is significantly higher than before 1980. Furthermore, much of the growth in university patents has tended to concentrate in a few utility classes, particularly those related to life sciences. In three of the key biomedical utility classes, universities' share of total patents increased from about 8 percent in the early 1970s to over 25 percent by the mid-1990s (National Science Foundation [NSF], 1998). Also, universities' adjusted gross licensing revenue has grown from $186 million going to 130 universities in 1991 to $862 million going to 190 universities in 1999 (Association of University Technology Managers [AUTM], 2000), with the preponderance of these sums reflecting activity in life sciences. An eightfold increase in university technology licensing offices from 1980 to 1995 is further evidence of increasing emphasis on the licensing of university discoveries (Mowery et al., 2001).

Contributing to the rise in patenting, particularly in genomics, is the intensification of defensive patenting. Overall, about a third of our industry respondents claimed to be increasing their patenting of gene sequences, assays, and other research tools as a response to the patenting of others to ensure freedom to operate (see also Henry et al., 2002).

Thus, we observe many patents (especially on research tools) owned by different parties with different agendas. In short, the patent landscape has genuinely become more complex – although not as complex as suggested by some. Nonetheless, conditions may indeed be conducive to a tragedy of the anticommons.

10.4.2 Preconditions for restricted access to upstream discoveries

Our second concern is that the restrictive assertion or licensing of patents on research tools – especially foundational upstream discoveries upon which subsequent research must build (such as transgenic mice, embryonic stem cells, or the knowledge of a potential drug target) – may undermine the advance of biomedical research. As suggested above, the key condition for this concern holds – namely that research tools are now commonly patented. One R&D manager, for example, has stated that "there has been a pronounced surge in patenting of research tools, previously more freely available in the public domain." Academic scientists we interviewed affirmed this view, observing a shift from a regime in which findings were more likely to be placed in the public domain with no IP protection.

Restricted access to upstream technology becomes a greater concern and more limiting on downstream research activity as the claims on the upstream patents are interpreted more broadly. The complaint about Human Genome Sciences (HGS) asserting its patent over the HIV receptor illustrates the concern that patentholders are able to exercise control over a broad area even when their own upstream invention is narrow and there is very little disclosed about the utility of the invention (Marshall, 2000a). At the time of the patent application, HGS knew only that they had found the gene for something that was a chemokine receptor. Later work published by NIH scientists detailed how this receptor (CCR5) worked with HIV, making this a very important drug target. Those who discovered the utility of this receptor for AIDS research and drug development filed patents, only to find that HGS's "latent" discovery had priority. The concern that arises is that knowledge of the reach of HGS's patent could have deterred subsequent research on exploring the role of the gene and the associated receptor.

Another way in which the absence of a clear written description may allow upstream patents to affect subsequent research directly is via "reach-through" patent claims (as distinct from license agreements that include royalties on the product discovered using a research tool). Here, the patent claims the target and any compound that acts on the target to produce the desired effect, without describing what those compounds are. If the patentholder is given broad rights to exclude others from pursuing research in this area, we could have the problem of no one in fact possessing the innovation, and greatly reduced incentives for non-patentholders to explore possible uses of the innovation.

Thus, there is a proliferation of patents on upstream discoveries and tools, and how those patents affect downstream discovery depends heavily on the breadth of claims. Although the United States Patent and Trademark Office (USPTO) has permitted broad claims to issue, there remains the question of how the courts will evaluate those claims.

10.4.3 *Evidence of an anticommons in biomedical research*

Given that the preconditions for an anticommons seem to exist, we turn to our findings on the incidence and nature of the different impediments to biomedical research that an anticommons may pose. These include breakdowns in negotiations over rights, royalty stacking, and "excessive" license fees.

10.4.3.1 Breakdowns

Perhaps the most extreme expression of an anticommons tragedy is the existence of multiple rights holders spawning a breakdown in negotiations over rights that leads to an R&D project's cessation. We find almost no evidence of such breakdowns. We asked respondents and searched the literature to identify cases in which projects were stopped because of an inability to obtain access to all the necessary IPR. In brief, respondents reported that negotiations over access to necessary IP from many rights holders rarely led to a project's cessation. Of the fifty-five respondents who addressed this issue (representing all three sectors), fifty-four could not point to a specific project stopped because of difficulties in getting agreement from multiple IP owners (the anticommons problem). For example, one respondent indicated that about a quarter of his firm's projects were terminated in the past year. Of these, none were terminated because of any difficulties with the in-licensing of tools. Instead, the key factors included pessimism about technical success and the size of the prospective market.[9]

A particular concern raised by Heller and Eisenberg (1998) and the NRC (1997, chap. 5) was the prospect that, by potentially increasing the number of patent rights corresponding to a single gene, patents on ESTs would proliferate the number of claimants to prospective drugs and increase the likelihood of bargaining breakdowns. Our respondents suggested that this has not occurred. The key concern was that patents on the partial sequence might give the patentholder rights to the whole gene or the associated protein, or at least that the patent might

---

[9] Numerous respondents reported that they did not initiate or had dropped projects if they learned that another firm had already acquired a proprietary position on a drug they were considering developing – that is, on the output of a drug discovery and testing process. But that is quite different from other firms having IP for the research tools – the inputs into the discovery process.

block later patents issued on the gene or the protein (as Doll [1998] of the USPTO suggested). Our respondents from industry and from the USPTO reflected the view of Genentech's Dennis Henner, however, who testified before Congress that EST patents do not dominate the full gene sequence patent, the protein, or the protein's use; these are separate inventions.[10] Also, although the existence of large numbers of EST patents may have had the potential to create anticommons problems, the new utility and written description guidelines implemented by the USPTO will now probably prevent many EST patents from issuing and will grant those that do issue only a narrow scope of claims. In addition, it is likely that already issued EST patents will be narrowly construed by the courts. Thus, the consensus is that the storm over ESTs has largely passed.

10.4.3.2 Royalty stacking
Another way in which multiple claimants on research tool IP may block drug discovery and development is the stacking of license fees and royalties to the point of overwhelming the commercial value of a prospective product. Most of our respondents reported that royalty stacking did not represent a significant or pervasive threat to ongoing R&D projects. One respondent said that, although stacking is a consideration, "I can't think of any example where someone said they did not develop a therapeutic because the royalty was not reasonable." We heard of only one instance in which a project was stopped because of royalty stacking. We were told, however, that, in this case, there were too many claimants to royalty percentages because of carelessness by a manager, who had given away royalty percentages without carefully accounting for prior agreements.[11] One of our other biotechnology respondents suggested, however, that "the royalty burden can become onerous," and that the stacking of royalties "comes up pretty regularly now" with the proliferation of IP. Even here, the respondent said that no projects had ever been stopped because of royalty stacking. Overall, about a half of our respondents complained about licensing costs for

---

[10] Testimony before House Judiciary Committee, 13 July 2000, htttp://www.house.gove/judiciary/henn0713.htm.

[11] We also had one respondent, an IP lawyer, who said that such cases where projects were stopped did exist, but client privilege prevented the respondent from giving details.

research tools, although nearly all of those concerned about licensing costs also went on to say that the research always went forward.

Royalty stacking does not represent a significant threat to ongoing R&D projects, for several reasons. First, and principally, the total of fees paid (as discussed below) typically does not push projects into a loss. Second, in the minority of cases in which the stacking of fees threatens a loss, compromises tend to be struck, often in the form of royalty offsets across the various IP holders. One respondent stated, "All are sensitive and aware of the stacking phenomenon so there is a basis for negotiation, so that you don't have excessive royalties." Finally, in the few cases in which such a problem could emerge, it also tends to be anticipated.[12] One firm executive we interviewed said he had a corporate-level committee that reviewed all such requests to make sure such problems do not occur.

10.4.3.3 Licensing fees for research tools

Although obtaining systematic data on the cost of patented research tools is difficult, half of our respondents provided enough information to allow us to approximate the range of such costs. The norm for total royalty payments for the various input technologies associated with a given drug development program is in the range of 1 to 5 percent of sales, and somewhat higher for exclusive licenses. Occasionally, royalty demands were 10 percent or higher, and these were described in such terms as "high" or "ridiculous." Firms (especially the large pharmaceutical firms) also license technologies – such as using a gene for screening or a vector or micro-arrays – for a fee ranging from $10,000 to $200,000. These fees (especially for genes) were often described (both by those buying and those selling such technologies) as small amounts that large pharmaceutical firms paid as insurance both to ensure freedom to operate and to avoid the cost of litigation. The cost of patented reagents could be two to four times as much as do-it-yourself versions (or, in the case of *Taq* polymerase, buying from an unlicensed vendor), although the overall cost to the project is generally small (at most a few percent).

---

[12] In response to a question as to whether his firm had ever had a case of a project being stopped for problems with royalty stacking, a biotechnology respondent stated: "No. It would be hard to find such a case, given the reality of how decisions are made. It is not a late-stage decision."

Large pharmaceutical firms have also been licensing access to genomic databases, and these database fees are often tens of millions of dollars and occasionally over $100 million (*Science*, 1997). In 1997, for access to its database, Incyte was reported to be charging $10 million to Upjohn and almost $16 million to Pfizer, as well as undisclosed amounts to eight other firms. These deals also include "low single-digit" royalties for the use of patented genes in drug development. Four pharmaceutical firms paid between $44 million and $90 million each to Millennium to access their data and research tools for identifying disease genes. In 1998 Bayer agreed to a deal in which they would pay up to $465 million to Millennium to have Millennium identify 225 new drug targets within five years (Malakoff and Service, 2001).

Overall, our respondents noted that, although these costs were higher than before the surge in research tool patents, they believed them to be within reason, largely because the productivity gains conferred by the licensed research tools were thought to be worth the price.

Our interviews suggested, however, that, although these costs were seen as manageable by large pharmaceuticals firms, and even by established biotech firms, small startup firms and university labs noted that such costs could be prohibitive, in effect making it impossible for them to license particular research tools. This sentiment was echoed by the university researchers we talked with, and was one justification for the "do-it-yourself" solution of making patented laboratory technology without paying royalties (Marshall, 1999b). Similarly, the manager of a small biotech startup told us that Incyte's licensing terms for access to their gene database amounted to several times the firm's whole annual budget. They were forced to rely on the public databases – a viable but second-best solution. One solution for universities has been the development of core facilities to share expensive resources, such as chip-making or high-throughput screening facilities.

Some firms (particularly genomics firms) holding rights over research tools did, however, offer discounted terms for university and government researchers. Celera, for example, licenses their database to firms for about $5 to $15 million per year and to university labs for about $7500 to $15,000 (Service, 2001). In 2000 Incyte began allowing single-gene searches of its database for free, with a charge of $3000 or more for ordering sequences or physical clones, making its database more accessible to small users (*Science*, 2000). Myriad also offers a discount rate (less than half the market rate) for academics doing NIH-funded research on breast cancer (Blanton, 2002).

In this section we have considered the costs of licensing research tool IP – but only the out-of-pocket, monetary costs. Costs can, however, also take non-monetary forms. The most prominent of these for university researchers are publication restrictions, which we did not examine.[13]

10.4.3.4 Projects not undertaken and broader determinants of R&D

Although the number of ongoing R&D projects stopped because of an anticommons problem is small, it is possible that firms avoid stacking and other difficulties in accessing IPR simply by not undertaking a project to begin with. As a practical matter, it is difficult to measure the extent to which projects were not started or redirected because of patent-related concerns. In brief, although redirecting projects to invent around research tool patents was common, it was relatively rare for firms to move to a new research area (perhaps a new disease, or even a very different way of approaching a disease) because of concerns over one or more research tool patents. Of the eleven industry respondents who did mention IP as a cause for redirecting their research, seven, however, were primarily concerned with IP on compounds, not on research tools.[14] An IP attorney with a large biotech firm suggested that patents on research tools were rarely determinative, reporting that, in the "scores of projects" that his firm considered undertaking over the years, he could remember only one where such patent rights dissuaded them from undertaking the project. Another biotech firm's

---

[13] Thursby and Thursby (1999) report that 44% of agreements to license university technologies to firms include publication delay clauses, with the average delay specified being almost 4 months. Also, Blumenthal et al. (1997) report that 20% of academic biomedical researchers have delayed research publication by 6 months or more, in part because of concerns about patents and commercialization. Thus a substantial fraction of university-industry agreements about the outputs of university research include delays of publication. We do not know whether these examples generalize to the case of agreements over the inputs to university research.

[14] And a large number (about a third of industry respondents) said that, when faced with rival patents on research tools, or even compounds, they were likely to go ahead with the research so long as they were able to develop their own IP that would protect their compounds (see below).

lawyer, while reporting that they had never stopped an ongoing project because of license stacking, stated that considerations of patents on both compounds and research tools did pre-empt projects:

We start very early on ... to assess the patent situation. When the patent situation looks too formidable, the project never gets off the ground ... Once you are well into development, you get patent issues, but not the show-stopper that you would identify early on.

Although we have no systematic data on projects never pursued, our findings on the absence of breakdowns is consistent with the notion that there are relatively few cases where otherwise commercially promising projects are not undertaken because of IP on research tools. Consider Heller's (1998) original article on the anticommons, which paints a vivid image of empty buildings in Moscow, unrented because the various owners and claim holders who could "veto" a rental arrangement were many and had trouble coming to agreement. Our analogue to an "empty building" is, of course, an R&D project that is stopped midway. However, if the argument that the proliferation of IP is generating an "anticommons" is correct, it follows that the rational anticipation of such difficulties would prevent the construction of some (or many) buildings. Likewise, some R&D projects may not be undertaken if firms anticipate difficulty in negotiating cost-effective access to the required IP. However, in the absence of any visible empty buildings (i.e. observed stopped projects), it is unlikely that the anticipation of breakdowns in negotiations or an excessive accumulation of claims (i.e. license stacking) prevented construction (i.e. undertaking the R&D project).[15]

Our interviews suggested that the main reasons why projects were not undertaken reflected considerations of technological opportunity,

---

[15] One version of this argument is that, if the anticommons problem were widespread but few stopped projects ("empty buildings") were observed, this could be due to "anticipation and redirection," whereby firms or university researchers redirect R&D towards projects for which they do not anticipate an anticommons problem. However, to do so with such high success (so that projects did not, in the end, get stopped), decision-makers would have to be very prescient about when they would, and when they would not, face such problems. This is unlikely, given the uncertainty of early-stage R&D (when researchers do not yet know what tools they might need) and given the lag in patent issuance (so that researchers do not even know which tools are patented – see Marshall, 2000a, and Merz et al., 2002). Both these factors lead to having substantially less than perfect information on potentially blocking patents.

demand, and internal resource constraints, with expected licensing fees or "tangles" of rights on tools playing a subordinate role, salient only for those projects that were commercially less viable. One industrial respondent affirmed that, although other considerations were key, royalty stacking could affect decisions at the margin: "I don't want to say a worthwhile therapeutic was not developed because of stacking problems. But, if we have two equally viable candidates, then we choose based on royalties." One biotechnology respondent was explicit, however, about the greater importance of expected demand and technological opportunity:

At the preclinical stage, you find you have ten candidates, and you can afford to continue work on three. The decision is a complex prediction based on the potential for technical success, the cost of manufacturing, the size of the market, what you can charge, what you need to put in for royalties. I am not familiar with royalty stacking being the deciding factor. The probability of technical success and the size of market are key.

This last remark also implies that the firm had more viable opportunities than it had the resources to pursue. Indeed, complaints about resource constraints as impediments to progress on promising research were more common than complaints about IP.

The notion that opportunities often exceed the ability of firms to pursue them suggests that, at least under some circumstances, the social cost of not pursuing projects because of IP considerations may not be as great as one might suppose. Indeed, four of our industry respondents expressed the view that redirection of research effort towards areas less encumbered by patents was not terribly costly for their firms or others because the technological opportunities in molecular biology and related fields were so rich and varied. As one biotech respondent put it, "There are lots of targets, lots of diseases." Some respondents have suggested that the value of targets has actually declined substantially because companies can't exploit all the targets they have, and so firms are more willing to license some of their targets, or abandon some of their patents and let the inventions shift to the public domain, because maintaining large portfolios of low-value patents is expensive. On the other hand, one can also argue that, even in the presence of rich opportunities, shifting may be costly to the extent that diminishing the number of firms trying to achieve some technical objective makes success less likely.

10.4.4 Evidence of restricted access to upstream discoveries and tools

Although biomedical research does not appear to be especially vulnerable to breakdowns over IP negotiations, restricted access to important research tools – especially foundational upstream discoveries – can potentially impede innovation in a field. Moreover, this has occurred in other settings as well (Merges and Nelson, 1990). Our question is whether the restrictions on access to such upstream discoveries, through, for example, exclusive licensing, has impeded biomedical innovation. As noted above, in contrast to the prospect of an anti-commons, this is not a problem of accessing multiple rights but one of accessing relatively few patents – perhaps even just a single patent – on a key tool or discovery.

In its report *Intellectual Property Rights and Research Tools in Molecular Biology*, the NRC (1997) provided a series of case studies on the uses of patents covering a small number of important research tools in molecular biology where the question of restricted access was considered. In the case of the Cohen–Boyer technology for recombinant DNA developed at Stanford University and the University of California – "arguably the defining technique of modern molecular biology" (NRC, 1997, p. 40) – the three patents were broadly licensed on a non-exclusive basis on a sliding scale, providing the basis for the creation of the biotechnology industry as we know it. The license was available for about $10,000 per year plus a royalty of 0.5 to 3 percent of sales (Hamilton, 1997). The two universities eventually had several hundred licensees, and the patent generated an estimated $200 million.

The second case was that of PCR technology, which "allows the specific and rapid amplification of DNA or targeted RNA sequences," and *Taq* polymerase, which is the enzyme used in the amplification. The technology was also key to subsequent innovation. It "had a profound impact on basic research not only because it makes many research tasks more efficient but also because it ... made feasible ... experimental approaches that were not possible before" (NRC, 1997, p. 43). In addition to being a discovery tool, the technology also provides a commercial product in the form of diagnostic tests. Developed by Cetus Corporation, the technology was sold to Roche in 1991 for $300 million. As the NRC (1997) reports, the controversy over the sale of the technology has been primarily over the amount of the licensing fees and the fees charged for the material (*Taq* polymerase) itself. Although Roche licensed the technology widely, particularly to the research community, they did charge high royalty rates on their licenses for diagnostic service applications. Also, small firms complained about Roche's fees for applications of the technology other than in diagnostics, which ranged between $100,000 and $500,000 initially with a royalty rate of 15 percent. The high price probably restricted access for some, especially small biotech firms.

The CellPro case, described in detail by Bar-Shalom and Cook-Deegan, 2002 (on which the following account draws primarily), also illustrates the potential for the owners of upstream patents to block development of cumulative systems technologies (see Merges and Nelson, 1994). Johns Hopkins University's Curt Civin discovered an antibody (My-10) that selectively binds to an antigen, CD34, found on stem cells but not on more differentiated cells. In 1990 Hopkins was awarded a patent that claims all antibodies that recognize CD34. Baxter obtained an exclusive license. The chief rival was CellPro, a company founded in 1989 based on two key technologies, one a method for using selectively binding antibodies to enrich bone marrow stem cells or deplete tumor cells, and the other an unpatented antibody, 12-8, that also binds to CD34, although in a different class of antibodies from Civin's My-10 and recognizing a different epitope (binding site) on CD34. CellPro combined these two discoveries with other innovations and know-how to produce a cell separator instrument for use in cancer therapies, particularly bone marrow transplants. Baxter offered CellPro a non-exclusive license for $750,000 plus a 16 percent royalty. CellPro felt this was uneconomic and, armed with a letter from outside counsel saying that CellPro's technology did not infringe and that the patent was probably invalid, decided to move forward with development and to sue to invalidate the patent. Although the jury ruled in CellPro's favor, the *Markman* decision reopened the case and the judge ruled for Baxter, assessing treble damages totaling $7.6 million, as well as $8 million in legal fees.[16] The court also ordered license terms similar to (though somewhat higher than) existing licenses, a royalty of over $1000 per machine.

---

[16] This provides an example of the scale for legal fees involved in such a case (see below).

CellPro lost the appeal and went bankrupt. Baxter allowed sales of CellPro's machine until its own instrument (which Baxter was developing all through this) received FDA approval. In the end, the technology did not prove to be widely effective, and more successful rival technologies were developed by others.

From our perspective, the main lesson of the CellPro case is that, to the degree that upstream patents are broadly interpreted, IP holders can use this broad claim to prevent others from engaging in the subsequent development needed to bring the patented technology to market. This is troubling when the patent owner cannot effectively develop the technology in a timely fashion, which was the case with Baxter, which was at least two years behind CellPro in bringing a product to market.[17]

Another case was that of the Harvard OncoMouse, licensed by Harvard exclusively to DuPont. The OncoMouse contained a recombinant activated oncogene sequence that permitted it to be employed both as an important model system for studying cancer and permitting early-stage testing of potential anticancer drugs. After years of negotiations NIH and DuPont finally signed a memo of understanding in January 2000 that, among other things, permitted the relatively unencumbered distribution of the technology from one academic institution to another, although under specific conditions.[18] Although this agreement was the cause of relief on the part of academic researchers, DuPont has more recently begun asserting its patent against selected institutions (Neighbour, 2002). The difficulty is that, although the initial press release suggested that these non-paying rights to use the OncoMouse covered non-profit recipients of NIH funding, the actual agreement stated that DuPont would make available similar rights to non-profit NIH grantees "under separate written agreements." Because most universities have not asked for those rights, they stand outside the

---

[17] Bar-Shalom and Cook-Deegan (2002) also suggest that royalty stacking may have made the technology economically infeasible. Johns Hopkins University had licensed to B-D, which in turn licensed to Baxter, which in turn licensed to others, with each taking a share of the rents. However, in this case, the license stacking was all on the same technology being passed from hand to hand. Thus this was not a tragedy of the anticommons, but one of a proliferation of middlemen.

[18] In 1998 NIH announced an agreement with DuPont covering cre-lox technology (see memorandum of understanding [MOU] at http://ott.od.nih.gov/pdfs/cre-lox.pdf). In January 2000 NIH announced an agreement covering OncoMouse (see MOU at http://ott.od.nih.gov/pdfs/oncomouse.pdf).

agreement, and DuPont has begun to approach some of them, claiming that they are infringing Harvard's patent rights and must take a license from DuPont. The difficulty is that these new license agreements, although also non-paying in principle, go well beyond the earlier understanding and make a series of stringent demands. Under the proposed agreement, for example, universities cannot use the technology in industry-sponsored research without the sponsor taking a commercial license, *notwithstanding the content or intent of the sponsored research* (Neighbour, 2002). It is unclear at this point, however, what success DuPont will have or how NIH and other institutions will respond.

The most visible recent controversy over access to IP covering a foundational biomedical discovery is the case of embryonic stem cell technology.[19] In brief, Geron funded the research of a University of Wisconsin developmental biologist, James Thompson, who in 1998 first isolated human embryonic stem cells and was issued a very broad patent. The Wisconsin Alumni Research Foundation (WARF), a university affiliate, held the patent and granted Geron exclusive rights to develop the cells into six tissue types that might be used to treat disease, as well as options to acquire the exclusive rights to others. Another beneficiary of Geron support, Johns Hopkins University, also provided Geron with exclusive licenses on stem cell technology. In August 2001 WARF sued Geron – which had been trying to expand its rights to include an additional twelve tissue types – to be able to offer licensing rights to Geron's competitors. In January 2002 a settlement was reached that narrowed Geron's exclusive commercial rights to the development of only three types of cells (neural, heart, and pancreatic), gave it only non-exclusive rights to develop treatments based on three other cell types (bone, blood, and cartilage), and removed its option to acquire exclusive rights over additional cell types. Geron and WARF also agreed to grant rights free of charge to academic and government scientists to use the stem cell patents for research but not for commercial purposes. Companies wishing to use the stem cells for research purposes would, however, have to license the patents. Thus it would appear that, although WARF would like to license the technology broadly, Geron retained control over key application areas of the technology and may well decide to pursue those applications itself.

---

[19] This paragraph is based on two articles in *The New York Times*: 14 August 2001, p. C2, and 10 January 2002, p. C11.

Indeed, David Greenwood, the chief executive officer and senior vice-president of Geron, noted that Geron did not have to allow others to develop products in the three areas in which it retained exclusive rights. It is unclear, however, whether Geron's now limited control of IPR blocks others' research on stem cell technology. According to one respondent, infringement of Geron's IP is commonplace. Moreover, scientific advances in both adult stem cell technology and the use of unfertilized eggs to spawn stem cells may weaken the constraints imposed by Geron's IP by broadening the access to the commercial development of non-infringing stem cell technology.

We have considered the question of research access to a small number of important upstream discoveries. Evidence from AUTM also suggests that, at least for licensing relationships between universities and small firms, access to relatively upstream discoveries – that is, the kind of discoveries that tend to originate from university labs – is commonly restricted. Specifically, in 1999 90 percent of licenses to startups were exclusive, whereas only 39 percent of licenses to large firms were exclusive (AUTM, 2000). Similarly, in their study of licensing practices for genetic inventions, Henry et al. (2002) report that 68 percent of licenses granted by university and public labs were exclusive, whereas only 27 percent of licenses granted by firms were exclusive. However, only a minority of university-based discoveries are patented to begin with. Henry et al. (2002), consistent with Mowery et al.'s prior (2001) results, find that only about 15 percent of university-based genetic discoveries are patented, with the vast majority going into the public domain without IP protection.

Even where universities employ restrictive licensing terms, however, it is not clear that such a practice diminishes follow-on discovery, at least when applied to smaller firms. One manager of a university-based startup suggested that exclusive licensing to smaller biotech firms may actually advance follow-on discovery:

The traditional way universities did this [technology transfer] would be to go license a large company. Those kinds of agreement [include a] ... minimal up-front [fee] and small royalty, 1–2 percent. What the experience has been, then, is often the large company will work on it for a while but if it doesn't look very promising, or they run into problems, which invariably they do ... since they haven't invested much in it, they don't have a whole lot of motivation to stick with it. So, most of these licensing agreements that universities have done ended up going nowhere. The idea the university had – and other universities are beginning to do this – is to create small companies like us, where the small company has every motivation to develop it because it is the only intellectual property that they have.

10.4.4.1 Restrictions on the use of targets

In our interviews, we heard widespread complaints from universities, biotechnology firms, and pharmaceutical firms over patentholders' assertion of exclusivity over an important class of research tools, namely "targets," which refers to any cell receptor, enzyme, or other protein implicated in a disease, thus representing a promising locus for drug intervention. Our respondents repeatedly complained about a firm excluding all others from exploiting its target (in the anticipation of doing so itself), or, similarly, a firm or university licensing the target exclusively. About one-third of respondents (representing all three types of respondents) voiced concerns over patents on gene targets (for example, the COX-2 enzyme patent, the CCR5 HIV receptor patent, and the hepatitis C protease patent).

Before considering the degree to which the assertion and licensing of IP on targets may be restricting their use in downstream research, we should recall that, to the extent that patents on targets do confer effective exclusivity, even over the ability of other firms to conduct research on a particular disease, this is the purpose of a patent: to allow temporary exclusivity. Responding to complaints about restricted access to their patented targets, a respondent from a pharmaceuticals firm stated: "Your competitors find out that you've filed against anything they might do. They complain, 'How can we do research?' I respond, 'It was not my intent for you to do research.'" Others also defended their rights to exclude rivals from their patented targets. More importantly, this right to assert exclusivity may confer a benefit in the form of increasing the incentives to do the research to discover the target to begin with, as well as incentives for follow-on investment to exploit the target. A key question, then, is whether those incentives can be protected while allowing reasonably broad access.[20]

---

[20] Related and important questions are: how much incentive do patents actually provide (i.e. how effective they are), and is this incentive necessary to bring forth the innovation, given the alternative means of capturing the rents from the innovation and given public subsidies for inventive activity (see Scherer, 2002, for a review of these issues)?

Patents on targets, if broad in scope and exploited on an exclusive basis, may preclude the benefits of different firms with distinctive capabilities and perceptions pursuing different approaches to the problem (see Nelson, 1961; Cohen and Klepper, 1992). For example, big pharmaceuticals firms have libraries of compounds that might affect the target. These libraries vary by firm and are either kept secret or patented. Thus, narrowing access to the target entails a social cost. The problem of "limited lines of attack" may be greater when exclusive access to a set of targets is held by a smaller firm with limited capabilities, and, as noted above, much of the university licensing of biomedical innovations to small firms is on an exclusive basis. Although perhaps biased, a scientist from a large pharmaceutical firm described the broader capabilities of the large pharmaceutical firm to develop the potential of a target: "[Once the target had been identified], then, the power of the pharmaceutical company comes into play. You put an army of fifty molecular biologists and one-third of the medicinal chemists at [the firm] on this single problem." In addition to the constraints imposed by firms' particular capabilities on the approaches taken to exploiting targets, interviews also indicated that there are also simply differences in firm strategies or expectations with regard to promising approaches.

Although limiting access to targets may well limit their exploitation, the question is how often this occurs. We do not have systematic data on what the frequency is. From interviews and secondary sources, however, we heard of a number of prominent examples of firms being accused of asserting exclusivity over (or allowing only limited access to) a target. One case that has garnered a lot of attention is Myriad and its patents on a breast cancer gene (BRCA1). Myriad has been accused of stifling research because it has been unwilling to license diagnostic use of its patents broadly (Blanton, 2002). Myriad counters that over a dozen institutions had been licensed to do tests and that "Myriad's position is to not require a research license for anybody," while reserving the right to decide whether particular uses are research or commercial (Bunk, 1999; Blanton, 2002). Myriad sent a letter threatening a lawsuit to the University of Pennsylvania to stop them from performing genetic tests, arguing that this was commercial infringement (see below). Chiron has also developed a reputation for aggressively enforcing its patents on research targets. Chiron has filed suits against four firms that were doing research on drugs that block the hepatitis C virus (HCV) protease (in addition to filing suits against three firms doing diagnostics), and some have claimed that these suits are deterring others from developing HCV drugs (Cohen, 1999). Chiron responded to this claim by pointing out that it had licensed its patent to five pharmaceutical companies for drug development work (as well as at least five firms for diagnostic testing) and that the firms being sued had refused a license on essentially the same terms, which included significant up-front payments as well as "reach-through" royalties on the drug.

Also mentioned by our respondents was the case of telomerase as a potential target for cancer drugs. One university scientist observed: "I've asked heads of discovery why they were not using telomerase as a target. The response was, 'intellectual property.'" A scientist from a biotech firm suggested that Geron, the key IP owner, had been stymied in pursuing this because of the complexity of the biology and had redirected their efforts towards stem cell research, which looked more promising. Upon investigation, we found that Geron had indeed established a substantial patent position, with fifty-six US patents related to telomerase. However, we also found that there is a great deal of research being done on telomerase in universities and that at least three other firms (Amgen, Novartis, and Boehringer Ingelheim) are reported to be pursuing telomerase as a target (Marx, 2002). In addition, Geron presented the results of three separate studies on telomerase-based anticancer projects at the April 2002 meetings of the American Association for Cancer Research. Furthermore, Geron has formed a number of non-exclusive licensing agreements for the exploitation of telomerase, typically with small biotech firms possessing complementary technologies. Thus, although we again see some evidence of researchers being excluded, we do not find a failure to exploit the target.

10.4.4.2 Costs and delays
In this section we consider the transactions costs associated with gaining access to one or multiple patents or responding to third-party assertions.[21] For instance, firms may avoid derailing in-house R&D projects, but only by engaging in long and costly negotiations or

---

[21] Here we are concerned only with social costs, not the transfers of rents reflected in licensing fees.

litigation with IP holders. Firms may also invent around or conduct the
R&D overseas, possibly at the cost of reducing R&D efficiency.
Finally, IP holders may have to invest in monitoring the use of their
IP, which, from a social welfare perspective, also constitutes a cost.
Over a third of respondents (representing all three sectors) noted that
dealing with research tool patents did cause delays and add to the cost
of research.

Litigation costs are likely to be a significant component of the social
costs of the assertion and licensing of patents in biomedicine.
Furthermore, biomedical patents are more likely to be litigated than
are patents on other technologies (Lanjouw and Schankerman, 2001).
Although estimates of litigation costs vary, estimates commonly ranged
between $1 and 10 million for each side (see, for example, the discussion of the CellPro case above, where attorney fees were $8 million).

In addition to these out-of-pocket expenses, we tried to estimate the
opportunity cost of engaging in patent litigation. Out of the sixteen
industry respondents who addressed this issue, all but one suggested
that litigation imposed a significant burden on the managers and
scientists involved. In terms of actual work time, estimates were usually
in terms of a few weeks over the course of a year for the individuals
involved. Respondents also underscored the time spent worrying about
the progress and outcome of the case.

About a third of our respondents addressed the question of negotiation delays or litigation, and nearly all of them felt that the process of
sifting through a large number of potentially relevant patents, and the
subsequent negotiations, were very time-consuming. One large pharmaceutical firm's attorney responsible for evaluating research tool IP
provided estimates for the time that attorneys were occupied with
evaluating the IP of third parties and the time associated with actual
negotiations, which implied a total of $2 million in annual expenses.
Another respondent from a large pharmaceuticals firm suggested that
the transactions costs associated with biotech IP in particular were
especially high. He gave the following metric: lawyers in the small
molecule division (of this firm) are responsible for about eight projects
each, whereas those in the biotech division can handle only about two
projects each, because of the greater complexity of dealing with input
technologies in biotech-based projects.

The question is: although they are perhaps high in absolute terms, do
these transactions costs represent a significant expense? The answer

depends on the firm. For large pharmaceutical firms, although the expense is by no means trivial, our respondents did not convey that it significantly affected their returns from drug development. For example, one executive responsible for biotech IP gave figures suggesting that the costs for evaluating and negotiating IP rights amounted to about one-thousandth of the firm's total R&D budget. This same figure (in absolute terms), however, could represent a significant burden for a small firm, especially one with limited access to capital.

Although our respondents suggested that IP reviews and negotiations are costly and time-consuming and that their complexity has increased, it is not clear whether these efforts have increased over the recent past. To address this question, we supplemented our interview data with data from the American Intellectual Property Law Association (AIPLA) and Biotechnology Industry Organization (BIO). The AIPLA's *Reports of the Economic Survey* (AIPLA, 1995; 1997; 2001) give the number of responding attorneys working in the area of biotechnology and the median percentage of effort dedicated to biotechnology by each respondent. Assuming that the AIPLA's data are representative (which they may not be, with only an 18 to 20 percent response rate), they suggest slightly more than a 10 percent increase in the number of attorneys working on biotech between 1995 and 2001 and a 25 percent jump in the amount of time (at least per the median) that each attorney commonly dedicates to biotechnology. Therefore, there is roughly a 33 percent increase in resources devoted to what one might broadly construe as the "transactions costs" of filing, enforcing, and contracting for patents. Ernst & Young LLP's annual *Biotechnology Industry Reports* suggest, however, that, in nominal terms, R&D expenditures by biotechnology firms have increased by over 80 percent during the 1994–2000 period. If we use an annual R&D cost deflator of 5 percent, then real R&D has increased by about 40 percent. Therefore, attorney activity per R&D dollar is unlikely to have increased significantly in the recent past. Even allowing for some increase in attorneys' hourly fees, these data suggest that the patenting of research tools has not itself dramatically increased demand for legal resources, and, by extension, that the transactions costs have not increased disproportionately.

10.4.4.3 Universities

There is particular concern among academic commentators about the effects of patenting on university research (Heller and Eisenberg, 1998;

Eisenberg, 2001; Barton, 2000; Cook-Deegan and McCormack, 2001). We find only limited support for the idea that negotiations over rights stymie pre-commercial research conducted in universities. Industrial respondents all claim that university researchers, to the extent that they are doing non-commercial work, are largely left alone. In fact, firms often welcome this research because it helps further develop knowledge of the patented technology. The university researchers among our respondents confirm this claim. Also, many of the firms interviewed expressed the view that the negative publicity that an aggressive assertion of rights against a university would entail was not worth it. One university technology transfer officer reported that the university would indeed receive letters of notification of infringement. The respondent indicated that the typical response was effectively to ignore such letters and inform the IP holder that the university was engaged in research, did not intend to threaten the firm's commercial interests, and would not cease its research. However, receiving such letters is not that common. For example, one respondent reported that in fifteen years as a university administrator, overseeing fifty faculty members, he had never had a case of a professor coming to him with a notification letter.

There is a major exception to this norm of leaving university researchers alone, and that is the case of clinical research based on diagnostic tests using patented technologies. Merz, Cho, and their colleagues have recently conducted several studies of the frequency with which clinical labs have been affected by patents on diagnostic tests. One study found that 25 percent of laboratory physicians reported abandoning a clinical test because of patents. They also reported royalty rates ranging from 9 percent for PCR to 75 percent for the human chorionic gonadotropin (hCG) patent. In a follow-up survey of 119 labs capable of performing hemochromatosis testing, they found that many had adopted the test immediately upon publication (Merz et al., 2002). When the patent was issued a year later, it was licensed exclusively to SmithKlineBeecham (SKB). Nearly all respondents in the Merz et al. study said they knew of the patent. About half had received letters from SKB. Twenty-six percent said they did not develop the genetic test for hemochromatosis, and another 4 percent said they abandoned the test, in part because of the patent. Much of the controversy around Myriad's use of its patent on BRCA1 revolves around this distinction between research and clinical practice.

Myriad allows licensees to do tests provided that no fees are charged and the tests are not used for clinical purposes. Myriad also provides reduced-fee diagnostic tests ($1200 versus $2680) for NIH-funded projects (Blanton, 2002). However, according to Myriad's Gregory Critchfield, "If you give test results back to patients, it crosses over the line, and it's no longer a simple research test. [It] is really a very bright line." (Blanton, 2002). On the other hand, Merz argues that "there is no clear line to be drawn between clinical testing and research testing, because the state of the art of genetic tests is such that much more clinical study is necessary to validate and extend the early discovery of a disease gene. Thus, the restriction of physicians from performing clinical testing will directly reduce the knowledge about these genes."[22] One of our respondents, a former medical school dean, noted that the fact that the universities charge for these tests complicates the matter, but that clinical work is critical for the research process. Cho et al. (2003) report that about a half of the diagnostic labs in their sample are also using the test results for clinical research, but that the sharing of test results within the clinical diagnostic communities is an important means of advancing clinical and scientific understanding of diseases.[23]

Thus, in some cases, firms are willing to assert their patents against universities that are doing diagnostic testing and charging a fee without licensing the patented tests, and at least some labs are stopping their testing as a result. However, the majority continue with the testing. So long as the university is not generating revenue based on the patented technology, universities appear to be largely left alone, although some firms will send letters.

10.5 Working solutions: overcoming the anticommons and restrictions on access

Notwithstanding concerns about the proliferation of IP on research inputs and about the ability of rights holders to limit access to upstream discoveries and promising research targets, the problem was generally considered to be manageable. Firms reported a variety of private

[22] Testimony before House Judiciary Committee, 13 July 2000, http://www.house.gove/judiciary/merz0713.htm.
[23] Personal communication, 24 January 2002.

strategies and institutional responses that limited the adverse effects of the changing IP landscape. Although negotiations over IP and licensing fees surely affect access, and sometimes choice of projects, our conclusion is that patents on research tools do not yet pose the threat to research projects that they might given the number of patents and the diversity of owners. In this section we review the private strategies adopted by firms and universities and the responses from government that allow research and commercialization to go forward despite the proliferation of biomedical intellectual property and claimants over the past decade or so.

One important reason why research tool patents tend not to interfere with research is that it is typically not that difficult to contract. Licensing is routine in the drug industry. Indeed, many of our responding firms suggested that, if a research tool was critical, they would buy access to it. Several companies that have patents on targets noted that, in addition to trying to develop their own therapeutics, they include the liberal and broad licensing of those targets to others as part of their business model, reflecting a belief on the part of some holders of target patents that by giving several firms a non-exclusive license they increase the chances that one will discover a useful drug. We also observe that most of what might be called "general-purpose" tools – tools that cut across numerous therapeutic and research applications that tend to be non-rival-in-use – tend to be licensed broadly. Thus, many of the more fundamental (general-purpose) research tools, such as genomics databases, DNA chips, recombinant DNA technology, PCR, etc., are made widely available through non-exclusive licenses. Incyte, for example, has licensed its genomics database to more than twenty pharmaceutical firms (which together account for about 75 percent of total private pharmaceutical R&D). They have also begun expanding their licensing program to include biotech firms and universities as well.[24] Similarly, *Taq* polymerase and thermal cyclers for PCR are available from a variety of authorized reagent and equipment vendors (Beck, 1998). Human Genome Sciences' semi-exclusive licensing of its databases to only about five firms reflects an exception to this pattern.

---

[24] Furthermore, Incyte's license requires users to "grant back" non-exclusive rights to the use of genes discovered from its database, providing freedom to operate to firms in the network and creating what Incyte refers to as an "IP Trust."

Liberal licensing practices are also encouraged to the extent that "inventing around" tool patents is feasible. Under such circumstances, patentholders are more willing to license on reasonable terms, assuming the prospective user does not invent around to begin with. The ability to invent around puts an upper limit on the value of the rival's patent. Indeed, our respondents frequently noted their ability to invent around a patent as one component in their suite of solutions to blocking patents. Firms have also occasionally developed technologies that, it was claimed, made it possible to circumvent a number of the patents in the field.[25] Although some respondents argued that target patents were often unassailable, others claimed that for many important diseases (AIDS, many types of cancer, etc.) there are likely to be multiple approaches to the metabolic pathways. One university scientist considered the issue of whether a patent on a target protein could confer exclusive rights to working on a disease:

I have never worked with a disease where one particular protein makes the only difference. A patent gets you exclusive rights to a class of drugs, but there may be other classes ... I could imagine a genetic disease where a single target was involved, but I don't think that the big medical problems fall into this case.

Aside from conventional methods for coming to terms, we find that firms have adopted a set of complementary strategies that create "working solutions" to address either a prospective anticommons (e.g. the need to license numerous tools) or a potentially blocking patent on one tool or discovery. These solutions include (in addition to licensing and inventing around) ignoring patents (sometimes invoking an informal research exemption), going offshore, creating public databases, and challenging patents in court. These working solutions thus combine to create a free space in the patent landscape that allows research projects to proceed relatively unencumbered.

---

[25] For example, Athersys, a biotech firm based in Cleveland, Ohio, advertises its RAGE technology, which uses automated techniques to create protein expression libraries (i.e. activate and express every gene and therefore produce every protein) without using any knowledge about the location and structure of the corresponding gene. The company's website reports that some established pharmaceutical firms (Bristol–Myers Squibb and Pfizer) had licensed this technology (http://www.athersys.com).

10.5.1 Infringement and the "research exemption"

One solution to restrictive patents on upstream inventions is simply to ignore some or all of them. Several respondents noted that infringement of research tool patents is often hard to detect, facilitating such behavior. Thus, if research tool patents have created a minefield, they are mines with fairly insensitive triggers.

University researchers have a reputation for routinely ignoring IPR in the course of their research (Seide and MacLeod, 1998). Respondents noted that many research tools were "do-it-yourself" technologies, and therefore they did not feel they should be required to pay royalties for the work. In fact, some strongly believed that these patented technologies were well known in the scientific community and therefore the patents were not valid (see, for example, Kornberg, 1995). University researchers will often invoke a "research exemption," although the legal research exemption as construed by the Court of Appeals for the Federal Circuit (CAFC) has been quite narrow.[26] Some reagent suppliers facilitate this practice by supplying "unlicensed" (and less expensive) materials, also invoking the research exemption. Promega, for example, sells *Taq* polymerase for about a half of what many licensed vendors charge, and asserts that many of its customers in university and government labs do not need a license under the experimental use exemption (Beck, 1998).

Many firms claim to be reluctant to enforce their patents against universities to the extent that the university is engaging in non-commercial research because of the low damage awards and bad publicity that suing a university would entail. For example, William Haseltine of HGS said that they were ready to give academics access to data and reagents related to their patented CCR5 HIV receptor: "We would not block anyone in the academic world from using this

---

[26] Building on *Roche Products, Inc. v. Bolar Pharm. Co.* (Federal Circuit 1984) and *Embrex v. Service Engineering Corp.*, the current standard of the CAFC for what qualifies for a research exemption includes uses of patented inventions "for amusement, to satisfy idle curiosity, or for strictly philosophical inquiry" (*Embrex*). Although there is some question about whether the term "philosophical inquiry" may actually refer to scientific inquiry (see Wegner, 2002), the 3 October 2002 decision of the CAFC in *Madey v. Duke*, discussed below, corroborates that the CAFC sustains a narrow interpretation of the research exemption.

for research purposes" (Marshall, 2000a). As one university technology transfer officer stated, "Asserting against a university doesn't
make sense. First, there are no damages. You cannot get injunctive
relief and/or damages. What have you gained? You've just made people
mad. Also, these firms are consumers of technology as well. No one will
talk to you if you sue. We all scratch each others' backs. You will
become an instant pariah if you sue a university." Similarly, from the
industry side, Leon Rosenberg of Bristol–Myers Squibb said, "Frankly,
we all know it is not good form to sue researchers in academic institutions and stifle their progress." (NRC, 1997, chap. 5, p. 3). These
quotations suggest that one limit on opportunism is being a member
of a community with the members being able to sanction overly
aggressive behavior (Rai, 1999). This vulnerability to such sanctions
is based on the need to buy as well as sell technology, or, perhaps
especially, to trade information informally. Indeed, there is a strong
interest in developing trusting relationships with university researchers
to encourage information sharing (for the general issue of trust and
information sharing, see Uzzi, 1996). A respondent at a biotech firm
put it this way:

We rely on lots of outside collaborations with academic labs. Our scientists
want to feel on good terms with the academic community. If you start suing,
it breaks down the good feeling. We give out our research tools for free,
frequently. All we ask is, if you invent anything that is directly related to the
tool, you allow us the freedom to practice.

We heard similar comments from those in universities, large pharmaceutical firms, and biotech firms.

One exception that may demonstrate the role of reciprocity in mitigating IP-based frictions is DuPont's recent aggressive assertion of its
exclusively licensed OncoMouse patent against universities that did
not follow the precise terms of a prior memo of understanding between
DuPont and NIH. In commenting on this behavior, Neighbour (2002)
muses as to why DuPont would do such a thing now that they are out of
the business of research in molecular biology. We would suggest that
that may be the explanation. They have now ceased to be a part of that
community and therefore have little to lose and revenue to gain when
they sacrifice the goodwill of that community. Thus, DuPont's behavior is consistent with the notion that a community of practice restrains
the aggressive assertion of IP.

Several respondents noted that they actually welcomed universities using their patented technologies because, if the university discovers a new use, the patentholder is best positioned to exploit the innovation. If the university becomes a competitor, however, firms feel they then have a right to assert their patents. As noted above, this is particularly evident in cases where university physicians use patent-protected discoveries as the basis for diagnostic tests (Merz et al., 2002; Blanton, 2002).

The infringement of research tool patents by firms also appears to be widespread. One-third of the industrial respondents (and all nine university or government lab respondents) acknowledged occasionally using patented research tools without a license, and most respondents suggested that infringement by others is widespread. The firms felt that much of their research would not yield commercially valuable discoveries, and thus they saw little need to spend money to secure the rights to use the input technology, particularly because it is very difficult to police such infractions. If the research looked promising, then they would get a license, if necessary.[27] Furthermore, at least a few industry respondents argued strongly that using a gene patent as a research tool did not infringe, or that infringement was limited to that experiment per se and did not extend to the product discovered (in part) by using the research tool – i.e. that the scope of research tool patent claims is quite limited. Because many of these patents are of debatable validity, they also felt that, if a license were not available, they could challenge the patent in court. Finally, not only is use of a patented research tool hard to detect but, because of the long drug development process, the six-year statute of limitations may expire before infringement is detected.

Consistent with this behavior, we also find that firms feel that it is not worth their while to assert their patents on all other firms that might be infringing. They may send a letter, offering terms, but will not aggressively pursue infringers on their marginal patents. Respondents pointed out that the cost of pursuing these cases greatly outweighs their

---

[27] If this is true, it suggests that, when firms do ask for a license, the patent owner ought to suspect that the tool has been useful in generating a valuable discovery. This knowledge may lead to asking for a high license fee. However, this urge is balanced by the recognition that this promising candidate still needs to get through risky clinical trials, and if the price is too high the buyer may chose one of his other promising candidates.

value in most instances: "The average suit costs millions of dollars. The target is worth $100,000. Even with treble damages, it doesn't pay to sue." There is an additional cost, and that is the risk of the patent being invalidated by the court. These firms have noted, however, that they will aggressively defend a patent central to the firm's competitive performance. Barring that, with reference to research tool patents, there is a sense that the industry practices "rational forbearance" (NRC, 1997, chap. 6).

Respondents also pointed out that patents are national but the research community is global. Thus, another means of avoiding research tool patents is to use the patented technology offshore. Although similar to the solution of ignoring the patent, in that it involves using patented technologies without securing the rights, this case differs in that firms are not violating the legal rights of the patent owner, at least not until there is a product developed and the firm tries to import the product. Furthermore, a district court decision in 2001 (*Bayer AG v. Housey Pharmaceuticals*) suggests that, even then, the drug maker may not be liable for infringement (see, for example, Maebius and Wegner, 2001).

In summary, by infringing (and informally invoking a research exemption), inventing around, going offshore, or invalidating patents in court, firms have been able to reduce the complexity of the patent landscape greatly. These strategies, combined with licensing when necessary, provide working solutions to the potential problem that an increasingly complicated patent landscape represents.

10.5.2 *Institutional responses by firms, the NIH, the USPTO, and courts*

In addition to these private responses to overcoming the barriers that patents may create, we have also observed firms (especially larger pharmaceutical firms), the courts, the NIH, and the USPTO under taking initiatives and policies that have had the effect (if not always the intent) of broadening and easing access to research tools. For example, with substantial public, private, and foundation support, public data-bases (e.g. GenBank or the Blueprint Worldwide Inc. venture to create a public "proteomics" database) and quasi-public databases (such as the Merck Gene Index and the SNP Consortium) have been created, making genomic information widely available. Similarly, Merck has sponsored an $8 million program to create 150 patent-free transgenic mice to be made available to the research community at cost, without patent or use restrictions. According to our respondents, these efforts partly represent an attempt by large pharmaceutical firms to undercut the genomics firms' business model by putting genomic and other related information into publicly available databases and then competing on the exploitation of this shared information to develop drug candidates (see Marshall, 1999a, 2001).[28] These initiatives represent a partial return to the time before the genomics revolution, when publicly funded university researchers produced a body of publicly available knowledge that was then used by pharmaceutical firms to help guide their search for drug candidates.

The NIH has also taken the lead in pressing for greater access to research tools. For example, since 1997 the NIH has negotiated with DuPont to provide more favorable terms for transgenic mice for the NIH and NIH-sponsored researchers (Marshall, 2000b). The NIH has also begun a "mouse initiative" to sequence the mouse genome and create transgenic mice. One of the conditions of funding is that grantees forgo patenting on this research. The NIH has also pushed for broader access to stem cells, as well as for a simplified, one-page material transfer agreement (MTA) without reach-through claims or publication restrictions. Scientific journals have also pushed for access to research materials. For example, biology journals have long made it a condition of publication that authors deposit sequences in public databases such as GenBank or Protein Data Bank (Walsh and Bayma, 1996). Similarly, when Celera published its human genome map findings, *Science*'s editors were able to gain for academics largely unrestricted access to Celera's proprietary database. Thus large institutional actors have been able to act as advocates for university researchers to increase their access to necessary research tools.

---

[28] For example, the firms in the SNP Consortium include Bayer, Bristol–Myers Squibb, Glaxo Wellcome, Hoechst Marion Roussel, Monsanto, Novartis, Pfizer, Roche, SmithKline Beecham, and Zeneca. Each firm has contributed $3 million, and Wellcome Trust has added another $14 million to the effort. Also, financed by IBM, Canada's MDS, Inc., and the Canadian government, the Blueprint Worldwide database could pose a threat to the joint effort of Myriad Genetics, Hitachi, and Oracle to launch a $185 million effort to map protein interactions, along with for-profit efforts by universities to market protein databases (*Wall Street Journal*, 2001).

Responding partly to concerns expressed by the NIH, universities, and large pharmaceutical firms, the USPTO has also adopted new policies that diminish the prospect of an anticommons. Specifically, in January 2001 the USPTO adopted new utility guidelines that have effectively raised the bar on the patentability of tools, particularly ESTs. These guidelines are designed to reduce the number of "invalid" patents (see Barton, 2000).

Some of our respondents suggested that recent court decisions have also mitigated potential problems due to research tool patents by limiting the scope of tool patents, or, in some cases, invalidating them. Thus, although patentholders have the right to sue for infringement, the perception is that they are increasingly likely to lose such a suit. Cockburn, Kortum, and Stern (2003) find that the CAFC went from upholding the plaintiff in about 60 percent of the cases to finding for the plaintiff in only 40 percent of the cases in recent years. One case that came up frequently among our respondents was *University of California v. Eli Lilly and Co.* As noted above, the University of California tried to argue that its patent on insulin, based on work on rats, covered Lilly's human-based bioengineered insulin production process. The CAFC ruled that the university did not in fact possess this claimed invention at the time of filing; therefore, the claim was not valid, and Lilly was not infringing. The case of *Rochester v. Searle* over COX-2 inhibitors is another example of the courts ruling against a research tool patentholder. The critical issue in the case was whether knowledge of a drug target allows one to claim ownership over specific classes of drugs (i.e. how broadly do initial discovery claims extend over future developments building on those discoveries?). The district court dismissed Rochester's complaint of infringement, and Rochester lost even upon appeal. Notwithstanding these decisions narrowing the scope of research tool patents, there is still, however, uncertainty about how the courts will handle research tool patents generally.

10.6 Discussion and conclusion

In this chapter we have considered two possible impacts of the patenting of research tools on biomedical research. First, we considered whether the existence of multiple research tool patents associated with a new product or process poses particular challenges for either research on or the commercialization of biomedical innovations.

Second, we examined whether restricted access to some upstream discovery – perhaps protected by only one patent – has significantly impeded subsequent innovation in the field. In brief, we find that the former issue – the "anticommons" – has not been especially problematic. The latter issue, of access, at least to foundational upstream discoveries, has not yet impeded biomedical innovation significantly, but our interviews and prior cases suggest that the prospect exists and ongoing scrutiny is warranted.

The patenting of research tools has made the patent landscape more complex. As suggested by Heller and Eisenberg (1998), our interviews confirmed that there are on average more patents and more patentholders than before involved in a given commercializable innovation in biomedicine, and many of these patents are on research tools. Despite this increased complexity, almost none of our respondents reported commercially or scientifically promising projects being stopped because of issues of access to IPR to research tools. Moreover, although we do not have comparably systematic evidence on projects never undertaken, our interviews suggest that IP on research tools, although sometimes impeding marginal projects, rarely precludes the pursuit of more promising projects. Why? Industrial and university researchers have been able to develop "working solutions" that allow their research to proceed. These working solutions combine taking licenses (i.e. successful contracting), inventing around patents, going offshore, the development and use of public databases and research tools, court challenges, and using the technology without a license (i.e. infringement), sometimes under an informal and typically self-proclaimed research exemption. In addition, the members of a research community (which includes both academic and commercial researchers) are somewhat reluctant to assert their IPR against one another if that means they will sacrifice the goodwill and information sharing that comes with membership in the community. Changes in the institutional environment, particularly new USPTO guidelines and some shift in the courts' views towards research tool patents, as well as pressure from powerful actors such as the NIH (stimulated perhaps by the early concerns articulating the anticommons problem), also appear to have reduced further the threat of breakdown. Finally, the very high technological opportunity in this industry means that firms can shift their research to areas less encumbered by intellectual property claims, and, therefore, the walling off of particular areas of research may not, under some circumstances, exact a high toll on social welfare.

Although stopped and stillborn projects are not evident, many of the working solutions to the IP complexity can impose social costs. Firms' circumvention of patents, the use of substitute research tools, inventing around or going offshore, although all privately rational strategies, constitute a social waste. Court challenges and even the contract negotiations themselves can also impose significant social costs. Litigation can be expensive and non-out-of-pocket costs, represented by the efforts devoted to the matter by researchers and management, can be substantial. Even when there is no court challenge, the negotiations can be long and complex and may impose costly delays. Disagreements can lead and have led to litigation, which is especially costly for small firms and universities. It is difficult to know, however, how much contracting costs in biomedicine reflect an enduring feature of IP in biomedicine and how much is transitional, arising from the uncertainty associated with the newness of the technology and uncertainties about the scope and validity of patent claims. Moreover, as new institutions (i.e. universities) and firms become owners of intellectual property, there is a costly period of adjustment as these new actors learn how to manage their IP effectively. The development of standard contracts and templates may be helpful in diminishing these adjustment costs, and funding agencies such as the NIH can play an important role in developing and encouraging the use of such standards.

The second issue that we examined is the impact on biomedical innovation of restricted access to research tools. In thinking about the issue of access, it is helpful to distinguish research tools along two dimensions. First, it is obviously of interest how essential or "foundational" a research tool is for subsequent innovation, both in the sense of whether the tool is key to subsequent research and in the sense of the breadth of innovation that might depend upon its use. Is the research tool a key building block for follow-on research on a specific approach to a specific disease, is the tool key to advance in a broad therapeutic area, or might its application even cut across a range of therapeutic and diagnostic domains?

A second dimension of interest is the degree to which a research tool is rival-in-use. By "rival-in-use" we mean research tools that are primarily used to develop innovations that will compete with one another in the market place. For instance, in the case of a receptor that is specific to a particular therapeutic approach to a disease, if one firm finds a compound that blocks the receptor, it undermines the ability of another firm to profit from its compound that blocks the same receptor. The defining feature of research tools that are not rival-in-use is that the use of the research tool by one firm will not typically reduce others' profits from using it. Such tools include PCR, micro-arrays, cre-lox, and combinatorial libraries. From a social welfare perspective, a research tool that is not rival-in-use is like a public good, in that it has a high fixed cost of development and zero or very low marginal cost in serving an additional user. Thus, the maximization of social welfare requires that the tool be made available to as large a set of users as possible.

We have observed that holders of IP on non-rival research tools often charge prices that permit broad access, at least among firms. In some of these cases, the IP holders have also charged higher prices to commercial clients and lower prices to university and other researchers who intend to use the tool largely for non-commercial purposes. From a social welfare perspective, such price discrimination expands the use of the tool and is indeed welfare-enhancing. There are, however, cases in which the IP holder cannot or does not develop a pricing strategy that allows low-value and academic projects access to the tool, as for instance in the case of DuPont's initial terms for the cre-lox technology or Affymetrix's initial terms for GeneChips. However, DuPont eventually bowed to pressure from the NIH (although, as noted above, the issue is not entirely settled), and Affymetrix developed a university pricing system that greatly increased access (while others developed do-it-yourself micro-arrays).[29]

The concern with regard to IP access tends to be the greatest when a research tool is rival-in-use and is potentially key to progress in one or more broad therapeutic areas. When a foundational research tool is rival-in-use, the IP holders often either attempt to develop the technology themselves or grant exclusive licenses. As suggested above, exclusive exploitation of a foundational discovery is unlikely to realize the

---

[29] We conjecture that it is exactly these non-rival-in-use technologies with many low-value uses that are likely to benefit from NIH intervention, if necessary, because there will be a large constituency of users who want access (including many researchers at the NIH itself), most of the research community uses will be low-value, and the cost to the patent owner of allowing these non-rival uses is low because the high-value uses are not necessarily affected.

full potential for building on that discovery, because no one firm can even conceive of all the different ways in which the discovery could be exploited, let alone actually do so. Geron's exclusive license for human embryonic stem cell technology shows how restrictions on access to an important, broadly useful rival-use technology can potentially retard its development. A more prosaic example is the pricing of licenses for diagnostic tests. Myriad's (and others') licensing practices show that, to the degree that a high price on a diagnostic test puts it out of the reach of clinics and hospitals involved in research that requires the test results, clinical research may be impeded, yielding long-term social costs.[30] The social welfare analysis of this situation is not however, straightforward. Even though knowledge, once developed, can be shared at little additional cost and may be best exploited through broad access, it does not follow that social welfare is maximized by mandating low-cost access if such access dampens the incentive to develop the research tool to begin with.

Many of the same kinds of "working solutions" that mitigate the prospect of an anticommons also apply to the issue of access for research. Our interviews suggest that a key "working solution," however, is probably infringement under the guise of a "research exemption." Firms and universities frequently ignore existing research tool patents, invoking a "research exemption" that is broader than the existing legal exemption and that is supported by norms of trust and exchange in the research community. As discussed above, such instances of possible infringement, especially on the part of universities, are tolerated by IP-holding firms, both for normative reasons and because of the high cost of enforcing rights through litigation, relative to the low pay-off for stopping a low-value infringement. One can rationalize the failure of the IP holder to monitor infringement aggressively as a form of price discrimination, and, as suggested above, economic theory suggests that such price discrimination can improve social welfare.[31]

---

[30] Here the difficulty is associated with the fact that the same activity that is rival-in-use (providing commercial diagnostic services) is also the (possibly non-rival-in-use) research use. The difficulty of separating these two activities in the American system of funding clinical research contributes to the problems associated with patents on diagnostic uses of genes.

[31] As long as the infringing uses do not reduce the value of the tool to the users with a high willingness to pay, such price discrimination is likely to be profitable privately as well.

There are two central questions to ask when considering the effects of a given research tool patent on the progress of biomedical research. The first has to do with the specifics of the biology in question: does current scientific knowledge provide us with many or few opportunities for modifying the biological system in question? As science progresses, we are likely to see an oscillation, with new discoveries opening promising but narrow short cuts, and further exploration of those discoveries uncovering a variety of lines of attack on the problem. Where there are many opportunities, the likelihood of a research tool patent impeding research is smaller. Here, again, the Geron case provides an illustration, with the recent development of alternatives to the use of embryonic cells for exploiting the promise of stem cells mitigating the restrictive impact of Geron's control over embryonic stem cell technology.

The second question has to do with the specifics of the legal rights in question, and has been highlighted by Merges and Nelson (1990) and Scotchmer (1991): does the scope of claims in this patent cover few or many of the research activities using this technology? As the USPTO and the courts become more familiar with a technology, uncertainty over the scope of patent claims should diminish. The outcome of the *Rochester v. Searle/Pharmacia* COX-2 case, for example, probably signals that research will proceed with somewhat reduced concern over upstream research tool patents, although one should then consider the impact of that decision on the incentives for developing that class of upstream discovery.

Through a combination of luck and appropriate institutional response, the United States appears to have avoided situations in which a single firm or organization, using its patents, has blocked research in one or more broad therapeutic areas. However, the danger remains that progress in a broad research area could be significantly impeded by a patentholder trying to reserve that area exclusively for itself. The question is whether something systematic needs to be done. One possibility that has been considered is a revision of the law providing for research exemptions to reflect the current norms and practices of the biomedical research community more accurately (see Rai, 1999; Ducor, 1997, 1999). One difficulty is that it is not easy to discern when research is commercial or non-commercial regardless of the kind of institution that is doing the research (see OECD, 2002). Thus it is not apparent that society would benefit from a policy response, as opposed to a continued reliance on the current ad hoc practices of de facto infringement under the informal rubric of the "research exemption." The viability of this latter approach may, however, be undermined by the recent (October 2002) CAFC decision in *Madey v. Duke*, which effectively narrows the research exemption to exclude, in essence, any use of IP in the course of university research. The effect of this decision is not to make the unauthorized use of others' IP in academic biomedical research illegal; such uses, as suggested above, were probably illegal already in the light of pre-Madey interpretations of the research exemption. Rather, this decision will focus attention on such practices, sensitizing both faculty and university administrations to the possible illegality of – and liability for – such uses of IP. This could well chill some of the "offending" biomedical research that is conducted in university settings. Given the importance of this informal exemption for allowing open science to proceed relatively unencumbered, this outcome would be unfortunate. Thus, policy-makers should consider an appropriate exemption for research intended for the public domain.

We cannot, therefore, rule out future problems resulting from patents currently under review, court decisions, new shifts in technology, or even assertions of patents on foundational discoveries. Therefore, we anticipate a continuing need for the active defense of open science. Yet the social system we observe has appeared to develop a robust combination of working solutions for dealing with these problems. Recent history suggests that these solutions can take time and expense to work out, and the results may not be optimal from either a private or social welfare perspective, but research generally moves forward. It should also be recalled that patents benefit biomedical innovation broadly by providing incentives that have called forth enormous investment in R&D (see Arora, Ceccagnoli, and Cohen, 2002), and that the research tools developed have increased the productivity of biomedical research (e.g. Henderson, Orsenigo, and Pisano, 1999).

Thus, our conclusion is that the biomedical enterprise seems to be succeeding, albeit with some difficulties, in developing an accommodation that incorporates both the need to provide strong incentives to conduct research and development and the need to maintain free space for discovery. Nonetheless, as technologies change and as court decisions such as *Madey v. Duke* emerge, these issues may need to be periodically revisited.

References

American Intellectual Property Law Association (1995), *Report of the Economic Survey*, American Intellectual Property Law Association, Washington, DC.

—— (1997), *Report of the Economic Survey*, American Intellectual Property Law Association, Washington, DC.

—— (2001), *Report of the Economic Survey*, American Intellectual Property Law Association, Washington, DC.

Arora, A. A., M. Ceccagnoli, and W. M. Cohen (2002), *R&D and the Patent Premium*, working paper, Carnegie Mellon University, Pittsburgh.

Arora, A. A., A. Fosfuri, and A. Gambardella (2001), *Markets for Technology*, MIT Press, Cambridge MA.

Association of University Technology Managers (2000), *AUTM Licensing Survey FY1999 Survey Summary*, Association of University Technology Managers, Northbrook, IL.

Bar-Shalom, A., and R. M. Cook-Deegan (2002), "Patents and innovation in cancer therapeutics: lessons from CellPro," *Milbank Quarterly*, 80, 637–76.

Barton, J. (2000), "Intellectual property rights: reforming the patent system," *Science*, 287, 1933–4.

Beck, S. (1998), "Do you have a license? Products licensed for PCR in research applications," *The Scientist*, 12 (12), 21.

Blanton, K. (2002), "Corporate takeover," *Boston Globe*, Magazine, online edition, 24 February.

Bunk, S. (1999), "Researchers feel threatened by disease gene patents," *The Scientist*, 13 (20), 7.

Cho, M. K., S. Illangasekare, M. A. Weaver, D. G. B. Leonard, and J. F. Merz (2003), "Effects of patents and licenses on the provision of clinical genetic testing services," *Journal of Molecular Diagnostics*, 5 (1), 3–8.

Cockburn, I., and R. M. Henderson (2000), "Publicly funded science and the productivity of the pharmaceutical industry," in A. Jaffe, J. Lerner, and S. Stern (eds.), *Innovation Policy and the Economy*, Vol. I, MIT Press, Cambridge, MA, 1–34.

Cockburn, I., R. Henderson, L. Orsenigo, and G. P. Pisano (2000), "Pharmaceuticals and biotechnology," in D. C. Mowery (ed.), *U.S. Industry in 2000: Studies in Competitive Performance*, National Academies Press, Washington, DC, 363–98.

Cockburn, I., S. Kortum, and S. Stern (2003), "Are all patent examiners equal? Examiners, patent characteristics, and litigation outcomes," in W. Cohen and S. Merrill (eds.), *Patents in the Knowledge-Based Economy*, National Academies Press, Washington, DC, 17–53.

Cohen, J. (1999), "Chiron stakes out its territory," *Science*, 285, 28.

Cohen, W. M., A. Goto, A. Nagata, R. R. Nelson, and J. P. Walsh (2002), "R&D spillovers, patents and the incentives to innovate in Japan and the United States," *Research Policy*, 31, 1349–67.

Cohen, W. M., and S. Klepper (1992), "The tradeoff between firm size and diversity for technological progress," *Journal of Small Business Economics*, 4, 1–14.

Cohen, W. M., R. R. Nelson, and J. P. Walsh (2000), *Protecting Their Intellectual Assets: Appropriability Conditions and why U.S. Manufacturing Firms Patent (or Not)*, Working Paper no. 7522, National Bureau of Economic Research, Cambridge, MA.

Cook-Deegan, R. M., and S. J. McCormack (2001), "Patents, secrecy, and DNA," *Science*, 293, 217.

Doll, J. J. (1998), "The patenting of DNA," *Science*, 280, 689–90.

Drews, J. (2000), "Drug discovery: a historical perspective," *Science*, 287, 1960–2.

Ducor, P. (1997), "Are patents and research compatible?," *Nature*, 387, 13–14.

(1999), "Research tool patents and the experimental use exemption," *Nature Biotechnology*, 17, 1027–8.

Eisenberg, R. S. (2001), "Bargaining over the transfer of proprietary research tools: is this market failing or emerging?," in R. C. Dreyfuss, D. L. Zimmerman, and H. First (eds.), *Expanding the Boundaries of Intellectual Property: Innovation Policy for the Knowledge Society*, Oxford University Press, Oxford, 223–50.

Evenson, R., and Y. Kislev (1973), "Research and productivity in wheat and maize," *Journal of Political Economy*, 81, 1309–29.

Gambardella, A. (1995), *Science and Innovation: The US Pharmaceutical Industry in the 1980s*, Cambridge University Press, Cambridge.

Hall, B. H., and R. H. Ziedonis (2001), "The patent paradox revisited: an empirical study of patenting in the U.S. semiconductor industry, 1979–1995," *RAND Journal of Economics*, 32 (1), 101–28.

Hamilton, J. O. (1997), "Stanford's DNA patent 'enforcer' Grolle closes the $200m book on Cohen–Boyer," *Signals Magazine*, online edition, 25 November.

Heller, M. A. (1998), "The tragedy of the anticommons: property in the transition from Marx to markets," *Harvard Law Review*, 111, 621–81.

Heller, M. A., and R. S. Eisenberg (1998), "Can patents deter innovation? The anticommons tragedy in biomedical research," *Science*, 280, 698–701.

Henderson, R. M., L. Orsenigo, and G. Pisano (1999), "The pharmaceutical industry and the revolution in molecular biology: exploring the

interactions between scientific, institutional, and organizational change," in D. C. Mowery and R. R. Nelson (eds.), *Sources of Industrial Leadership: Studies of Seven Industries*, Cambridge University Press, Cambridge, 267–311.

Hicks, D., T. Breitzman, D. Olivastro, and K. Hamilton (2001), "The changing composition of innovative activity in the US – a portrait based on patent analysis," *Research Policy*, 30 (4), 681–704.

Henry, M. R., M. K Cho, M. A. Weaver, and J. F. Merz (2002), "DNA patenting and licensing," *Science*, 297, 1279.

Jewkes, J., D. Sawers, and R. Stillerman (1958), *The Sources of Invention*, Macmillan, London.

Kornberg, A. (1995), *The Golden Helix*, University Science Books, Sausalito, CA.

Lanjouw, J. O., and M. Schankerman (2001), *Enforcing Intellectual Property Rights*, Working Paper no. 8656, National Bureau of Economic Research, Cambridge, MA.

Levin, R. (1982), "The semiconductor industry," in R. R. Nelson (ed.), *Government and Technical Progress: A Cross-Industry Analysis*, Pergamon Press, New York, 9–100.

Levin, R., A. Klevorick, R. R. Nelson, and S. G. Winter (1987), "Appropriating the returns from industrial R&D," *Brookings Papers on Economic Activity*, 3, 783–820.

Maebius, S., and H. Wegner (2001), "Research methods patents: a territoriality loophole," *National Law Journal*, 24 December, C3–C4.

Malakoff, D., and R. F. Service (2001), "Genomania meets the bottom line," *Science*, 291, 1193–203.

Mansfield, E. (1986), "Patents and innovation: an empirical study," *Management Science*, 32, 173–81.

Marshall, E. (1999a), "Drug firms to create public database of genetic mutations," *Science*, 284, 406–7.

(1999b), "Do-it-yourself gene watching," *Science*, 286, 444–7.

(2000a), "Patent on HIV receptor provokes an outcry," *Science*, 287, 1375–7.

(2000b), "Property claims: a deluge of patents creates legal hassles for research," *Science*, 288, 255–7.

(2001), "Bermuda rules: community spirit, with teeth," *Science*, 291, 1192.

Marx, J. (2002), "Chromosome end game draws a crowd," *Science*, 295, 2348–51.

Merges, R. P. (1994), "Intellectual property rights and bargaining breakdown: the case of blocking patents," *Tennessee Law Review*, 62 (1), 74–106.

Merges, R. P., and R. R. Nelson (1990), "On the complex economics of patent scope," *Columbia Law Review*, 90 (4), 839–916.

(1994), "On limiting or encouraging rivalry in technical progress: the effect of patent scope decisions," *Journal of Economic Behavior and Organization*, 25, 1–24.

Merz, J. F., D. G. Kriss, D. D. G. Leonard, and M. K. Cho (2002), "Diagnostic testing fails the test," *Nature*, 415, 577–9.

Mowery, D. C., R. R. Nelson, B. Sampat, and A. Ziedonis (2001), "The growth of patents and licensing by U.S. universities," *Research Policy*, 30, 99–119.

National Research Council (1997), *Intellectual Property Rights and Research Tools in Molecular Biology*, National Academies Press, Washington, DC.

National Science Foundation (1998), *Science and Engineering Indicators*, Government Printing Office, Washington, DC.

Neighbour, A. (2002), "Presentation to the National Cancer Policy Board," Institute of Medicine, Washington, DC, 23 April.

Nelson, R. R. (1961), "Uncertainty, learning, and the economics of parallel research and development efforts," *Review of Economics and Statistics*, 43, 351–64.

(1982), "The role of knowledge in R&D efficiency," *Quarterly Journal of Economics*, 97, 453–70.

Organisation for Economic Co-operation and Development (2002), *Genetic Inventions, IPRS and Licensing Practices*, report by the Working Party on Biotechnology, Directorate for Science, Technology and Industry, Organisation for Economic Co-operation and Development, Paris.

Rai, A. K. (1999), "Regulating scientific research: intellectual property rights and the norms of science," *Northwestern University Law Review*, 94 (1), 77–152.

Scherer, F. M. (2002), "The economics of human gene patents," *Academic Medicine*, 77, 1348–66.

Scherer, F. M., S. Herzstein, Jr., A. Dreyfoos, W. Whitney, O. Bachmann, C. Pesek, C. Scott, T. Kelly, and J. Galvin (1959), *Patents and the Corporation: A Report on Industrial Technology Under Changing Public Policy*, 2nd edn., Graduate School of Business Administration, Harvard University, Boston.

Science (1997), "Genomics' wheelers and dealers," 275, 774–5.

(2000), "A cheaper way to buy genomic data," *Science*, 288, 223.

Scotchmer, S. (1991), "Standing on the shoulders of giants: cumulative research and the patent law," *Journal of Economic Perspectives*, 5 (1), 29–41.

Seide, R. K., and J. M. MacLeod (1998), "Comment on Heller and Eisenberg," *ScienceOnline*, http://www.sciencemag.org/feature/data/980465/seide.shl.

Service, R. F. (2001), "Can data banks tally profits?," *Science*, 291, 1203.

Shapiro, C. (2000), "Navigating the patent thicket: cross-licenses, patent pools, and standard-setting," in A. Jaffe, J. Lerner, and S. Stern (eds.), *Innovation Policy and the Economy*, Vol. I, MIT Press, Cambridge, MA, 119–50.

Thursby, J. G., and M. C. Thursby (1999), *Purdue Licensing Survey: A Summary of Results*, unpublished manuscript, Krannert Graduate School of Management, Purdue University, West Lafayette, IN.

Uzzi, B. (1996), "The sources and consequences of embeddedness for the economic performance of organizations: the network effect," *American Sociological Review*, 61 (4), 674–98.

Wall Street Journal (2001), "IBM and others are financing a public database on proteins," online edition, 30 May.

Walsh, J. P., and T. Bayma (1996), "Computer networks and scientific work," *Social Studies of Science*, 26, 661–703.

Wegner, H. C. (2002), *The Right to Experiment with a Patented Invention*, paper presented to a meeting of the Bar Association of the District of Columbia, Patent, Trademark and Copyright Section, Washington, DC, 10 December.

Whyte, W. F. (1984), *Learning from the Field*, Sage Publications, Beverly Hills.

11 | Upstream patents and public health: the case of genetic testing for breast cancer

FABIENNE ORSI, CHRISTINE SEVILLA,
AND BENJAMIN CORIAT

11.1 Introduction

The aim of this chapter is to explore the medical and health care implications of the growing tendency to grant patents on human genes. Taking the genetic testing of breast cancer as an example, we show how the granting of these large-scope patents, covering upstream scientific information, is likely to have very damaging consequences as far downstream as the health care system itself.

In the second section we explain why the granting of this new type of "upstream" patent, exemplified by patents on genes, is so significant. We show that the major problem posed by these patents, which benefit from an extremely large scope, including "future" and "potential" applications, is that they are likely to give rise to widespread monopolies. This will impede the design and marketing of products developed using the same genetic information, with serious consequences not only for the future of research but also for the whole chain of medical practice.

In the third section we use the case of genetic testing for breast cancer, linked to the BRCA1 and BRCA2 genes, to illustrate the issues raised by these new patents. In the case of BRCA1 and BRCA2, the private American firm Myriad Genetics has been granted several patents, providing it with a monopoly over the diagnostic and therapeutic activities based on these genes. We begin by describing how the granting of these patents enabled the firm to build a monopoly over every aspect of diagnostic activity in the United States. We then show how the extension to Europe of American patents (through the European Patent Office [EPO], which has adapted its doctrine to converge with that of the US Patent and Trademark Office), has provoked unprecedented problems in Europe, creating serious tensions within current medical practices.

11.2 The patentability of genes: the origins and significance of a major change in doctrine

The introduction of gene patenting represented a major divergence between the rules on patentability and the economic foundations of patent law. This is one of the keys to understanding the significance of the entry of human genes into the field of patenting, and why it has provoked such fierce debate. To explore this point, we shall start by running briefly through the classic economic foundations of patent law. We can then show clearly how the new American doctrine on gene patentability has caused such a break between the principles and practice of patent law.

11.2.1 The economic foundations of patent law

Traditional economic analysis sets out several principles that determine the foundations of patent law (classically defined as the exclusive but temporary right to enjoy the proceeds of an invention – including the right to prevent competitors from using it).

Ever since the seminal article written by Kenneth Arrow in 1962, if not earlier, it has been recognized that an economy composed of private, decentralized agents in competition is constantly threatened with underinvestment in research. This is due to the indivisible nature of the good "information" – including the products of research. Granting inventors a patent – in other words, a "temporary monopoly" to exploit their inventions – is intended to provide a sufficient incentive for private firms to invest in research activities, by making up for the shortcomings of the market. Fundamentally, therefore, the purpose of patents is to compensate for so-called "market failures," while at the same time curbing monopolies and restrictive or discriminatory practices,[1] which would deprive the public of the benefits of the inventions. So, the "optimal" patent must create the right balance between two opposing requirements: the encouragement of innovation, on the one hand, and its diffusion at a reasonable cost, on the other.

Finding the right balance is a particularly sensitive issue in the field of medical innovations, because access to a great number of these innovations has such huge implications for public health. As Scherer and Watal

---

[1] What competition law formalizes as "abuse of a dominant position."

(2001, p. 4) have emphasized in the specific case of drug patents, the central question is "*how to balance the desire to make new drugs affordable to all who need them, and yet retain strong incentives for inventing and developing new and better treatments.*" In countries such as France, where the social security system guarantees the financial resources to satisfy the demand for medical products and services, the question takes on a particular dimension, as the monopoly price provided by the patent is covered by a series of "transfers," to the benefit of the producers of medical products. This creates particular tensions over the sharing of the social costs of the patent protection.

Thus, according to economic theory, all patenting systems should be governed by considerations of social welfare. And this goes well beyond the case of human health, where the welfare issue is of particular importance. While guaranteeing the incentive to innovate, such systems must limit the social cost of the protection given to innovators by restricting the rights conferred on patentees.[2]

Another key question has to be addressed, quite apart from the quality and importance of the protection granted to patentees. This concerns the definition of "patentable objects" – in other words, the "frontier" that separates information and knowledge which can be patented from that which cannot. On a purely theoretical level, the search for this frontier has stimulated, particularly in the United States, certain observations of crucial importance concerning the status of basic research. Following on from the work of Nelson (1959), Arrow, setting out a principle that would subsequently be adopted and developed by other authors,[3] has stressed the need to distinguish basic research from other research activities. He argues that, because it occupies a very "upstream" position in the R&D process, the specific purpose of basic research is to provide *common knowledge bases* – in other words, *multiple-use inputs* – for other research activities. The results of basic research are characterized by the fact that they can

---

[2] Note that all patenting systems demand something in return. The inventor must reveal the contents of his invention, so that society can benefit from the new knowledge and other players can develop it further or invent around it. In accordance with this principle, patenting systems have always required a written description of the invention as a condition for the granting of the patent.

[3] Here we refer to current work by "law and economics" specialists, such as Heller and Eisenberg (1998), Rai (2001), and Rai and Eisenberg (2003). See also Nelson (2003).

be used only for future advances in research or for the development of new products. Consequently, as any private appropriation of the results of basic research would work against the fruitful development of innovation, by impeding their use, Arrow contends that all researchers should have free access to these results, in the interests of public welfare.

In this approach, long recognized as the authority in the matter, the patent is seen *as a constituent element of a frontier between upstream and downstream research activities.* Only patents on downstream research products are considered capable of playing a positive role in the encouragement of innovation. It is important to note that this frontier principle also explains why basic research is described as the product of an "open science" type of organization (see Dasgupta and David, 1994, in line with Merton, 1973). One of the specific characteristics of this mode of research organization is the rule that discoveries must be publicized, since the research itself has been mainly publicly funded.

At the end of this section we emphasize the coherence of this economic perspective with the legal governance of patent rights up to the 1980s. Before we do so, however, we present another, radically different economic view of the role of patents in the promotion of innovation, proposed by Kitch in 1977. Kitch considers the patent to be a kind of *"prospecting right."* Countering the predominant doctrines of the time, he argues that only *"extended"* (i.e. large-scope) patents delivered high enough *"upstream"* and, granted to a *"unique operator,"* can provide sufficient incentive for the fruitful development of innovation (Kitch, 1977).

Kitch's thesis, diametrically opposed to that of Arrow, is founded on the idea that, just like the rights granted to mineowners, a patent should confer the exclusive right not only to "exploit" discoveries by developing industrial applications for them (the veins known to exist, to continue the mining metaphor) but also to "explore" the possibilities offered by a given scientific breakthrough (the as yet unknown veins present in the territory of the mine). For Kitch, patents must provide "prospects," which means that they have to possess at least two essential characteristics: (i) they must be of a large scope, and (ii) they must be delivered very far upstream. An additional argument put forward by Kitch, crucial to this debate, is that the exploitation of these patents must be placed "under one unique command" – that of the patentholder. According to Kitch, this is the only way that the exploitation (and exploration) of the potentials of the discovery can be coordinated efficiently. His argument is that this situation is "optimal" for collective welfare, because the granting of exclusive control over future applications of the invention is the only means of reducing the many transaction costs involved in the circulation and exploitation of the information contained in the patent. Thus, conferring exclusive rights on the patentee is the best way to limit the effects and costs of duplicate research, as the exchange and coordination of information between competitors is managed by the holder of the exclusive rights. For Kitch, the "prospecting" function of the patent enables each firm to announce its field of investigation, signaling to competitors that they should either give up this path of research or coordinate their activity with the exclusive rights holder. The patentee can thus choose either to develop the future "prospects" himself or to coordinate the R&D involved by licensing other firms in possession of other resources or information. Finally, the author contends, this type of patenting system "lowers the cost for the owner of technological information of contracting with other firms possessing complementary information and resources" (p. 277).

In his article, Kitch does not even refer to the work of Arrow, let alone discuss it. This is all the more surprising, and important for the subject of this paper, since the thesis of Kitch stands in total opposition to that of Arrow, even while claiming to adopt the same levels of analysis. In both cases, the reasoning is constructed from the perspective of *social welfare*. Likewise, it is worth noting that Kitch never discusses the relative efficiency of upstream patents compared with the free circulation of information within the community of innovators (researchers and innovators alike) that prevails in the world of open science. The virtues of upstream patents are examined only in the light of the advantages (and disadvantages) of trade secrets from the point of view of individual firms.

At the time of its publication, Kitch's article aroused little interest among either economists or specialists in patent law. And yet, as we shall see, his thesis now lies at the heart of the debate on the patentability of genes and the upstream results of biopharmaceutical research. Major discontinuities have also occurred in connection with the activities subject to patenting. As we discuss below, genes are paradigmatic examples.

Up until the 1980s patent law in Continental Europe was based on an essential distinction between "discoveries" and "inventions," in accordance with the "frontier" principle explained above. Only inventions were considered valid subjects of patenting.[4] Formally, this distinction is specific to European patent law. Nevertheless, in the United States, where the distinction is formally absent or irrelevant, other legal considerations led to the same practical end result. In Anglo-Saxon law, the "frontier" principle was established by the fact that an object could be patented only if the "practical or commercial utility" of the invention had been proved. In principle, this excluded scientific discoveries from the field of patentability (Orsi, 2002). In the United States (and other common law countries), the essential criterion of patentability is that of the recognized "utility" of the invention, and this characteristic is not defined by patent law but by jurisprudence. The doctrine of utility was established during the 1960s by a number of legal rulings. The first of these, the *Brenner v. Manson* case at the US Supreme Court,[5] stated that a patent could only be granted if it could be proved that the invention was *"operable and capable to use."* The judges specified that *"a patent is not a hunting license ... it is not a reward for the search, but compensation for its successful conclusion."* This idea that patents are not *hunting licenses* is worth noting. It means that the American doctrine of the time rejected Kitch's thesis that patents should be considered as rights to exploit "deposits" of future discoveries. Here, the court based its decision on a relatively narrow conception of patents, as rights that should be conferred only to compensate specific inventions of proven utility. As Eisenberg has pointed out, this ruling aimed explicitly to exclude scientific discoveries, which were considered *not to satisfy* the criterion of utility, to the extent that they can be transformed into useful – and therefore patentable – results only after additional work has been carried out by others, leading, in the words of the court, to their "successful conclusion" (Eisenberg, 1987). Finally, this doctrine clearly established that *scientific discoveries cannot satisfy the criterion of utility because they are considered "basic*

---

[4] Provided, however, that they met the traditional criteria of patentability. To be patentable, an invention had to be new, entail an inventive activity, and be open to industrial application.

[5] Supreme Court of the United States (1966), *Brenner v. Manson*, 388 U.S. 519, available at http://www.law.cornell.edu.

*tools" of science and technology and as such they are too far removed
from the "world of commerce"* (Eisenberg, 1987).

11.2.2 Genetic patentability: breaking with the traditional economic view

The patenting of genes represented a major break with the prevailing
doctrine and a large step towards the development of a new regime of
intellectual property rights over living matter.[6] Established during the
1980s under the impetus of a major Supreme Court ruling, one of the
first consequences of this new IPR regime was the granting of numerous
patents covering a large number of living organisms, including multi-
cellular organisms such as genetically modified mammals.

However, the policy of the USPTO towards the patenting of genes
remained relatively strict during the whole of the 1980s. In strict application of the criterion of utility defined by jurisprudence, the USPTO
rejected most requests for patents on genetic material[7] (Eisenberg,
2000). At the beginning of the 1990s, for example, the office turned
down a request from the National Institute of Health, a US government
agency, for patents on several partial gene sequences, on the grounds that
the information submitted for patent lacked "practical utility."[8] In this
case, the office adopted a particularly firm and clear position, emphasizing that the gene sequences in question were basic tools of research, and
as such could not satisfy the criterion of utility (Harnett, 1994).

The situation changed in 1995, when a major ruling by the Court of
Appeals for the Federal Circuit considerably loosened the criterion of

---

[6] For a detailed presentation of the conditions of the emergence of this new regime,
see Orsi (2002). The establishment of this regime formed part of a more general
movement towards the strengthening and expansion of IPR in the United States;
on this point, see Coriat and Orsi (2002).

[7] However, during this decade the USPTO did grant many patents on protein-
coding genes, the therapeutic interest of which had been demonstrated in the
requests. The patents covered not only the proteins themselves but also the coding
DNA and their various practical applications. Patents were granted, for example,
for the insulin-, interferon- and erythropoietin-coding genes at the same time as
for the proteins themselves.

[8] This patent request aroused fierce controversy within the international scientific
community, because the action of the NIH went completely against the rules for
the free circulation of knowledge that had been established for the Human
Genome Project. The NIH had itself been active in the promotion of these rules.
For a detailed account of this controversy, see Académie des Sciences (1992).

utility. This ruling led the USPTO to modify its policy on the patent-ability of genes. Following the Appeal Court ruling, the USPTO published new utility examination guidelines recommending less strict application of the criterion of utility, notably for genetic material (USPTO, 2001).

The granting of patents on genes linked to certain illnesses shows the serious practical consequences of this loosening in the criterion of utility. In this case, these patents are delivered before any practical application of the genes has been demonstrated. *The patentholder simply reveals the existence of a defective gene, associating it with commercial uses which are no more than potential and which require further research.* A good example is the patent granted on a gene linked to susceptibility to breast cancer,[9] the BRCA1 gene. The patent, co-owned by the University of Utah, the NIH, and the firm Myriad Genetics, covered firstly the methods and materials used to isolate the gene (notably the use of micro-organisms and DNA fragments as probes), then the *complete and mutated gene sequences* (germinal and somatic mutations) associated with susceptibility to breast and ovarian cancer, *the protein associated with these mutations*, and, lastly, *gene and protein applications* in diagnosis and treatment (Cassier and Gaudillière, 2000a). In other words, property rights over the gene itself and its applications were conferred even before the research had been carried out to prove that the BRCA1 gene had a real, practical application.

The final result of these changes is that patent protection now covers not only downstream products, derived from the use of scientific knowledge and of proven practical utility, but also the "input," which may give rise to future inventions. The ground covered by patent protection has expanded such that it leaves the knowledge base itself (the "discovery") open to industrial development. Until this happened, IPR appeared to delimit the frontier between the complementary domains of basic knowledge and commercial exploitation of this knowledge. Now this basic knowledge has become a central object of the system of appropriation, and its exploitation is the exclusive right of the patentholder. *Therefore, we now find Kitch's thesis of "prospecting rights" at the very heart of the new rules on patentability* – rules that now apply to the upstream results of biomedical research.

---

[9] This gene has also been linked to susceptibility to ovarian cancer.

What is remarkable is that this transformation has occurred without the slightest sign of any corresponding change in the domain of the theory and analysis of IPR.

Let us return to the claim that the granting of an exclusive right constitutes a powerful incentive to innovate. As Heller and Eisenberg (1998) have pointed out, the essential question is not whether a firm possessing exclusive patent rights over "generic" knowledge is encouraged to innovate or not; the essential question is whether this situation is "optimum" *from the perspective of social welfare* – the perspective adopted by both Arrow and Kitch. Heller and Eisenberg's argument concerning the "anticommons tragedy" is a direct extension of Arrow's reasoning, applied to the specific case of gene patentability. The authors contend that the coordination costs entailed by this new type of patent would be so high as to render exploitation of the patent impossible, leading to what the authors call the *anticommons tragedy*. This terms describes a situation in which several players each possess a fraction of a resource, and consequently have the right to exclude the others from the exploitation of their fraction, with the result that no one can obtain effective use of the resource in its entirety. When applied to genes, the "anticommons" concept represents the way in which knowledge of the human genome has been chopped up into a multitude of patents held by different actors, threatening to increase coordination problems and access costs to such a magnitude that the development of subsequent research may be seriously impeded. Consequently, *"a proliferation of intellectual property rights upstream may be stifling lifesaving innovations further downstream in the course of research and development"* (Heller and Eisenberg, 1998, p. 698).

As for the argument that the empowerment of one exclusive coordinator favours the "economy of duplication," Rai (2001) observes that "this argument – that multiple research paths can be pursued under broad upstream rights through multiple tailored licences – is the weak link that ultimately undermines Kitch's chain of reasoning." Rai's counter-argument is founded, in particular, on the case studies from Merges and Nelson (1994), who show that, in the case of cumulative innovations, large-scope patents on basic inventions have not led to the establishment of licenses and have penalized subsequent development. As Rai notes, the scale of the failure of such a model depends on the nature of the industry involved, and even more strongly on the scope of the upstream patent granted.

Finally, a growing number of authors concur that patents on the results of upstream research represent a serious threat to the progress of biomedical research and innovation. Recent studies have stressed the need for a return to the principle of free access to the results of basic research (Nelson, 2003; Rai and Eisenberg, 2003). Otherwise the currently established US doctrine may well prove to have extremely grave consequences, not only for research but also for the whole organization of public health systems.

In order to illustrate these points, let us examine the implications of granting patents on breast cancer genes, and of the strategy developed in and around these patents by Myriad Genetics.

11.3 BRCA genes and genetic testing for breast cancer: from patent to monopoly

The legal situation concerning the BRCA1 and BRCA2 genes, linked to a genetic susceptibility to breast cancer, has attracted a great deal of attention. There are several reasons for this. First, although only 5 percent of cases are of a hereditary form, breast cancer is nevertheless the most widespread type of cancer among women in the West. In addition, as we shall see, these genes have been used to develop what is called "predictive" genetic testing of the possible future development of breast cancer. This explains the economic importance and emotional charge attached to this case. Lastly, although the full effects of the exploitation of this patent in the United States are yet to appear, the resulting monopoly is far enough advanced to be able to carry out a first analysis of the type of institutional organization to which it gives rise. Likewise, although the consequences of the extension of this patent to Europe are as yet impossible to predict, the many points of controversy and conflict it will provoke can already be clearly identified.

Obviously, the case of the BRCA genes is just one specific example of the granting of patents on genes. Nevertheless, for the reasons just given, it provides us with a clear vision of the stakes, tensions, and risks – in terms of health care – created by this type of patenting.[10]

---

[10] Meanwhile, the very question of their patentability is still a subject of controversy in Europe. See the debates on the European directive on Legal Protection and Biotechnological Inventions (Directive 98/44/CE of the European Parliament and Council of 6 July 1998 *Journal Officiel des Communautés Européennes*, no. L213, 30 July).

We first describe the strategy adopted by Myriad Genetics in the United States for the exploitation of its patents, and next explore an extension of this strategy on current French practices.

11.3.1 The strategy of Myriad Genetics in the United States

As we have already seen in the case of the BRCA1 gene, the patents granted by the USPTO on breast cancer genes are an example of the new generation of large-scope upstream patents, covering not only the DNA sequence of the genes, and therefore any reproduction or by-product, but also all diagnostic and therapeutic applications, without any limitations as to the associated techniques of diagnosis or treatment.

The private genomics company Myriad Genetics was founded in Salt Lake City by a researcher from the University of Utah, after the localization of the BRCA1 gene by a team from the Breast Cancer Linkage Consortium in 1990 (Cassier and Gaudillière, 2000a). In 1994 Myriad Genetics, engaged in research into genes linked to common illnesses, succeeded in identifying the BRCA1 gene, enabling it to obtain several patents from the USPTO (Cassier and Gaudillière, 2000a). In fact, several distinct research bodies were granted US patents on the BRCA1 gene, and then on the BRCA2 gene, which was identified in 1995. This quickly created friction between the different patent-holders, leading, most notably, to a lawsuit for counterfeiting, between OncorMed, the first laboratory to market a BRCA1/2 test in the United States in 1996 (Brower, 1997; Kaufert, 2000), and Myriad Genetics. This resulted in an agreement in 1998 through which the Utah firm obtained all the patents on the two genes held, or to be obtained in the future, by OncorMed (Cassier and Gaudillière, 2000a).

At this point, things took a particular industrial turn. Boosted by all the patents it had acquired, Myriad Genetics set out to exercise to the full its monopolistic rights in the United States over every single diagnostic and therapeutic use of the BRCA1 and BRCA2 genes.

It agreed to give a few licences for *therapeutic* uses, notably to the Eli Lilly pharmaceutical group for BRCA1, *but chose to develop the diagnostic applications – which appeared to be immediately realizable and highly profitable – itself*. For this purpose, it created a subsidiary, Myriad Genetic Laboratories.

The first step taken by Myriad Genetics in its strategy to establish a monopoly was to take control over the whole activity of testing in the

United States (Cassier and Gaudillière, 2000b). Until 1996 BRCA1/2 tests had been carried out only in a research context, without charge for the patient. The tests were proposed by public genetics laboratories and also marketed by private laboratories such as Myriad Genetics, OncorMed, the Genetics and IVF Institute, and the genetic diagnosis laboratory of the University of Pennsylvania (Williams-Jones, 2002). Each laboratory used its own preferred gene analysis technique. But all this was to change in 1996, when Myriad Genetics succeeded in getting its rights recognized by the USPTO. It then set out systematically to sue or threaten to sue all its competitors, such as OncorMed or the University of Pennsylvania, until it had eliminated all its rivals in the market of BRCA1/2 tests (Williams-Jones, 2002).

In practical terms, this meant that all the blood samples required for the genetic tests had to be taken by the prescribing doctors *using exclusively the collection and dispatch kit supplied on order by Myriad Genetic Laboratories*,[11] and then sent to the firm in Salt Lake City for analysis. To start with, Myriad Genetics centralized all BRCA1/2 testing[12] in its Utah facilities (Cassier and Gaudillière, 2000a). Then, in 2001, it signed an agreement with the firm LabCorp, a specialist in biological analysis with a huge network of forty-seven laboratories and 24,000 employees. This contract of exclusivity for the United States allowed LabCorp to carry out tests of family members and frequent mutations in return for the marketing of all Myriad Genetics predictive tests through LabCorp's distribution network of 600 salespeople and its clientele of 200,000 doctors.

However, the first research into family mutations, entailing complete analysis of the two genes, was still carried out exclusively in Salt Lake City, by the technique of direct sequencing[13] (Cassier and Gaudillière, 2000a; Williams-Jones, 2002). Each patient was billed more than $2500 for this analysis, most often reimbursed by their health insurance (Wadman, 2001). In 1999 390 health insurance companies had already agreed to reimburse the BRCA1/2 tests (Williams-Jones, 2002). The fact that such a large number of companies agreed to

---

[11] See http://www.myriadtests.com/provider/orderform.htm.
[12] This is a first search for mutation in the family, consisting of a complete analysis of the two genes or a test of frequent mutations among founding populations, and a test for the mutation already identified among family members.
[13] This is a technique for the precise determination of sequence, based on the use of robots connected to analytical software.

meet the costs of the tests was the result of the active strategy adopted by Myriad Genetics. The company was quick to seek not only to increase the number of prescriptions for tests – notably by financing, from 1999 onwards, a training program for doctors proposed by the American Medical Association (AMA) – but also to ensure that the demand for tests was solvent, by negotiating agreements with American insurance companies. The signing of agreements with different health maintenance organizations (HMOs), such as Harvard Pilgrim Health Care, Aetna US Healthcare and Kaiser Permanente, provides a clear example of this twin strategy (Cassier and Gaudillière, 2000b). These agreements enabled Myriad Genetics to guarantee not only the payment of these expensive analyses but also their use by many health plan professionals. For example, by signing an agreement with Harvard Pilgrim Health Care in July 2000 the company gained access to a network of 25,000 health professionals, including 18,000 general practitioners and organ specialists caring for 1 million people insured with Harvard Pilgrim in three different states – Massachusetts, Maine, and New Hampshire.

11.3.2 The extension of the new doctrine to Europe: tensions and contradictions with current practices

The practice developed in the United States soon clashed with European rules and practices. In Europe, contrary to the type of organization, de facto and partly de jure imposed by Myriad Genetics in the United States, the BRCA1/2 tests are still carried out by *laboratories* essentially located in establishments forming part of the *public health system*. Originally offered mainly in a research context, these tests have been progressively introduced into common clinical practice (Cassier and Gaudillière, 2000b).

In France, the tests are mostly prescribed by specialist consultants, and carried out by more than twelve different laboratories, usually located in anticancer centres (CRLCCs) and regional university hospitals (CHRUs) (Sevilla et al., 2004). Interestingly, at present, these tests are not explicitly covered by public health care, and the laboratories generally find internal ways to cover the costs. Up until 2002 this emerging new activity was financed out of the budgets of the different establishments and by public funds donated by charities and research funds. This situation began to change in 2002, following a call for tender from the Direction des Hôpitaux et de l'Organisation des Soins[14] (Department of Hospitals and Health Care Organisation), allocating an operating budget to laboratories, on certain conditions, to finance their whole activity of genetic susceptibility tests, notably on the BRCA1/2 genes. However, the major threat to a "universal" use of upstream genetic knowledge basically stems from the request by Myriad Genetics for the extension of its patent rights to Europe, made in 1995, followed by the EPO granting the company three patents in 2001, even if contested (Butler and Goodman, 2001; Benowitz, 2003).

If these patents are ultimately granted by the EPO and if no compulsory licensing clause is allowed,[15] Myriad Genetics will be able to impose a testing process similar to that established in the United States. It has already announced its intention to establish it in Canada, Australia, the United Kingdom, Switzerland, Germany, and Austria. If the monopoly demanded by the firm is not challenged, we can expect to see a dramatic reorganization of the French testing process, threatening almost all the laboratories carrying out BRCA1/2 analyses in France. The first complete searches would be carried out exclusively in Salt Lake City by Myriad Genetic Laboratories, and tests for frequent mutations or for mutations already identified in the family would be carried out in France only by a French laboratory licensed by the patentholder. Ultimately, this would imply a national health system paying uncheckable bills to a foreign, profit-maximizing entity (Morgan et al., 2003).

Let us consider in greater detail the implications of such institutional changes. First, European patents would give Myriad Genetics the power to impose its specific practices on the whole chain of diagnostic operations, as it has already done in the United States. This includes the modes of supply of tests, the price of analyses, the choice of licensed laboratories, the sites in which the analyses are carried out, and the dispatch of blood samples abroad. Above all, the analytic techniques used for the tests are entirely up to the discretion of the firm, even if the firm's choice goes against the principles of economic efficiency. As we have seen, the firm uses direct sequencing of the two genes for the first complete search for family mutations. However, it has been

[14] Circular DHOS/OPRC no. 454, 14 August 2002.
[15] According to article L.613–16 of the intellectual property code.

demonstrated that this is the most expensive technique, and its effectiveness is matched by at least one alternative, the cost of which is at least 30 percent lower (Sevilla et al., 2002; Sevilla et al., 2003).[16]

Second, there is an issue in the fact that the pressure to increase demand could be applied through several distinct channels. An obvious means of increasing demand is the widening of the circle of prescribing doctors. Currently, in France, the doctors who prescribe tests are consultants in oncogenetics, of whom there were forty in 2001, spread over thirty-four establishments, essentially in CRLCCs and CHRUs (Sevilla et al., 2004). However, if Myriad Genetics intensifies the diffusion of information about these tests to health professionals and the general public, it is possible that general practitioners and organ specialists will also be led to prescribe these analyses. This could result in an exaggerated increase in prescriptions, particularly given the pressure exerted on doctors either through the company's marketing activities (Wazana, 2000; Armstrong et al., 2002), or by patients (Findlay, 2001) whose demands may be based on a misconception of the risk (Press et al., 2001; Huiart et al., 2002) – a misconception that could be accentuated by the diffusion of simplistic and alarmist commercial information (Gollust, Hull, and Wilfond, 2002; Wolfe, 2002). After all, in the French case the public health code does allow a prescription to be made by any doctor, on condition that his patient is already ill.[17] However, as long as these tests remain outside the social security listing of reimbursable acts (being financed since 2002 by an operating budget allocated to French laboratories by the Ministry of Health[18]), the success of the firm's efforts to boost demand is likely to remain limited by the threshold of solvency.

Obviously, the universal care system is likely to come under strain as soon as it is hit by the above pressures, which – let it be noted – happen ultimately to be profit-driven ones, and, in that respect, stemming from a monopolistic firm.

---

[16] Cost-effectiveness analysis comparing direct sequencing with the nineteen possible alternative techniques for the first search for BRCA1 gene mutations shows clearly that, although direct sequencing is the most effective method of analysis, with 100 percent sensitivity, it is also the most expensive ($1200). At least five alternative cost-effective strategies exist, with costs between 30 percent and 90 percent below those of direct sequencing and a rate of false negatives ranging between 2 percent and 13 percent.

[17] Article R.145–15–5 of the public health code.

[18] Circular DHOS/OPRC no. 454, 14 August 2002.

Conversely, let us explore the impact of such a monopoly on *the organization of research itself*. As we have already pointed out, one of the constraints imposed by the company is the dispatch of blood samples to the United States for a first complete search for family mutations. The analyses for frequent or known mutations are then carried out by a licensed laboratory in the patient's own country.

In practice, this policy represents nothing less than the *establishment of an international, rights-depriving DNA bank*. It also means the loss, for any world laboratory, of the practice of searching for mutations on these two genes. In this fast-evolving field this can only result in biologists and technicians losing their expertise in the techniques of analysis and, above all, in the interpretation of results. In fact, this might be even worse than the anticommons situation described by Heller and Eisenberg (1998), creating a situation of irreversibility that would threaten firstly the pursuit of research in this fundamental domain and subsequently the whole health care system.

11.4 Conclusions

One of the central implications of the patentability of genes is granting upstream patents with a very large scope, including the protection of virtual, future inventions that are not described in the request for patent itself.

In fact, gene patents – a radical break with the doctrines and practices governing IPR until the 1980s – seriously threaten both the future of research and the whole public health system.

By examining the impact and effective consequences of gene patentability through the textbook example of patents granted on the BRCA1 and BRCA2 genes, we have tried to demonstrate the huge changes that accompany the new doctrine. Here the story of Myriad Genetics highlights how, through its acquisition of patents on these genes, this firm has been able to impose a whole set of unprecedented monopolistic practices in the United States. The extension of such practices to Europe in general, and more specifically France, would result in the health system becoming responsible for the reimbursement of the patentholder with the de facto right to extract as much rent as possible from any public health system. Such a company could exploit its monopoly, not only to impose its mode of carrying out the tests but also to dictate its prices, despite the fact that the strategy of analysis chosen by the firm might be far from the most cost-effective.

Together, further upstream, the establishment of a rights-depriving database (formed thanks to the protocols imposed in this example by the monopolistic company itself) represents a serious threat to the development of future research.

References

Académie des Sciences (1992), *La brevetabilité du génome*, Report no. 32, Editions Techniques et Documents, Paris.

Armstrong, K., J. Stopfer, K. Calzone, G. Fitzgerald, J. Coyne, and B. Weber (2002), "What does my doctor think? Preferences for knowing the doctor's opinion among women considering clinical testing for BRCA1/2 mutations," *Genetic Testing*, 6, 115–18.

Arrow, K. (1962), "Economic welfare and allocation of resources for inventions," in R. R. Nelson (ed.), *The Rate and Direction of Inventive Activity*, Princeton University Press, Princeton, NJ, 609–25.

Benowitz, S. (2003), "European groups oppose Myriad's latest patent on BRCA1," *Journal of the National Cancer Institute*, 95, 8–9.

Brower, V. (1997), "Testing, testing . . . testing?," *Nature Medicine*, 3, 131–2.

Butler, D., and S. Goodman (2001), "French researchers take a stand against cancer gene patent," *Nature*, 413, 95–6.

Cassier, M., and J. P. Gaudillière (2000a), "Le génome: bien privé ou bien commun?," *Biofutur*, 204, 26–30.

(2000b), "Recherche, médecine et marché: la génétique du cancer du sein," *Sciences Sociales et Santé*, 18, 29–49.

Coriat, B., and F. Orsi (2002), "Establishing a new intellectual property rights regime in the United States: origins, content, problems," *Research Policy*, 31 (8–9), 1491–507.

Dasgupta, P., and P. A. David (1994), "Toward a new economics of science", *Research Policy*, 23 (5), 487–521.

Eisenberg, R. S. (1987), "Property rights and the norms of science in biotechnology research," *Yale Law Journal*, 97 (2), 177–231.

(2000), "Analyze this: a law and economics agenda for patent system", *Vanderbilt Law Review*, 53 (6), 2081–98.

Findlay, S. D. (2001), "Direct-to-consumer promotion of prescription drugs: economic implications for patients, payers and providers," *Pharmacoeconomics*, 19, 109–19.

Gollust, S. E., S. C. Hull, and B. S. Wilfond (2002), "Limitations of direct-to-consumer advertising for clinical genetic testing," *Journal of the American Medical Association*, 288, 1762–7.

Harnett, C. J. (1994), "The Human Genome Project and the downside of federal technology transfer," *Risk: Health, Safety and Environment*, Spring.

Heller, M. A., and R. S. Eisenberg (1998), "Can patent deter innovation? The anticommons tragedy in biomedical research," *Science*, 280, 698–701.

Huiart, L., F. Eisinger, D. Stoppa-Lyonnet, C. Lasset, C. Nogues, P. Vennin, H. Sobol, and C. Julian-Reynier (2002), "Effects of genetic consultation on perception of a family risk of breast/ovarian cancer and determinants of inaccurate perception after the consultation," *Journal of Clinical Epidemiology*, 55, 665–75.

Kaufert, P. A. (2000), "Health policy and the new genetics," *Social Science and Medicine*, 51, 821–9.

Kitch, T. (1977), "The nature and function of the patent system," *Journal of Law and Economics*, 20, 265–90.

Merges, R. P., and R. R. Nelson (1994), "On limiting or encouraging rivalry in technical progress: the effect of patent scope decisions," *Journal of Economic Behavior and Organization*, 25, 1–24.

Merton, R. (1973), *The Sociology of Science: Theoretical and Empirical Investigation*, University of Chicago Press, Chicago.

Morgan, S., J. Hurley, F. Miller, and M. Giacomini (2003), "Predictive genetic tests and health system costs," *Canadian Medical Association Journal*, 168, 989–91.

Nelson, R. R. (1959) "The simple economics of basic scientific research," *Journal of Political Economy*, 67, 297–306.

  (2003), *The Market Economy and the Scientific Commons*, working paper, School of International and Public Affairs, Columbia University, New York.

Orsi, F. (2002), "La constitution d'un nouveau droit de propriété intellectuelle sur le vivant aux Etats-Unis: origine et signification économique d'un dépassement de frontière," *Revue d'Economie Industrielle*, 99 (2), 65–86.

Press, N. A., Y. Yasui, S. Reynolds, S. J. Durfy, and W. Burke (2001), "Women's interest in genetic testing for breast cancer susceptibility may be based on unrealistic expectations," *American Journal of Medical Genetics*, 99, 99–110.

Rai, A. K. (2001), "Fostering cumulative innovation in the biopharmaceutical industry: the role of patents and antitrust," *Berkeley Technology Law Journal*, 16, 813–53.

Rai, A. K., and R. S. Eisenberg (2003), "Bayh–Dole reform and the progress of biomedicine," *Law and Contemporary Problems*, 66, 289–313.

Scherer, F. M., and J. Watal (2001), *Post-Trips Options for Access to Patented Medicines in Developing Countries*, Working Paper WG4, Center for Metropolitan History, London.

Sevilla, C., P. Bourret, C. Noguès, J. P. Moatti, H. Sobol, Groupe Génétique et Cancer, and C. Julian-Reynier (2004), "L'offre de tests de prédisposition génétique au cancer du sein ou de l'ovaire en France," *Médecine/Sciences*, 20, 788–92.

Sevilla, C., C. Julian-Reynier, F. Eisinger, D. Stoppa-Lyonnet, B. Bressac de Paillerets, H. Sobol, and J. P. Moatti (2003), "The impact of gene patents on the cost-effective delivery of care: the case of BRCA1 genetic testing," *International Journal of Technology Assessment in Health Care*, 19, 287–300.

Sevilla, C., J. P. Moatti, C. Julian-Reynier, F. Eisinger, D. Stoppa-Lyonnet, B. Bressac de Paillerets, and H. Sobol (2002), "Testing for BRCA1 mutations: a cost-effectiveness analysis," *European Journal of Human Genetics*, 10, 599–606.

USPTO (2001), "Utility examination guidelines," *Federal Register*, 66, 1092–9.

Wadman, M. (2001), "Testing time for gene patent as Europe rebels," *Nature*, 413, 443.

Wazana, A. (2000), "Physicians and the pharmaceutical industry: is a gift ever just a gift?," *Journal of the American Medical Association*, 283, 373–80.

Williams-Jones, B. (2002), "History of a gene patent: tracing the development and application of commercial BRCA testing," *Health Law Journal*, 10, 121–44.

Wolfe, S. M. (2002), "Direct-to-consumer advertising – education or emotion promotion?," *New England Journal of Medicine*, 346, 524–6.

12 Competition, regulation, and intellectual property management in genetically modified foods: evidence from survey data

PIERRE REGIBEAU AND
KATHARINE ROCKETT

12.1 Introduction

Genetically modified food represents a unique opportunity to trace a new technology from its inception. It thereby provides a rich example within which we can examine the management of a new technology, as well as public policy towards new technologies. This chapter presents a series of hypotheses regarding industry structure, regulation, and patent policy towards GM food crops. We have designed and implemented a survey focusing on the innovations that contribute towards the production of new GM plant varieties to allow us to gather information on the plausibility of these hypotheses. This study summarizes the support (or lack of support) we found for them in our responses.

While our sample size is small, a number of suggestive findings emerge. First, we investigate the competitive structure of GM food. The industry involves a long vertical chain, moving from innovations to approved crops, to cultivation, processing, distribution, and – finally – retail. We investigate the structure of the first two stages of this chain. We find some support for our hypothesis that GM food is perceived, at these stages, as a separate industry from the traditional food sector. This suggests that the relatively high concentration ratios measured for GM food are reflective of the true concentration level of this industry. Further, they are well over the levels that generally trigger antitrust scrutiny. Second, for rapidly changing or new markets, it has been suggested that "innovation market" structure be evaluated as a

We would like to thank three anonymous readers for their helpful comments and to acknowledge the support of Economic and Social Research Council grant no. L145251003 for this project.

prospective measure of product market structure. We use the survey's support for patents as a measure of market power to propose and evaluate weighted patent counts as an implementation of the innovation market concept. We find that this implementation tends to undervalue patents that are very important in our market, so that the prospective measure of market share tends to understate actual levels of concentration, quite drastically in the case of highly important patents.

As we move from patent portfolios to products that are actually approved for commercialization, we expect to see higher concentration. After all, this is a new industry, and product approvals take time. Indeed, we postulate that regulatory approval could have a concentrating influence on this industry by erecting a significant and persistent entry barrier. Our survey responses lend general support to this hypothesis. In fact, entities owning patents in this area appear to use licensing to delegate regulatory costs. We suggest that a two-tier structure in this industry results, with "design" firms providing technology on license to larger "hub" firms, which concentrate patents, obtain regulatory approval, and manufacture. More stringent GM regulations are likely to increase this concentration. In fact, responses indicate that most of the entry barrier is not due to learning economies, as the regulatory burden is not reported to fall greatly as entities accumulate submissions. This, in turn, indicates that the concentrating influence of regulation is likely to be a long-term phenomenon in this field.

Third, we investigate whether patents actually elicit research in this area through their reward function, as is, indeed, the presumption of most economic models of the patent system. Patents and their associated revenues are not cited in our sample as the primary cause for the *specific* research into GM food being undertaken, even for entities that continue to be active in this field. Indeed, our respondents indicated that other incentives generated their research in this area. On the face of it, this would suggest that strengthening patent protection would be unlikely to generate more research spending by the "reward" route. However, respondents also said that, whatever the reason that was cited for the research to have taken place, patents are revenue generators, indicating that patents do fulfil, somewhat, the function of raising the returns to research-based firms. Additionally, our respondents indicated that patents do appear to be viewed as valuable "defensive" tools for raising the cost of imitation. In this sense, the salient function of the patent system may be closer to creating incentives to disclose innovations in the form of patents than to directing research by creating a reward. In fact, most patent systems have a dual function, providing not just a reward to elicit research but also an incentive to disclose innovations by creating property rights that protect innovations from imitation, although the latter function tends to be the less studied in the economics literature.

Finally, we investigate the role of litigation in this industry, and in particular its substitutability with licensing. We find that this substitution does, in fact, appear to be occurring. We received mixed support, however, for Lanjouw and Schankerman's (2001) hypothesis that smaller entities may face an entry barrier of higher litigation costs than larger firms that have inventories of "spare" technologies that can be traded in order to stave off litigation. In our sample, smaller entities are small corporations – not individuals, as in their study – and do not seem to be suffering a lack of choice of using licensing arrangements to avoid litigation. In fact, they seem to be considering a wider range of strategies than this strict substitution. This may indicate that the Lanjouw–Schankerman entry concerns may apply more strictly to individuals than to smaller entities in general. We received even more limited support for Lerner's (1995) contention that high litigation costs could cause research to be redirected away from litigious areas. Instead, we received more support for the view that firms with high litigation costs may exploit their technologies by licensing them to firms with lower litigation costs, while leaving the area of research unchanged. In fact, if one recasts the Lerner model with an option of licensing, this could easily be the result, as we will detail later. This would imply, in turn, that litigation costs are not a barrier to entry into research areas at the "design" level of this industry, even in highly competitive research areas, even though they may be a barrier to entering the "hub" group.

We are not the first to use survey data to gather information on technology management. A recent series of papers[1] use survey data to frame further empirical work on various technologies to address questions that are similar to ours. Our study differs in that it is focused on GM food and its unique issues (such as the patentability of genes), as

---

[1] See Hall and Ziedonis (2001), Ziedonis (2000), Sakakibara and Branstetter (2001), and Cohen, Nelson, and Walsh (2000) for some examples.

well as being quite recent. The latter allows us to check some of the hypotheses that have been put forward in the last few years regarding the role of litigation in intellectual property strategy. While supporting empirical work using a large data set would be desirable, there are several impediments to this in the area of our work. First, because GM animal and medical research involves very different competitor groups, both at the product market and the research stages, our work is focused solely on GM plants, limiting the amount of data potentially available for econometric analysis.[2] Second, genetically modified food is still, even at the time of writing, a new technology, on which relatively little information is available even without the further area restrictions we have imposed. Finally, GM food research often represents only a portion of the overall research projects of the firms involved in the area. Using more general statistics on these research programs could, then, be misleading to the extent that GM food represents a small percentage of their overall research expenditure and so is not necessarily determinate in overall strategy. Our preference, then, is to present the survey on its own as suggestive of areas for future probing as the industry develops.

The chapter will now proceed to describe our hypotheses. Next, we briefly discuss the survey.[3] We then present our findings in more detail. The last section of the chapter presents some concluding remarks.

12.2 Our hypotheses

As a first step towards discussing policy towards firms in the area of GM food, we need to define the scope of the industry. In particular, we need to know whether this industry should be considered, for the purposes of analyzing behavior, as separate from the traditional plant breeding industry, other areas of biotechnology, or any other group. In terms of our survey respondents, we need to know from them which entities they consider to be their competitors.

The definition of a market is a relatively well-explored concept in the area of competition policy and, in particular, merger policy. Mergers in concentrated industries are reviewed for their impact on welfare by

---

[2] Our scope includes genetic modifications to plants directly incorporated into food, and genetic modification *techniques* applicable to plant-based food.

[3] The survey itself and a brief summary of the specifics of the methodology can be found in the working paper version of this chapter; see Regibeau and Rockett (2005).

competition authorities. Since concentration will tend to increase the more narrowly a market is defined, the definition of the market is crucial to determining whether this review will take place and on what terms. Clearly, too, the types of behavior that could be expected of firms in a relatively concentrated industry would differ from those in a relatively fragmented industry.

At the heart of most concepts of market definition is the idea of substitutability. In merger policy, this is reflected in some measures of price elasticity of demand, or predicted price responses in the face of a hypothetical increase in industry concentration. This type of exercise is particularly difficult to conduct for emerging industries, as information on prices and products may not be readily available and substantial change in supply or demand conditions may occur in a short period of time, resulting in traditional analysis giving only a fleeting glimpse of the industry. Simply waiting to see how the market will develop as it matures is not always an option, as merger activity in the early days of an industry can be quite frequent. GM food is a prime example of this, with extremely high merger activity in the late 1990s[4] and moderate levels still continuing. As a result, the guidelines of the Competition Commission (2003) allow it to consider how rivalry may be expected to develop over time, relying on survey and other "soft" data in cases where hard information is limited.

The 1995 antitrust guidelines for licensing intellectual property, used in the United States, extend the concept of market to "innovation markets," where such a market consists of "the research and development directed to particular new or improved goods and processes ... and close substitutes."[5] Innovation markets give us an idea of the prospective universe of firms that will be product market competitors in the future, and hence are relevant to rapidly developing new fields, such as GM food. In this sense, we need information from our respondents not only on their product market competition but on the competitors they face in research.

We used the survey as an instrument to elicit this information by asking our respondents which entities they viewed broadly as product market competitors, as well as how many products competed directly

---

[4] Counting the activity of the top producers in this area only, over the last ten years there have been close to fifty mergers.

[5] 1995 guidelines, section 3.2.3, as quoted in Scotchmer (2004).

with their GM products.[6] Hence, we had questions set to investigate the following hypothesis.

Hypothesis 1: GM food constitutes a separate industry from non-GM food

Once the market has been defined, we need some measure of market power in order to discuss the structure of the industry as well as public policy towards it. Market shares in output markets are a standard measure, but current market shares in an emerging market may not be reflective of even near-term structure. An alternative would be to build an IP-based market structure measure as a means of implementing the innovation market concept. In theoretical models of patent design, patents are often taken as synonymous with monopoly power in output markets.[7] If we take this at face value, then it suggests that patent counts in sub-fields (such as corn or soy) could be used as a proxy for market power derived from the innovation market. There are several problems with using patent counts to reflect monopoly power, however. First, in many markets, survey evidence indicates that market power comes from factors other than patents, most notably learning by doing, secrecy, and sales efforts.[8] It should be noted, however, that the evidence varies considerably by industry, with drugs standing out as one where patents are viewed as highly effective. Before proceeding with a patent count, then, we had to check with our participants that the standard conception from theoretical models that patents are important determinants of market power holds for this industry.

Hypothesis 2: Patents are an important "prospective" indicator of market power in this industry

Second, pure patent counts do not correct for the difference in relative importance of patents in generating market power or for their cross-effects in building patent portfolios that effectively create market power. Most patents, whatever the field, are relatively

---

[6] A companion test would measure the substitutability of GM and non-GM crops more directly. This has been discussed, but no definitive measure of cross-elasticity has been developed, to our knowledge. For more information on the usage and pricing of GM and non-GM crops, see Fernandez-Cornejo and McBride (2002).

[7] For a summary of theoretical models of patent design, see Scotchmer (2004).

[8] See Levin et al. (1987) and Cohen, Nelson, and Walsh (2000) for relevant survey data.

unimportant. This is true for GM food as well. For example, if one were to use forward citations by other patents as a measure of importance of patents, for all GM plant patents in the United States granted as at the end of 2000, 25 percent still had zero citations four years later and 53 percent had two or fewer. The equivalent percentages for our respondent sample are 37 percent with no forward citations at all four years on, and 80 percent with two or fewer. To address the patent count/market power link, Lanjouw and Schankerman (2004) develop an index of patent characteristics that can be correlated with economic value for firms and so can be taken as a measure of the relative commercial importance of patented technology, following on work by others[9] who have used forward citations as a measure of importance. For our work, forward citations will be retained as a primary measure of importance, as this is the part of the Lanjouw–Schankerman index that has the most salience in the related industry of biotechnology, and as the other aspects of the index have less compelling justification in our sample.[10] For some examples where actual market shares are available several years later, weighted patent share measures are computed and compared with actual market shares as a means of discussing how accurate patent shares are in predicting market power for this industry. If they are accurate, then an implication is that the concentration of weighted patent portfolios can be an important candidate tool for evaluating future market power.

Finally, in the field of GM food, where regulatory approval for commercialization is required, patent counts – even weighted ones – do not account for the concentrating influence of the regulatory process. Where regulatory approval is very costly, this influence can be significant, with the result that patent counts should be taken – at best – as a probable lower bound on the true "prospective" level of concentration in GM food. We postulate two possible mechanisms by

[9] For example, see Trajtenberg (1990). Also, see Hall, Jaffe, and Trajtenberg (2005) for discussion.

[10] The number of claims in the patent is the other part of the index that receives considerable weight for biotechnology in Lanjouw and Schankerman's work. In fact, the number of claims has increased considerably in the field of GM food over the last twenty years, but at the same time each individual claim has become far less cited. As a result, the number of claims *alone* would appear to overstate the importance of patents with multiple claims. See Regibeau and Rockett (2004) for more discussion of the number of claims as a measure of importance in the field of GM plants.

which regulation could concentrate this industry. First, firms with extensive experience with similar regulatory approval processes might have a higher level of expertise than those with little experience. This "learning" effect could give firms with more experience an edge early on in this industry, but might disappear over time as more firms gained similar portfolios of experience. The second proposed mechanism is simpler: the cost of regulatory approval could serve as an entry cost to the industry, reducing the number of firms that this industry could support in the "downstream" stage after regulatory approval. This barrier could be expected to persist over time. Together, these two mechanisms could also lead to extensive licensing, as smaller patentholders attempt to avoid regulatory approval by selling their technology to their larger, and more experienced, rivals. Hence, we expect the industry to concentrate around firms with the "complementary skill" of gaining regulatory approval, whereas a second tier of "design" firms produces patents as a final output for license to the firms with regulatory ability. Hall and Ziedonis (2001) note a similar "two-tier" structure emerging in the semiconductor industry in the 1980s, with design firms not undertaking the (large and growing) cost of manufacturing facilities. Summarizing, we have questions aimed at probing the following two hypotheses.

Hypothesis 3: Regulation has a concentrating influence on GM foods, due to the combined effects of expertise and (fixed) entry cost

Hypothesis 4: Industry participants delegate regulatory approval by means of licensing agreements to a core of firms, resulting in increasing industry concentration at the approval stage and a "two-tiered" industry structure

Within this two-tiered industry structure, we investigate the role of patents in generating research in the next few hypotheses. Above, we looked at whether patents were a good measure of market power. In the hypothesis below, we ask whether this is the *reason* why research in GM food occurs. This is certainly the presumption in many theoretical models of the effect of patents on R&D activity.[11] Sakakibara and Branstetter (2001), Hall and Ziedonis (2001), and Bessen and Maskin (2000) all point out that one might expect that, to the extent that patents generate a reward to research, a stronger patent right would elicit more

---

[11] See Scotchmer (2004) for a summary of this link in the literature.

R&D. In fact, this connection is not supported empirically in any of these papers, and is theoretically ambiguous even if patents are functioning effectively as "rewards."[12] Delving more deeply into the reward function of patents, survey results in Levin et al. (1987) and Cohen, Nelson, and Walsh (2000) suggest that managers do not find patents to be the most important generators of research "rewards." An exception to this is drugs, where patents are found to be viewed as important to generating innovative efforts. To the extent that GM food, as a sub-field of biotechnology, may behave in a similar way to drugs, it might be expected that patents would play a larger role here as well.

Hypothesis 5: Patents elicit GM food research through the "reward" that monopoly power creates

On the other hand, if patents do not stimulate research by means of their reward function, it begs the question of the role of patents in generating scientific progress. Theoretical models of optimal patent design as well as legal work on the patent system[13] point to an alternative role for patents – the "disclosure" function – whereby the grant of exclusionary property rights on innovative results raises the incentive to disclose those results by reducing the ability of others to appropriate the gains of innovation. Such disclosure is crucial, the argument goes, for furthering scientific progress that builds on earlier results. If patents are, indeed, having such a function, then they should be claimed to raise the barriers to imitation significantly. Hence, we can investigate this function within our survey population. In particular, if the response is positive, but the reward function response is negative, it indicates that an important avenue for future research on optimal patent policy would be to investigate more thoroughly the implications of the disclosure function of patents, as it may be the more salient aspect of the patent system for some populations.

Hypothesis 6: Patents function to increase incentives for the disclosure of research results by raising the cost of imitation

[12] Some models of research indicate that stronger patents imply increased research expenditure, some do not. See Sakakibara and Branstetter (2001) and references therein. Intuitively, increasing patent strength may have more to do with the mixture of imitative and non-imitative research than the absolute level of total research expenditure, encouraging the latter and discouraging the former.

[13] See Scotchmer and Green (1990) and Matutes, Regibeau, and Rockett (1996) for a discussion of the legal and economic aspects of the disclosure function.

We are not concerned solely with the amount of research conducted by firms but also with the type of research. A particular concern that is unique to genetic engineering is the patentability of genes and its effect on research priorities. Harhoff, Regibeau, and Rockett (2001) argue that patents on pure genetic material may not be as socially desirable as patents that require genetic material to be embedded in a particular application. The intuition as to why pure gene patents may best be patentable is that, in a setting where innovation is sequential (so that later innovations build on the work of earlier innovations), the social value of the initial innovation should include the net social value of subsequent innovations. Second-generation inventors should, then, be limited to recovering their costs, but all the remaining surplus should be channeled to first-generation inventors. As a practical matter, this would require broad patents on first innovations. In the case of GM, we could conceive of such broad patents as patents on genes regardless of their applications. Harhoff, Regibeau, and Rockett (2001) argue that such a policy would be socially undesirable, however, because it directs research towards output that has relatively little social value: by raising the "prize" for the pure gene, it draws work away from applications of that gene, which is where more social value is generated. For example, herbicide-resistant crops are an application of GM technology with a value to farmers that a specific gene by itself would not have.

With our survey, we can obtain views on whether pure gene patents would, indeed, have an effect on the direction of research. Hence, we have our next hypothesis.

Hypothesis 7: Pure gene patents would cause research in this industry to be redirected away from applications and towards pure genetic material

The research area of genetic manipulation potentially has a very wide set of applications. Given this, we could ask whether firms tend to choose to "race" with each other on similar types of innovation or whether they choose very different trajectories to avoid each other. Clearly, the R&D race literature does not give firm predictions on this point: firms may rush to the same area despite competition because an application appears particularly profitable, or they may avoid each other in order to reduce the expense incurred by racing.[14] Lerner

---

[14] See Scotchmer (2004) for a brief summary of some of the literature on patent races.

(1995), in a study of new biotechnology firms in the United States, finds some evidence that firms avoid areas where other researchers have a "head start," but is able to go further to show that firms with a high cost of litigation tend to avoid research areas populated by other firms – particularly ones with lower litigation costs. Hence, litigation costs appear to be "scaring" some firms away from certain research areas in biotechnology. This type of result might be expected to come through in our sample, as GM food is a highly litigious area. Although US patent litigation takes a long time to develop and work its way through the courts, one can observe from European data that patent oppositions in Europe are extremely high in GM food, at 25 percent. This is three times the opposition rate for biotechnology and pharmaceuticals, a seemingly similar field (see Harhoff, Regibeau, and Rockett, 2001). The level of experience with some form of patent litigation in our sample is similarly high, with 23 percent of our participants having experienced litigation involving the patents that form the basis of our survey.

While Lerner empirically verifies litigation's link to research trajectory for biotechnology as a whole, the theory rests on several assumptions that may not hold for our sample. In particular, it must be the case that litigation costs cannot be delegated by means of licensing agreements. Consider a simple conception of the Lerner model. Litigation cost is a characteristic of the firm, independent of the area of future research to the extent that it depends on size and past litigation experience, whatever the technology field. Hence, the population of firms has two types: low- and high-litigation-cost firms. These firms may choose their area of research. Areas of research are of two types: areas where other firms hold patents ("populated") and areas where no other firms are working ("empty"). Assume that, if an area is populated, an existing patentholder will litigate any new firm patenting in the area. If the area is empty, no litigation will follow patenting. Suppose further that litigation is always successful. A prospective researcher chooses the area of research, then, by computing the profit of entering the populated area net of the litigation fee and comparing it to the profit of entering the empty area with no deductions. As long as the profit of the populated area is not high enough to overcome the cost of litigation, the empty area will always be chosen. For firms with higher litigation costs, the range of profits for which the empty area is chosen is larger. Hence, one can conclude that empirical work should indicate a stronger tendency for firms that are small or have no litigation experience to patent in "empty" fields.

Now, suppose that we consider delegating the litigation to a licensee with lower litigation costs. Suppose also that the licensor would change research area in the absence of a licensing agreement (so that the profits from changing field exceed those of staying in the populated area and shouldering the litigation cost). Hence, for licensing to occur, the licensor must earn more from the licensing agreement from the alternative of changing field. Suppose further that, at the lower litigation cost, research in the populated area is more profitable, taking into account litigation, than research in the empty field. This means that litigation expense is the determining element of the decision – our case of interest. In this case, there will be a surplus left over after the minimum license fee that is necessary to induce participation has been paid to the licensor. Still, the licensee will accept this delegation only if the revenues from the patent net of the (low) litigation fees exceed the profits it would make by refusing the license. As long as this alternative profit is small, then the licensee will always accept, and research should be conducted in the populated area. Therefore, no change in research area occurs due to the litigation expense. Rather, firms organize litigation "efficiently" by delegating to the low-cost firm whilst remaining in the populated area of research.[15]

Other considerations could be added as well, but the main point is that if licensing can be used as a tool to delegate the litigation process it separates the research from the litigation decision, with the result that, while litigation and licensing strategy would be expected to be heavily related, litigation and research strategy might not. At the very least, it could weaken the effects pointed out by Lerner (1995). We investigate, then, the research decision and the role that licensing and litigation play in that decision with the following hypothesis.

Hypothesis 8: Licensing is used to delegate litigation costs in this industry

[15] If patents tend to be complementary, being used within areas of research to build patent portfolios or being used together to generate the technology underlying a particular product, then the argument can be made stronger that the licensor would wish to remain in the populated field. This is because a blocking patent that is key to the strength of a patent portfolio owned by the licensee could result in a better bargaining position for the licensor in the negotiations.

Related to this, we consider other effects that the threat of litigation has on the marketing strategy of our respondents. In particular, Hall and Ziedonis (2001), as well as Lanjouw and Schankerman (2001), have suggested that litigation should rise as the cost of alternative strategies to resolve patent infringements (specifically, licensing and out-of-court settlement) rise. These alternative strategies should be more available to firms that have technologies available "off the shelf" to use as bargaining chips in cross-licensing agreements, so creating entry barriers for smaller entities. This leads us to our last hypothesis.

Hypothesis 9: Litigation is a higher entry barrier for smaller entities in this industry than for larger entities

12.3 The survey

We used patents from the USPTO and the EPO websites to construct a universe of granted patents up to the end of 2000 in the area of GM food based on plants. We screened each patent by carefully reading the claims in order to classify the patent as applying or not applying to GM food. This method was much more accurate than the standard method of using patent class as a criterion for selection, as there is no single patent class for that that adequately captures GM food, and GM food alone.[16] Our procedure resulted in a universe of 141 entities that were listed as assignees for patents on GM food. This was our list of possible survey recipients worldwide. We attempted to contact each of these entities in order to obtain agreement to respond to the survey. This was not always possible. For example, we were able to obtain no contact information for individuals holding patents in this area. In all, ninety-six of the entities we were able to contact agreed to participate and received a survey. Of these, we have collected twenty-six completed surveys. Clearly, we could not compel entities to respond, so our sample does not necessarily represent a random selection. Our sample does, however, represent a diverse cross-section.

---

[16] For example, the 800 class groups all plant patents, whatever the technology that generates them. Class 435 relates more closely to genetic manipulation but includes plants, animals, and some medical applications.

Table 12.1 Characteristics of the sample

	United States	Canada	Europe	Other
University	9	–	1	–
Small firms	4	1	–	
Large firms	–	–	5	1
Government agencies	–	2	1	2

While we divided our responses from firms into those under 1000 employees ("small") and those over 1000 ("large") employees, the actual responses fell into a "small entity" group composed of fewer than 200 employees and a "large entity" group composed of over 2500 employees. As a result, small firms refer to *very* small firms in our sample, while large firms are very large.

In each case we asked to speak with a person who was knowledge-able about the questions we asked. As our questions touched upon a number of different policies (research strategy, patenting, litigation, product market competition and so on), there were times when we had to contact a number of different people within the same enterprise to obtain informed responses. Further, we attempted to have questions that could be "matched" to detect inconsistencies in the answers, possibly related to misrepresentations reflecting particular agendas supported by survey participants, as well as to balance responses from respondents with different characteristics (and so, perhaps, opposing agendas). Hence, our discussion of the hypotheses often relies on taking several responses together to form an overall narrative rather than on single responses. We also supplemented the paper version of the survey with telephone contact, which was aimed both at making sure that the survey represented the true rather than the biased views of the respondents, and at clarifying any confusions. Overall, we detected relatively few contradictions that would lead one to be suspicious of the veracity of the answers; however, the responses represent at best the *perceptions* of *some* industry participants, which are undoubtedly colored by the type of organization that the respondent represented and the experience that each organization had in this field. Finally, we collected the data under a confidentiality agreement that allows us only to release aggregated responses in the discussion that follows.

12.4 Our findings

12.4.1 *Industry definition and patents as a measure of market power (hypotheses 1 and 2)*

Our first hypothesis was that GM food represented a different industry from others, specifically from traditional breeding and other biotechnology. We asked respondents how many firms were viewed as important product market competitors and, in a separate question, research competitors. Respondents were also invited, but not obliged, to name competitors in each market. Overall, just under a half (46 percent) cited at least one competitor in the relevant product market. The named competitors were drawn overwhelmingly from the same set of large research and manufacturing firms.[17] This set provided little support for the hypothesis that competitors were drawn from traditional breeders or biotechnology firms in areas other than agricultural applications. Furthermore, those naming competitors listed the same research competitors as product market competitors, suggesting that the "invention" market is comprised of, for all intents and purposes, the same group as the final product market.[18] Indeed, attention seemed to be focused on a core group of large firms (the "hub"), with a fringe of entities (the "spokes"), largely in separate applications, working on development only and not viewing each other as competitors.[19]

Those not listing competitors fell into two groups: those that said definitively that they had no important competitors, and those that did not know the use to which their technology was being put.[20] Interestingly, half the universities named firms as important research

---

[17] Of the competitors that were named, two out of seventeen could be considered small. One of these small firms had significant investment by very large partner firms that were members of the group consistently named by others as the competitor set.

[18] In fact, even those that *did not* name competitors indicated that the research competitors were generally the same as the product market competitors.

[19] Only one firm listed another set of small firms as working in a similar area.

[20] Note that these were all universities or governmental bodies with process technologies on license to other entities. Those not knowing the use to which the patent had been put were not necessarily those charging a zero license fee. One could perhaps infer that these were cases where a standard licensing contract was negotiated, based on a royalty plus a fee that was designed to cover the cost of the research so that the actual use to which the technology was put would not be vital to knowing how to set up the license.

competitors, and those naming firms listed the same group of large firms as the others.[21] This could suggest two things. First, it could suggest that universities could be considered as part of the "spokes" of this industry, contributing technology by license to the core group. Second, it could suggest that this is an area where there is considerable similarity between the research agendas of firms and universities (or government bodies). In fact, while Henderson, Jaffe, and Trajtenberg (1998) find that university research is more "general" and more "important"[22] than corporate research across a wide variety of fields, Conti, Regibeau, and Rockett (2003) find empirically that there is much less difference in the field of genetically modified plants. This survey, then, does not provide evidence to contradict the relative similarity of university and corporate work in the area of GM food.[23]

Our respondents appear to view themselves as truly having some monopoly position. When asked whether the product in which their GM technology was embedded (either produced by themselves or by another firm under license) had a direct product market competitor produced by another entity, the *unanimous* response was in the negative.[24] Clearly, then, each of these respondents derives at least some degree of market power either directly or indirectly from the technology. The research competitors cited on the surveys should, then, be viewed as working in the same *general* research area, but perhaps not developing precisely the same applications as the respondents. It is also likely that, taking these results together, while the precise research results may be unique, at least some members of the hub group have enough know-how to pursue competing research if they so desire. This suggests that, without patent protection, an exclusive hold on the technology would not be secured.

---

[21] Universities also listed the same firms in the product and research markets when they were named, and, even when they were not named, generally noted that the two groups were the same.

[22] For a discussion of these measures and of university patenting, see not just Henderson, Jaffe, and Trajtenberg (1998) but also Trajtenberg, Henderson, and Jaffe (1997).

[23] As an aside, there was no systematic difference between the views of private and public universities in any of our survey responses.

[24] We verified this by requesting an estimate of the price elasticity of demand, obtaining estimates that were consistent with a considerable degree of monopoly power in the output market overall.

We pursued this issue by attempting to determine more directly whether *patents* were the source of this market power. When asked by how much a lack of patentability would affect profits, those companies with products currently on the market responded, on average, that profits would drop less than 10 percent. A problem with interpreting this response, however, is that it does not correct for the importance of the patents: if most patents are, indeed, "unimportant" then the bulk of them would tend to have very little effect on the profits of firms. In fact, the survey response distribution of patent importance – as measured by forward citations by other patents – has a median number of net citations received in the first four years of the patent's life of one (compared to zero for all GM plant patents in the United States through the end of 2000) but none over three, missing the (slightly less than) 10 percent of the (US) GM plant patents that have more than three net citations in the first four years.[25] Hence, while our survey does not reflect a less important set of patents than the broader population (as reflected in the medians), our responses do tend to reflect patents that are "run-of-the-mill" rather than those that are very important, since we do not have any observations from the "right-hand tail" of citations distribution. Still, it is significant that even these "run-of-the-mill" patents can account for a 10 percent effect on profits for this group. We follow up, then, with a question more directed at the role of patents in creating market power by asking by how much the cost and time of imitation increased due to the patent. Here, firms generally responded positively, with an average rise in the cost of imitation of approximately 42 percent and an average increase in the time of imitation of 28 percent.[26] There was wide variance in these numbers, ranging from

[25] As our survey tends to reflect more recently granted patents than the overall population of GM plant patents, we choose to measure importance as the citations during the first four years of life rather than the entire lifetime citation pattern, as this latter measure would systematically favour earlier patents. "Net citations" refers to citations net of self-citations. If the citations were to include self-citations, the point would remain the same, however, as the percentages differ only slightly.

[26] As a comparison, Mansfield (1986) finds that most imitation costs (69 percent of his sample) rise less than 20 percent due to patenting in a study of thirty-three products across a variety of industries. Eighteen percent of his sample have an increase of 100 percent or more. For our sample, and focusing on for-profit firms, we have 35 percent listing no increase in imitation cost, one-quarter listing a 25 percent increase, another quarter listing a 50 percent increase, and 15 percent listing a prohibitive increase. This is a somewhat more favorable

zero to "prohibitively expensive" in terms of time or money. Still, in light of this, perhaps the way to interpret the responses is that the patent is "doing its job" in creating a relatively effective barrier to imitation, but the distribution of patents in our survey reflects the general distribution of patents in that most are, at best, modest sources of revenue. This suggests that patent counts may be used to measure market power, but that they should certainly be weighted by importance if they are to reflect actual or potential commercial strength in the market.

Suppose that we were to implement weighted patent shares, then, as a measure of prospective market power for this industry to evaluate a merger. For example, we could take all patents granted in the industry (or for the crop in question) up to four years before the proposed merger date and then track the citations measured during the first four years of the life of each of those patents (relative to the total population of patents that might have cited it) as a measure of its importance. Each merging entity's market share would, then, be reflected in its share of the total patents, with each patent weighted by its importance relative to the average importance for the entire population. As an example, if we were to do this as of 2000 we would obtain a prospective estimate of US market shares – based on US patents – for corn of only 23 percent for Monsanto when we do not apply the weighting but 43 percent when the weighting is applied. For soy, we obtain figures of 17 percent for Monsanto when we do not apply the weighting and 28 percent when we do.[27] Clearly, correcting for importance makes a huge difference to these figures.[28]

Does this represent an accurate prospective measure of actual market shares? In order to establish this, we need to let some time pass. While it is a little difficult to get agreement as to Monsanto's US market share and how to measure it,[29] a rough estimate is that Monsanto's current

distribution than Mansfield's, but holds for an industry where we might expect patents to be more powerful.

[27] This calculation is based only on patents that specifically make claims applying to the crop in question.

[28] Normally, one would want to take into account, in a weighting measure, the expiration date of patents. In the case at hand, however, the patents are sufficiently recent for their expiration dates not to be a major factor entering into the weighting.

[29] Market shares can be calculated by trait or by seed. The corn figure is the percentage of the genetically modified crop planted accounted for by

actual market shares in the United States for corn and soy are, respectively, 49 percent (up very slightly from 43 percent each year since 2000) and 90 percent (down very slightly from virtually the entire market each year from 2000). These are both higher than our "prospective" measures, above. While the corn figure is not very different from the weighted patent estimate, the soy figure is quite far off. The explanation for this large difference may be the strength of certain soy patents that are underestimated by the citations count: some patents now owned by Monsanto underlie almost any GM soy currently grown. In other words, to produce in this field, a license of some basic technology is very important, leading to the high share of Monsanto technology in this crop. Subsequent research (which would be reflected in the citations count) has not yet done justice to the importance of the Monsanto patent portfolio in the current soy market. This would generally be true of citation-based measures; extremes of market importance would be relatively poorly measured even by weighted citations.[30]

Hence, the corn and soy figures illustrate an advantage and a disadvantage of taking the innovation market information into account as a prospective market share measure: when technologies are quite powerful, their influence may be underestimated in weighted patent counts; on the other hand, in the moderate ranges, the measure can perform relatively well. Still, if we take the patent count calculation seriously, it would suggest several things. First, in terms of soy, one

Monsanto's brand of GM seed, including licensing. The second is the percentage of U.S. planted acreage with Monsanto traits, even though the seed may not be sold by Monsanto. Clearly, this is heavily influenced by licensing activity as well. Even these estimates vary substantially depending on the source, however, leading to their designation as "rough." Estimates of market shares can be found in Casale (2004) and McMahon (2004).

[30] Figures for other firms in corn are: 13 percent, weighted and unweighted, for Du Pont; 15 percent, weighted and unweighted, for AgrEvo; and 4 percent weighted, but 17 percent unweighted, for Syngenta. Dow's market share, weighted and unweighted, is negligible based on IP measures. This can be compared to reported market shares in corn of 40 percent for Du Pont (through Pioneer) and 15 percent for Syngenta. (Others were unavailable.) The Du Pont figure, unfortunately, includes a non-GM component, and so is not directly comparable to the IP shares. For soy, the only other company to hold a significant patent portfolio is Du Pont, with an unweighted and weighted share of 33 percent. All figures are computed using the most recent acquisitions in order to attribute patents to firms, but do not take into account any other marketing arrangements between firms.

would be concerned about increased power by a company such as Monsanto in the seed industry, as a powerful patent position in seed-making technology as well as seed production facilities would raise the possibility of foreclosure.[31] This supports some recent decisions to enforce the licensing of technology as being a part of acquisitions in this field (see Hayenga, 2003, for a summary of some recent decisions). Second, in terms of corn, our "prospective" measure is quite close to the actual market share measure, suggesting that no near-term changes would normally be expected based on the measured technology position. Still, the absolute measure is quite high, suggesting that further consolidation might be a subject of concern. For soy, the intellectual property is not nearly as concentrated as the market, suggesting that concentration figures should drop from their current extremes.

Hence, we have found support for relatively high concentration in this industry and for the use of weighted patent counts as a prospective measure of market power. We have also found support for a "hub and spoke" pattern, with a relatively small number of large firms being the product and research market focus, and many smaller design firms providing technology to the large firms but not competing head to head with each other. Now we turn, in the discussion of the subsequent hypotheses, to the roles of regulatory approval, patents, and litigation in generating this structure.

12.4.2 *The role of regulation (hypotheses 3 and 4)*

We asked several questions to investigate the role of regulation in this industry, and its effect on competition (specifically, on industry concentration). First, we asked entities to map out the regulatory process to which they had had to submit their GM technologies. Next, we asked about the level of costs (relative to profits) of this procedure, the relative burden that regulation imposed in terms of increasing the time to market of the technology, the change in cost of obtaining regulatory approval as the number of submissions increased, and the effect of a change in regulatory approval cost on licensing strategy.

---

[31] Monsanto has acquired a number of seed producers in recent years. Most recently, it has acquired Seminis (February 2005). Seminis does not currently have a strong position in soy, however.

Most entities in our sample were not directly involved in regulatory approval: only 15 percent of our respondents had, themselves, submitted output to a regulatory process beyond patenting. While this might appear surprising in such a highly regulated area, it has ready explanations. First, some entities stated that they were participating in process technologies (such as technologies to induce a certain trait in any plant), so that regulation applied to the plant that was modified rather than to the process itself. Others (a further 15 percent of the sample) were using the patents in question merely to strengthen existing patent portfolios rather than to contribute directly to the creation of a product, so regulatory approval for output based on the patent was not being contemplated.

More importantly, three-quarters (77 percent) of the sample were obtaining revenues from the patents by means of licensing agreements. Many licensors commented that all regulatory costs were handled by the licensee and, as such, regulatory costs were not of direct concern to them. Hence, delegating regulatory approval by means of the license was a regular practice in our sample. We pursued this by asking about the effects that a change in regulatory cost would have on licensing strategy. Some, including both firms and other entities, indicated that they had no realistic alternative to licensing; manufacturing was not an option for them. Hence, they claimed, whatever the cost of regulatory approval, they would continue to license equally intensively as long as their research costs were covered. This says two things. First, the design firms were delegating both regulatory and manufacturing costs through the license, seemingly considering both a barrier to entry. More interestingly, it appears that the main point of the license was to recoup the cost of research, leaving a large enough surplus to the licensee that even further increases in regulatory burden would not be likely to affect the propensity to license, even though it might affect licensee margins.

The regulation that applied to the firms that actually had participated directly in regulatory approval varied substantially, with some reporting a very light and short burden, and others reporting a complex process taking up to ten years. Similarly, the reduction in profits due to the regulatory process varied from (virtually) zero to 75 percent, with the higher figure associated with a complex and lengthy procedure involving approval by a variety of agencies. Hence, the degree of potential entry deterrence that regulation creates was very uneven across our sample, when one considers those firms that actually submitted products to regulators. Not surprisingly, the degree of reported regulatory burden was correlated with the response to the question of how much a 10 percent increase in regulatory cost would affect licensing activity: the higher the entity's reported regulatory burden, the larger the increase in licensing out that the entity said that it would conduct in response to a further increase in regulatory cost. Similarly, a larger propensity to license in was reported by the larger firms in the sample as a response to larger regulatory cost. This lends support to the comments made by current licensors that licensing is a standard tool to delegate regulatory approval in this industry, resulting in increasing concentration at the approval stage as the regulatory burden increases.

In terms of learning effects, the responses that cited multiple use of the regulatory system reported relatively fast learning, with all learning reported as having occurred after a single regulatory experience, and a percentage reduction in the cost of regulatory approval upon increased submissions that was relatively modest (on the order of 10 percent). While this would reflect a relatively modest learning curve, it is the response of entities that actually chose to submit products themselves; the firms that perceived the process as too onerous – and as a result licensed out – would not have reported a response to this or other regulatory questions. Still, this response does not lend strong support to the need to help firms *learn* about the regulatory process in order to promote entry.

In sum, there was some support for the hypothesis that regulation is a concentrating influence in this industry, and that it may discourage entry into the approval stage even if it does not hamper entry into generating new technologies at the research stage. Learning effects, while they are present, appear to be relatively mild and to occur with relatively few submissions. Hence, we appear to have some support for regulation's playing a role in generating a two-tier structure in this industry for those technologies subject to a large regulatory hurdle, but also a suggestion that this is *not* a short-term phenomenon that could be overcome by sponsored learning.

12.4.3 Patent policy in GM (hypotheses 5, 6, and 7)

We asked several questions to elicit the role that patents were playing in this industry. These included questions about whether patents were a way of stimulating research in this field, the precise nature of the protection that patents afforded the researchers, the opinion of the respondents on whether genes alone should be protected, and, more concretely, whether the respondents would change their research plans if patents on genes alone were made available. Conversely, we also asked whether the lack of patents on GM technology would decrease research in this area.

The overwhelming majority of respondents, firms and universities included, thought that their research spending in this area would not have decreased if patents had been *un*available on GM innovations. Unsurprisingly, no universities found patenting to be relevant to their research decision. More surprisingly, few of the companies did. Even when we looked only at the firms that were continuing to receive patents regularly in this area (in other words, those that listed further patents in the area of GM food that they had received since the end of 2000), the responses did not give a greater weight to patents in terms of stimulating research. One reason for this is that some patentholders – including firms – reported receiving compensation in ways other than direct patent "rewards." Specifically, research funds received from an external source before research was undertaken were cited in several instances.[32] As long as research costs were covered, these entities noted that this funding determined whether the research occurred and what the subject of the research was: the patent position did not have a *direct* effect. For these firms, any income from licensing the patent was viewed as "icing on the cake," but was not determinate. In this sense, the design firms seemed "one step removed" from any incentives that patents might create. A second reason was that just over a half of the companies commented in the margin that GM was not their main line of research. Instead, one can infer from the comments that the GM patents were either offshoots of other research programs or a way of keeping abreast of a new developing field, but not a main, strategic area of development. In this sense, patentable innovations in this area were "fortuitous" for these firms, rather than planned revenue generators: they contributed to the existence of the firm, but their revenues were not determinate to its research focus.

---

[32] These included direct grants and general corporate funding (such as venture capital).

In order to verify the role of patents, we asked several more questions regarding the reasons for patents to have value in this area. While universities continued to view patents as unimportant to their research agendas, almost a half of the relevant firms reported that profits would fall if GM innovations were *not* patentable. In other words, while the research would have occurred in the absence of patents, the patents were still, in fact, contributing currently to profits. This implies that patents must be raising the overall returns to the entity, and so increasing the returns to research as an activity, even if they are not the reason that particular research trajectories are chosen. Hence, we received some support for the view that patents do create a "reward," but that this reward is not what determined whether the research supporting these patents had been chosen.

As mentioned earlier, patents were viewed as contributing significantly to increased imitation cost or time by our respondents. One interpretation of this apparent imitation barrier is that patents are functioning as relatively effective property rights, mapping out areas of exclusivity for their owners. In turn, this implies that the role of the patent in increasing the incentives to disclose research progress may be important. For some of our sample the disclosure function was noted explicitly as the reason to patent: universities and non-profits stated that the patent was being used as a way of "getting information out." Further, these entities indicated that their answers to the effect on imitation cost and time were affected by their eagerness to disclose: they felt that the patent would not increase imitation cost partly because they intended to *facilitate* future research in the area. For firms, the imitation cost and time effect was much higher, suggesting that the patents' exclusionary rights potentially had "bite" in terms of increasing appropriability even if universities and non-profits did not choose to take advantage of this.[33] Interestingly, this disclosure function of IP protection has received little attention in the economics literature, despite the weight given to it by this population.

Finally, we asked all respondents to give an opinion on whether patents in this area should be available on genes alone, and, if they

---

[33] Perhaps surprisingly, the one group that appeared to regard patents as an important means of eliciting research in this area was the governmental category. On the other hand, other responses indicated that the income from the patents was not a primary concern to this group. Perhaps this has something to do with the compensation systems in these agencies.

were, what difference this would make to their research. Consistent with our other responses on the effect of patents on research trajectory, we obtained exclusively negative responses to the first question and only a single positive answer to the second. As there are ethical and legal ramifications to granting patents on genes alone, these responses may have been motivated by purely non-economic considerations. Unfortunately, no respondents chose to elaborate on their reasons. Still, we are left to conclude that our survey indicates no support for the view that change in patent "breadth" would have an effect on the direction of research, as reported by those currently in the field.[34]

Our conclusions regarding patent policy from this section are, then, mixed. While patents clearly play a role in generating revenues for the entities involved, it is not clear that these revenues are the reason that the specific research occurred in the first place, even for private firms and those continuing to work in the field. Hence, while a role for the "reward" function of patents exists for the sample in that profits would generally fall in the absence of patents, the more salient function of patents that came through in the survey was to create property rights that effectively raised the barrier to imitation. In this sense, the relatively under-studied role of patents in encouraging the disclosure of results came through strongly in the sample. Perhaps the responses reflected the view of many of our respondents that they were "one step removed" from the patent system as a means of generating research. Consistent with other responses on the lack of patentability's effect on research priorities, all but one of our respondents felt that pure gene patents would make little difference to the type of research they are doing.

12.4.4 The roles of litigation and licensing (hypotheses 8 and 9)

A final area of concern was the enforcement of patent rights, and the role of litigation in IP management. First, we investigated the views of participants on the cost burden of litigation. Second, we explored

---

[34] A single firm did stand out by responding positively to this question. This firm also viewed patents as more important in directing research in general than other respondents.

strategies of reducing or avoiding this cost. While we asked directly about licensing as a strategy for avoiding litigation, we also obtained confirming evidence (to be discussed below) that licensing was a rather standard response to potential or actual litigation. In order to obtain information about the relative bargaining positions of the firms performing the licensing, we asked about the percentage of profits that the licensor attempted to recover from any licensing out and the structure of the licensing contracts.

We had understood before starting our work that licensing activity was extremely high in GM food. As we mentioned earlier, this was confirmed in our sample. For our participants, licensing contracts have primarily an up-front fixed fee plus royalty structure, although a minority of our respondents (27 percent) had a percentage of their earnings coming from either pure royalty or pure fixed-fee contracts. On average, just over a half (51 percent) of the contracts issued by these firms were exclusive. Covering the cost of patenting was noted by several respondents as the reason for the fixed fee. Finally, the percentage of profits that entities attempted to recover was either quite high for the for-profit firms (averaging 62 percent) or zero for the non-profits, universities, and other entities. In the latter case, presumably licensing attempted to cover the costs of the patent only, even when a royalty was used as part of the contract.

Based on our survey, it appears that licensing plays a series of roles in this industry. As mentioned earlier, it is clearly a major source of revenue for a number of our respondents, constituting 100 percent of revenues in a number of cases. Further, these respondents anticipated a significant share (more than 80 percent) of the profits generated by the technology, indicating a relatively strong bargaining position in this industry for licensors. Licensing also plays a role in avoiding litigation, however, ranking as the strategy of choice to respond to litigation for 42 percent of our respondents. Cross-licensing came in as the strategy of choice to avoid litigation for 30 percent of our respondents. Six respondents cited various other strategies, including conducting new R&D and threatening litigation to avoid litigation. When asked directly whether a rise in the cost of enforcing the patent would result in an increase in licensing activity, 42 percent responded positively. This appears to reflect a strong view that licensing is a standard response to litigation, and, further, that the higher the litigation costs the more licensing (out and in) one might expect.

The ranking of cross-licensing and licensing to avoid litigation did not appear to depend on the size of the entity, with small entities actually listing cross-licensing as the preferred response more frequently than large entities. The larger firms had far more litigation experience surrounding the patents in question in our survey, although they also tended to have older patents so that their exposure per patent was longer. Further, it is not the more important patents in the sample that are the more litigated patents. In fact, the more litigated ones tend to be somewhat less important than the average[35] and have almost exactly the mean number of backwards citations. Hence, one cannot conclude that our sample represents a case where more important patents, patents with a longer genealogy, or patents belonging to smaller entities tend to attract more litigation. In this sense, our sample does not reflect the concerns of Lanjouw and Schankerman that certain types of patents or patentholders tend to attract litigation.

Several comments are in order here, however. First, our sample does not include individuals. These are contrasted with corporations in the Lanjouw and Schankerman paper to make the point that smaller entities may be involved more frequently in litigation as opposed to licensing arrangements. Our responses would suggest that perhaps this sort of bias may apply more to individuals than to even quite small corporations, as even the small firms in our sample appeared to be "playing the licensing game" as much as the big players. On the other hand, and more in line with Lanjouw and Schankerman's work, it was exclusively small entities in our sample that listed litigation as a response to the threat of litigation, albeit sometimes qualified by specifying that a partner would have to be found in order to bring litigation.

The fact that litigation and cross-licensing are both listed relatively frequently as preferred strategies for smaller firms leaves us with something of a puzzle, for which our survey does not provide a ready answer. The explanation for this may be that size in terms of employees does not necessarily correspond with size in terms of stock of technology. It could be that those that have little to trade are those volunteering litigation while those that have much to trade are not. It could

---

[35] Again, importance is measured by net forward citations in the first four years of the patent's life, corrected for the size of the potential citing population at the time of grant.

also suggest that litigation and cross-licensing are viewed as complements, by some of these respondents, rather than as substitutes.
Clearly, this is a very different conception of the relation between
these two instruments. It may also be that cross-licensing has a cost
that is not much lower than that of litigation when one takes into
account the implicit cost of licensing when it was *not* optimal to do
so in the absence of the possibility of litigation. In this sense, cross-
licensing and litigation could be alternatives between which some
respondents were indifferent. The indifference could lead to the ambiguous ranking of these alternatives. Third, our smaller entities appeared
to be considering a larger list of alternatives than just litigation and
cross-licensing alone: partnering along with litigation may be one way
of addressing the Lanjouw–Schankerman concerns in a highly litigious
industry. In fact, Harhoff et al. (2001) find that some litigation is more
broadly based in GM food than in other parts of biotechnology.
Finally, to the extent that higher litigation rates in this industry reflect
less selection about who gets litigated, it may simply be that there is less
selection of the type that Lanjouw and Schankerman observe occurring
for this group. In short, our responses indicated that further study of
this issue would be important, as the interaction between licensing and
litigation could be fairly complex.

Our respondents stated *unanimously* that a change in litigation cost
would not affect research focus. The one that indicated that a rise in
litigation cost would affect research indicated that it would need to
work harder to create a stronger patent thicket, not that it would
change research focus. In this sense, we received no confirmation in
our sample for Lerner's theory of a link between litigation cost and
research trajectory. This is somewhat surprising given the litigiousness
of the area. Further, for those respondents listing infringement or
invalidity suits as part of the total cost of maintaining their patents in
this area, these costs averaged a significant 20 percent of the total costs
of patenting (including research costs). In addition, those firms that
noted that they no longer worked in the GM food area were exclusively
large, whereas those that noted that they were continuing to work in
the area were exclusively small, with a mix of litigation experience in
both groups. This is not consistent with high litigation cost driving
firms into other areas of research: it should be the small firms (those
with a high cost of litigation) that are leaving. There may be an
explanation for this that is consistent with Lerner's results, however.

As was mentioned above, our respondents reported a strong linkage between litigation cost and licensing activity. In fact, a number of respondents specifically noted that licensees were handling all the litigation costs. As our section on hypotheses stated, once licensing is introduced, the research–litigation link can be broken. In fact, this may be precisely what is occurring for this sample.

Hence, it appears that licensing is used to delegate many costs in this industry: litigation costs, regulatory costs, and manufacturing costs. While this has implications for a relatively concentrated "hub" at the center of the industry, it also implies that firms that view themselves as "inefficient" at litigation or regulatory approval can participate at the research stage and license the technology rather than avoid the research area altogether.

12.5 Conclusions

Our survey responses have suggested several conclusions about the GM food industry. First, the industry appears to be separate and highly concentrated, comprising, on the one hand, a small hub that conducts research, regulatory approval, and manufacturing and, on the other, a large number of spokes focusing on technology provision to the hub. The concentration levels for both innovation and current market provision of approved GM food crops appears high compared to normal triggers for the scrutiny of further merger activity. Licensing is undertaken to delegate regulatory and litigation costs to the hub, resulting in the "two-tier" structure of the industry. Regulatory costs appear to be barriers to entry to the hub, but not to be primarily due to learning economies. In this sense, they are persistent concentrating influences, not likely to be largely affected by sponsored learning for industry participants. Patents, while clearly generating income for industry participants, do not appear to be directing research through their reward function. In particular, we did not receive much support for the idea that targeted changes in patent scope – and, in particular, patents on pure genes – would affect research trajectories in this field. Finally, litigation appears not to affect research trajectory heavily in our sample, perhaps due to the interaction between litigation, research, and licensing strategies. Clearly, as our sample is small and not necessarily random, all these conclusions must be qualified as suggestive rather than definitive.

Our responses also suggested several areas for future work. First, weighted patent counts fared relatively well to predict moderate concentration levels, but fared less well in predicting the extremes of concentration for this industry. The contribution of weighted patent counts to the implementation of the innovation market concept to measure prospective concentration in an industry could well be evaluated across a wider set of industries and a wider time period. Second, our responses point clearly to an important interaction between licensing and litigation. The nature of this interaction is not, however, completely clear. These alternatives are not necessarily substitutes in all cases, licensing is not necessarily the only alternative considered to litigation, and the implicit cost of licensing compared to that of litigation is also difficult to judge. Third, our responses pointed to a relatively important role in intellectual property strategy for disclosure, and perhaps less emphasis on reward. This bears more investigation, as the disclosure function is the less studied of these two functions of patents.

References

Bessen, J., and E. Maskin (2000), *Sequential Innovation, Patents, and Imitation*, Working Paper no. 00–01 Massachusetts Institute of Technology, Cambridge, MA.

Casale, C. (2004), *Monsanto: Momentum in the Field*, available at http//:www.Monsanto.com/Monsanto/content/investor/financial/presentations/2004/monmouth2.pdf [accessed 12 January 2005].

Cohen, W. M., R. R. Nelson, and J. P. Walsh (2000), *Protecting Their Intellectual Assets: Appropriability Conditions and Why U.S. Manufacturing Firms Patent (or Not)*, Working Paper no. 7552, National Bureau of Economic Research, Cambridge, MA.

Competition Commission (2003), *Merger References: Competition Commission Guidelines*, Her Majesty's Stationery office, London.

Conti, M., P. Regibeau, and K. Rockett (2003), *How Basic is Patented University Research? The Case of GM Crops*, Discussion Paper no. 558, University of Essex, Colchester.

Fernandez-Cornejo, J., and W. McBride (2002), *Adoption of Bioengineered Crops*, Agricultural Economic Report no. AER810, Economic Research Service, United States Department of Agriculture, Washington, DC.

Hall, B., A. Jaffe, and M. Trajtenberg (2005), "Market value and patent citations," *RAND Journal of Economics*, 36 (1), 16–38.

Hall, B., and R. Ziedonis (2001), "The patent paradox revisited: an empirical study of patenting in the U.S. semiconductor industry 1979–1995," *RAND Journal of Economics*, 32 (1), 101–28.

Harhoff, D., P. Regibeau, and K. Rockett (2001), "Some simple economics of GM food," *Economic Policy*, 33, 265–99.

Hayenga, M. (2003), "Structural change in the biotech seed and chemical industrial complex," *AgBioForum: The Journal of Agrobiotechnology Management and Economics*, 1 (2), 43–55.

Henderson, R., A. Jaffe, and M. Trajtenberg (1998), "Universities as a source of commercial technology: a detailed analysis of university patenting, 1965–1988," *Review of Economics and Statistics*, 80, 119–27.

Lanjouw, J., and M. Schankerman (2001), "Characteristics of patent litigation: a window on competition," *RAND Journal of Economics*, 32 (1), 129–51.

(2004), "Patent quality and research productivity: measuring innovation with multiple indicators," *Economic Journal*, 114, 441–65.

Lerner, J. (1995), "Patenting in the shadow of competitors," *Journal of Law and Economics*, 38 (2), 463–96.

Levin, R., A. Klevorick, R. R. Nelson, and S. G. Winter (1987), "Appropriating the returns from industrial R&D," *Brookings Papers on Economic Activity*, 3, 783–820.

Mansfield, E. (1986), "Patents and innovation: an empirical study," *Management Science*, 32, 173–81.

Matutes, C., P. Regibeau, and K. Rockett (1996), "Optimal patent design and the diffusion of innovations," *RAND Journal of Economics*, 27 (1), 60–83.

McMahon, K. (2004), *Seed Tech Titans*, available at http://apply-mag.com/mag/farming_seed_tech_titans_2/ [accessed 12 January 2005].

Regibeau, P., and K. Rockett (2004), *Are More Important Patents Approved More Slowly and Should They Be?*, Mimeo, University of Essex, Colchester.

(2005), *Competition, Regulation, and Intellectual Property Management in Genetically Modified Foods: Evidence from Survey Data*, Discussion Paper no. 591, University of Essex, Colchester.

Sakakibara, M., and L. Branstetter (2001), "Do stronger patents induce more innovation? Evidence from the 1988 Japanese patent law reforms," *RAND Journal of Economics*, 32 (1), 77–100.

Scotchmer, S. (2004), *Innovation and Incentives*, MIT Press, Cambridge, MA.

Scotchmer, S., and J. Green (1990), "Novelty and disclosure in patent law," *RAND Journal of Economics*, 21 (1), 131–46.

Trajtenberg, M. (1990), "A penny for your quotes: patent citations and the value of innovations," *RAND Journal of Economics*, 20 (1), 172–87.

Trajtenberg, M., R. Henderson, and A. Jaffe (1997), "University versus corporate patents: a window on the basicness of invention," *Economics of Innovation and New Technology*, 5, 19–50.

Ziedonis, R. (2000), *Patent Protection and Firm Strategy in the Semiconductor Industry*, Ph.D. dissertation, Walter A. Haas School of Business, University of California, Berkeley.

13 | Governance, policy, and industry strategies: pharmaceuticals and agro-biotechnology

JOYCE TAIT, JOANNA CHATAWAY,
CATHERINE LYALL, AND
DAVID WIELD

13.1 New approaches to the governance of innovation

The emergence of new fundamental knowledge in life sciences is leading to new, "breakthrough" technologies, completely new product ranges, and new innovation trajectories. Some breakthrough technologies are potentially disruptive (Spinardi and Williams, 2005), in the sense that they step outside existing paradigms, requiring a major shift in product types and in their place in the market. Examples include GM crops in agro-biotechnology, and stem cells and pharmaco-genetics in the pharmaceutical industry. The shift of emphasis from innovation based on chemical knowledge to biotechnology-based pathways creates turbulence in companies' internal product development strategies, changes the balance of competitiveness among companies, and opens up new areas of regulatory uncertainty. At the same time, governance processes are becoming more complex and are placing greater constraints and uncertainties on companies that have made major investments in the development of new types of product over long periods of time.

This chapter is based on a series of research projects[1] we have conducted to study innovation in pharmaceutical and agro-biotechnology companies and the interactions between:

- science, technology, and innovation strategies in multinational companies (MNCs);

[1] Policy Influences on Technology for Agriculture (PITA), European Union 4th Framework Programme, Targeted Socio-Economic Research Programme (TSER) (http://technology.open.ac.uk/cts/pita/); the SUPRA seminars: Best Practice in Evidence-Based Policy, UK Economic and Social Research Council (http://www.supra.ed.ac.uk); *New Modes of Governance*, eds. C. Lyall and J. Tait (eds.), (2005) Ashgate Publishing; Innogen: Centre for Social and Economic Research on Innovation in Genomics (ESRC) (http://www.innogen.ac.uk).

- policy development, risk regulation, and governance; and
- public and stakeholder attitudes and concerns.

The background for this research has been the emergence of new governance structures and policy processes in Europe and North America (Giddens, 1999; Cabinet Office, 1999a; Commission of the European Communities, 2001). This change has involved a move away from the previous *government* agenda (a top–down legislative approach led by the institutions of the state, which tends to regulate the behavior of people and institutions in quite detailed and compartmentalized ways) towards *governance* (which attempts to set the parameters of the system within which people and institutions behave, with the assumption that incentives and self-regulation can achieve the desired outcomes) (Lyall and Tait, 2005).

Globalization is one of the main drivers of these new policy approaches, and the resulting rapidity of technological change presents challenges for the promotion of science and innovation and for its management and control. We argue that medical and agricultural developments such as stem cells, genetic data banks, and GM crops are evolving more rapidly than the relevant policy and regulatory systems, and that new types of product are crossing the boundaries of traditional regulatory systems. At the same time, the increasingly globally based organization of research and innovation, and the truly global reach of the companies that contribute both to scientific discoveries and to the commercial exploitation of this knowledge, are adding to the pressures for new modes of governance for science and technology.

The development of this new governance agenda has been debated in various academic disciplines, including institutional economics, international relations, organizational studies, development studies, political science, and public administration (Stoker, 1998). Sloat (2002) has noted that the term "governance" seems to be applied to "everything from corporations to rural society," but most commentators accept that "governance" is no longer a synonym for "government," the significant difference being in the processes adopted for achieving policy aims.

This policy revolution involves concepts such as policy integration ("joined-up" policy-making), an emphasis on "evidence-based" policy, the use of standards and guidelines linked to policy evaluation, the encouragement of openness, stakeholder and public engagement and consultation, and the avoidance of unnecessary regulatory burdens for industry. In theory, this should encourage a much freer flow of ideas across governments and government departments and from one level of government to another, focusing on ideas that can contribute to an effective system of governance rather than on the ideology that generated the ideas, again demonstrating the need for better integration across policy domains.

These ideas have been explored and tested much more thoroughly in policy contexts such as social security and education. However, they are increasingly influential in the governance of science and innovation, although in these cases there is much less guidance for, and experience of, their effective implementation (Lyall and Tait, 2005). The following sections describe the aspects of the new governance approach that are most relevant to innovation in life sciences.

13.1.1 Networks and stakeholder engagement

The increased role for non-government actors in policy-making, often under the heading of public or stakeholder engagement, implies an increasingly complex set of state–society relationships dominated by networks rather than hierarchies, so that the role of government is increasingly one of coordination and steering (Bache, 2003). The boundaries between and within public and private sectors have become blurred, and the role of the state has changed from being the main provider of policy to one of facilitating interactions across a wide range of interests. This perspective thus focuses on the coordination of multiple actors and institutions to debate, define, and achieve policy goals in complex political arenas (Sloat, 2002).

Highlighting collaborative public–private efforts and cooperative rather than adversarial policy strategies implies that there is now less reliance on coercive policy instruments and more on subtle techniques involving a restructuring of state institutions, creating new institutional forms that operate at a distance from control by the political elite.

However, this is one area where tensions are emerging in the translation of new governance ideas to industry sectors such as pharmaceuticals and agro-biotechnology. Top–down, *government*-style regulation is still necessary in such cases to control the safety and efficacy of new products, and there is little guidance yet on how to merge this requirement with the more horizontal *governance* tradition without creating tensions among stakeholders and, in some cases, unrealistic expectations.

13.1.2 *Integrated policy approaches*

The emphasis on integrated policy by the UK government is reflected in the "Modernising Government" initiative (Cabinet Office, 2000, 1999b), with the aim to improve policy formulation and implementation in areas that cut across the policy boundaries of traditional government departments. This requirement comes partly from a perceived need to remove the contradictions and inefficiencies caused when policies or regulations emerging from different government departments or different levels of government (regional, national, international) are inconsistent or provide incompatible signals to policy targets. Policy integration is also needed to deal with the complexity and uncertainty associated with many decisions concerning innovation.

Policy integration across functional boundaries can take place vertically (linking international/national or national/regional levels of government) and/or horizontally (linking different functional departments at the same governance level – for example, those responsible for the promotion and for the regulation of innovation). Generally, vertical integration is easier to achieve than horizontal integration because of the still-hierarchical relationship among policy players. Horizontally, policy approaches for innovation are compartmentalized in different departments and agencies that compete for power rather than cooperate to tackle policy issues (Cooke, Boekholt, and Todtling, 2000, p. 142).

Effective policy integration would imply that science-related policies ought to be crucial components of new governance initiatives, but we find little evidence of this, and the "Modernising Government" agenda concentrates almost entirely on social policy (social welfare, crime, health, and education). However, there have been attempts to provide an integrated framework of tax, subsidy, regulation, trade, patent and regional policy for the development of science-based industries in the United Kingdom, particularly in respect of the horizontal coordination of agencies' priorities (DTI, 1998).

13.1.3 *Governance and the evidence base*

One of the key requirements of the new governance approach is that policies or regulatory actions should be evidence-based, in several senses of the term. They should be based on evidence that policy intervention is required in a particular area, the specific policy instruments adopted and the threshold levels for regulatory standards should be based on evidence that they are likely to work as intended, and monitoring should be undertaken to check that policies are indeed working as intended. These points are reflected in the focus of modern governance on "what works" (Davies, Nutley, and Smith, 2000).

However, as the degree of stakeholder and public engagement in policy development increases, as outlined above, the evidence base for new policy and regulatory decisions is more likely to be contested by some stakeholder groups, increasing the timescale and the cost of policy-making, and in extreme cases – such as GM crops in Europe – making particular products ungovernable, at least in the short term. There are, therefore, unresolved and generally unacknowledged tensions in the expectation that policy-makers will simultaneously engage with a wider range of stakeholders and also increasingly base their decisions on evidence rather than advocacy.

Command-and-control-style govern*ment*, backed up by sanctions and penalties, will continue to be needed to regulate the safety and reliability of a wide range of products, processes, and facilities. Nevertheless, a common tactic among the diverse groups and networks of public stakeholders that tend to engage with policy decisions on drugs, pesticides, and new life-science-based products is to challenge data put forward as evidence, and for each "side" in a controversy to give selective attention to evidence that fits with its own objectives. Such challenges downgrade the value of research findings as evidence to support decision-making, and policy-makers are finding it increasingly difficult to reach decisions that are widely accepted.

There has always been uncertainty surrounding the evidence used as a basis for decision-making for science and innovation, but under the "government"-based approach this was often resolved only after a product had been approved and on the market for a period of time. In the case of pesticides, defects that emerge after a product has been in use for some time have been dealt with by withdrawing the product and replacing it with an alternative, or by altering its formulation or pattern of use (Tait and Levidow, 1992). This has allowed a great deal of scientific, organizational, and policy learning about products, their regulation, and their safe use. Similarly, the emergence of evidence about side effects in approved drugs has, until recently, been treated as an inevitable part of the learning process in the long-term trajectory of the development of better, more effective drugs.

Increasingly in such cases, public stakeholder groups invoke the precautionary principle in an attempt to ensure that potentially defective products do not reach the market place and so are not tested in use. They also tend to challenge the evidence put forward by companies in support of product registration, often imputing unethical motivations. Indeed, many of the complicating factors involved in the application of the governance agenda to innovative life science products arise from complex interactions between the still-necessary govern*ment*-based regulation and control and the main themes of the govern*ance* agenda, particularly the contrasting and sometimes incompatible requirements for policy decisions to be evidence-based and, at the same time, to involve a greater degree of stakeholder engagement.

13.2 Categorizing policy and regulatory signals

The pharmaceutical and agro-biotechnology industry sectors, although increasingly subject to more governance-based influences, still operate against a background of intense and very restrictive regulation and control at national and global levels. In our research on these industry sectors, we have proposed a range of different policy categories as a basis for analyzing the effectiveness of different policy approaches and their impacts on policy targets in different contexts. Elsewhere (Lyall and Tait, 2005), we have discussed the development of governance approaches to science, technology, and innovation from the perspectives of policy-makers and public groups rather than industry and scientists.

All classifications reflect the purposes and perspectives of those who develop them, and our approach here derives from a systemic, interdisciplinary perspective on policy analysis with the purpose of contributing to the understanding and improvement of regulatory processes as they impact on their target groups.

It is not appropriate to attempt to develop a single set of policy categories that would encapsulate all analytical needs and cover all the circumstances relevant to the life science industries. This would have required oversimplification of our data and premature loss of the richness of understanding of the industry and its management contained within them. The approaches outlined here reflect the interactions between a traditional govern*ment* approach and the new govern*ance* agenda, among policy-makers, innovation communities, the markets for products, and public interest groups.

13.2.1 Enabling, constraining, discriminating, and indiscriminate policy instruments

Bemelmans-Videc, Rist, and Vedung (2003, p. 250) have developed an approach to regulatory policy that classifies instruments according to their mode of operation and also the nature of their impact on the policy targets, where regulations are described as "sticks" and economic instruments as "carrots" (figure 13.1).

Superficially, this approach seems relevant to the government/ governance distinction among policy types, with regulation (sticks) corresponding to government-based approaches and economic means (carrots) being relevant to the governance agenda. However, there is some semantic confusion in this framework, in that, as figure 13.1 shows, both regulations and economic means are described as being either "affirmative" or "negative," implying that each can act as either carrots or sticks.

In our research on agro-biotechnology companies (Tait, Chataway, and Wield, 2000), managers described their responses to different regulatory systems in Europe and the United States in a way that implied the need for a more subtle categorization of regulatory instruments than the above, to recognize the continuing need to regulate pesticides and drugs *as products*, but at the same time to guide policy-makers and regulators towards a more effective, more governance-based approach to the design of policy instruments.

We are proposing an alternative categorization of policy instruments that reflects these perspectives from company managers in responding to policy and regulatory initiatives, emphasizing and supporting the

Policy instruments	Affirmative and negative variants
Regulation (the stick)	Affirmative (prescriptions)
	Negative (proscriptions)
Economic means (the carrot)	Affirmative (subsidies, grants, in-kind services)
	Negative (taxes, fees, physical obstacles)

Figure 13.1. Policy carrots and sticks
Source: Adapted from Bemelmans-Videc, Rist, and Vedung (2003).

	Enabling	Constraining
Discriminating	**Box 1** US Food Quality Protection Act (FQPA), 1996 Pesticide Review (EU Pesticide Directive 91/414/EEC) Common Agricultural Policy (CAP) reform	**Box 2** Pesticide Registration and Review (EU Pesticide Directive 91/414/EEC) US FQPA
Indiscriminate	**Box 3** EU CAP (in the past)	**Box 4** EU Drinking Water Directive (80/778/EEC) CAP reform

Figure 13.2. Pesticide policy dimensions from the perspective of the agrochemical industry

effectiveness and efficiency of regulatory instruments (Chataway, Tait, and Wield, 2004; Tait and Chataway, forthcoming).

We categorize policies and regulations according to whether they are perceived as *enabling* or *constraining* by industry managers (equivalent to "affirmative" or "negative" in the classification of Bemelmans-Videc, Rist, and Vedung). This can have a major impact on their effectiveness and on the cost of implementation. Our second basis for classification is whether regulations and policies are seen as *indiscriminate* or as *discriminating* among products. Indiscriminate policies are usually much less effective than intended, or can even have negative, counter-intuitive effects on the regulatory target.[2] As explained below, an appropriate basis for discrimination among pesticides, for example, would be one that takes account of the toxicological profiles of different pesticides for human health and the environment.

Figure 13.2 shows where we would place several pesticide regulatory and policy instruments on these two dimensions – enabling/constraining and discriminating/indiscriminate. Note that these are not fundamental, immutable categories. They relate to the perceptions of

---

[2] In some cases, policies that discriminate among products on a basis that is inappropriate to the policy aim can have negative effects similar to those of indiscriminate policies.

managers, and, even in the same company, one manager may see a particular regulatory policy as constraining while another sees it as enabling. The impact of this perception on manager and company behavior is the important criterion, and such distinctions can even be the deciding factor in determining which companies survive and which do not in complex and turbulent regulatory environments.

This mutability of categories depending on perceptions could be seen as a weakness of this approach. On the other hand, it could be used constructively as a means to bring together the strict regulatory requirements of the government approach and the consideration of stakeholder attitudes and wishes required by the governance approach. In areas such as pesticide regulation, the views of environmental groups, as stakeholders, usually impact on policy decisions about whether and how to regulate in a particular area in a fairly confrontational manner, as for example with the recent EU Chemicals Directive (REACH). We are suggesting here that a more subtle, joint approach that engaged stakeholders (e.g. from industry and environmental groups) in discussions about mutually acceptable or, indeed, beneficial forms of regulation could be one means of reconciling the need for strict regulation with stakeholder engagement.

13.2.1.1 US Food Quality Protection Act

The US Food Quality Protection Act (FQPA) of 1996, according to some of our agro-biotechnology industry interviewees, fundamentally changed the way that companies respond to signals from the Environmental Protection Agency (EPA) in the regulation of pesticides (Tait and Chataway, 2000). The new safety standard – reasonable certainty of no harm – that is required to be applied to all pesticides used on food crops is linked to a system that expedites the approval of safer pesticides (see http://www.epa.gov/oppfead1/fqpa) – i.e. pesticides that can meet this requirement can be submitted for registration by companies on a "fast-track" basis. Such instruments selectively *enable* some companies to gain an advantage over others and can in a very short space of time alter the behavior of a whole industry sector.

The FQPA was reported by some managers in our interviews to have had a major impact on pesticide development strategies such that any product without environmental and health-related benefits over others was unlikely to be registered by the EPA in a reasonable timescale. For the companies we interviewed, the FQPA was thus enabling and discriminating (figure 13.2, box 1). However, for those companies that did not have appropriate products in their pipeline, it was perceived as constraining (box 2).

We have tried, so far unsuccessfully, to find evidence from policy decisions by the EPA that the FQPA is indeed operating so as to improve the environment and health-related properties of new pesticides being registered. However, the FQPA seems to be a regulatory instrument that *works* in the governance sense, although this seems to be a marginal outcome from legislation that was intended primarily to remove the outdated "Delaney Clause" from US legislation.[3] Improved environmental performance by pesticides does not seem to have been an important enough policy aim for the EPA to be collecting evidence of its success (Yogendra, 2004).

13.2.1.2 Pesticide registration and review

Most pesticide registration systems, unlike the FQPA, do not discriminate on the basis of whether particular chemicals could be regarded as "safer" than currently available products. Any product capable of passing the regulatory process is deemed "safe enough." Thus, they do discriminate among chemicals on the basis of their safety to the environment and to human health, although only in the sense of meeting a minimum standard, rather than through a "ratchet-type" system that tries continually to improve the standard. They also act as a constraint on rather than an incentive to the behavior of companies, in that they do not encourage companies to compete with one another on the basis of continually improving the safety of products (figure 13.2, box 2).

The term "pesticide review" applies to chemicals approved in previous years where new information or new analytic techniques may require reconsideration of this approval. It discriminates among chemicals, in that it denies "approved" status to some chemicals currently on the market that are deemed to be more damaging to health or the environment and opens up new market niches for less damaging products. This mechanism is constraining for companies with products

---

[3] The Delaney Clause, included in Section 409 of the Federal Food, Drug, and Cosmetic Act, 1958, applied a zero-risk standard for chemicals, including pesticides, that might be present in processed foods. It prohibited any chemical found to induce cancer in humans or animals, regardless of the dose level required. Such "zero-risk" standards prove impossible to implement in practice.

that are no longer approved (figure 13.2, box 2), but it can also be enabling for companies that respond constructively to the newly opened market niches for alternative and safer products (box 1). A recent example of this type of effect operating in the pharmaceutical industry sector is the impact of the withdrawal of Merck's Vioxx pain relief drug in creating a major marketing opportunity for Mobic, produced by Abbott Laboratories (Bowe, 2005).

13.2.1.3 Common Agricultural Policy

Throughout much of its life the CAP has acted as an incentive to farmers to use inputs such as pesticides on their crops, with no discrimination between products.[4] The policy has thus been enabling as far as industry managers are concerned, providing a ready market for all pesticides with no discrimination on the basis of the quality of the chemicals (figure 13.2, box 3). Where CAP reforms are directed indiscriminately to reducing the levels of pesticide use, this is likely to discourage farmers from using the more expensive new pesticides that are less damaging to people and the environment, and therefore discourage innovation in the agrochemical industry (box 4). Where reform discriminates in favour of more benign chemicals, this will create new market niches and stimulate industry innovation (box 1).

13.2.1.4 The European Drinking Water Directive

The European Union's Drinking Water Directive (80/778/EEC) set a very low limit on the permitted level of contamination of drinking water by pesticides, which was indiscriminate in that it did not distinguish among chemicals on the basis of toxicity. As a result, the screening systems set up by agrochemical companies to detect promising new pesticides automatically reject any chemical with a tendency to be readily mobilized in soils (figure 13.2, box 4). This is not to suggest that soil mobility is an inappropriate criterion to use in these

---

[4] This is explained by the detailed implementation of the CAP in the 1980s – for example, for vegetable crops on which high levels of pesticides are used. An "intervention price" was set, and if market prices fell below this level crops were purchased by the state. If perishable, they were destroyed. However, in order to qualify for the intervention price, the crop had to be free from blemishes and pest damage. Farmers were thus encouraged to use pesticides to ensure that, if prices fell too low, their crops would be blemish-free and thus would qualify for the intervention price, even if the produce had never appeared on the market.

circumstances, but a relatively non-toxic chemical with a high mobility in soil may not present an environmental problem and, indeed, may be safer in many other respects than one that is more toxic but less mobile. Thus, a more appropriate and discriminating set of criteria would be a combination of relative toxicity, persistence, and mobility in soils.

Zeneca Agrochemicals were able to give us an example of this effect in practice (Tait and Chataway, 2000). Their strobilurin fungicides are widely regarded as very safe products. One of this group of chemicals was the first product to be registered under the fast-track system set up by the US Food Quality Protection Act, and the company was awarded the UK Queen's Award for Technological Achievement in 1999 in recognition of the qualities of this product. However, this class of chemicals narrowly escaped being rejected at an early stage of product development because of its mobility in soils, and hence the danger of falling foul of the EU Drinking Water Directive.

13.2.2 Product and process-based regulatory systems

13.2.2.1 Pesticides

The regulatory systems that have developed over the years for pesticides can be described as "reactive," whereby the industry and its products are controlled by a system set up *in response to* quantitative evidence of adverse impacts that have arisen in earlier generations of products. New products are screened to ensure that they do not give rise to similar hazards. The regulatory system is built up slowly in a piecemeal fashion, as products in use exhibit different, often unexpected, hazards, and decisions about the need for – and level of – regulation are taken in relation to the relevant costs and benefits (Tait and Levidow, 1992).

The regulation of pesticides thus occurs through a learning process involving a reaction to new information or the removal of previous uncertainties, followed by the prevention of any recurrence of unacceptable side effects. The system is largely in the hands of experts, who advise governments on the need for the introduction of new regulations or the withdrawal of approval for a chemical. Similar procedures are in place for the approval of drugs in both Europe and the United States.

This inexorable intensification of regulation for pesticides and drugs is an example of old-style *government*, and it has continued throughout the period, when many other regulatory and policy areas were being subjected to deregulation initiatives as part of a switch to more *governance*-based systems. For both pesticides and drugs the overall form of the regulatory system has remained, and the constraints it has placed on industry have been gradually extended and intensified.

13.2.2.2 GM crops

Approaches to the regulation of drugs and pesticides have thus remained govern*ment*-based in both the United States and the European Union. However, in the 1980s, when policy-makers began to consider how to regulate GM crops (a new technology with potentially disruptive properties for the agrochemical industry sector), radical differences of approach began to emerge between the United States and the European Union in proposals for the regulation of this new technology.

The European approach came under attack from policy-makers and regulators in the United States and OECD (Office of Science and Technology Policy [OSTP], 1990; OECD, 1986) for being *process-based* – i.e. for regulating GM crops on the basis of the process whereby they were developed (genetic modification) – rather than as products with a particular intended use. The OECD/US *product-based* approach thus proposed to regulate GM crops through the existing regulatory systems for pesticides, foods, and seeds (operated by the EPA, the US Department of Agriculture [USDA], and the FDA). The European Union, on the other hand, set about developing an entirely new regulatory system to cover all GM organisms, including new crops, considering genetic modification to be a radical departure from previous technology (pesticides) with potentially different and unpredictable properties.

This fundamental disagreement over how GM crops should be categorized and regulated has rumbled on throughout the 1990s and underlies the current US challenge to the European Union through the World Trade Organisation (Murphy and Yanacopulos, forthcoming).

The process-based approach was closely related to the emergence of the precautionary principle in European legislation. The companies developing GM crops were those that had previously produced pesticides such as the organochlorines, which had caused widespread public controversy. It is therefore not surprising that they became the first industry sector to be subjected systematically to a more precautionary approach to its regulation. An early policy document produced under the auspices of the European Commission (Mantegazzini, 1986) and a report from the UK Royal Commission on Environmental Pollution (1989), among many other documents produced around this time, confirmed the precautionary nature of the proposed systems being developed in Europe.

The term "process-based" was generally used by those who wished to oppose the EU precautionary approach to regulation and to advocate "product-based" approaches as an alternative.

In the context of the shift from government-based to governance-based policy and regulatory systems, the precautionary approach, although it could not in any sense be described as part of a deregulatory agenda, has thus allowed many other aspects of governance-based approaches to gain access to decisions on GM crop regulatory systems, particularly the much stronger role of public consultation and stakeholder dialogue in government decision-making on science and technology issues (Agriculture and Environment Biotechnology Commission, 2003).

The precautionary principle itself is not necessarily at fault here; indeed, the European Commission has developed guidelines on its implementation that should be an improvement on the current situation (Commission of the European Communities, 2000). However, the precautionary principle has also encouraged stakeholders to bring arguments based on ethics and values into risk debates in ways that were not previously possible, and, in the United Kingdom at least, environmental and other non-governmental organizations (NGOs) have been able to dictate the terms of the debate to an unprecedented degree (Tait, 2001).

13.2.2.3 Drugs and stem cells

There is currently broad transatlantic consensus over the regulation of drugs, as has been the case for pesticides. However, the potential for disputes, as in the agro-biotechnology sector, is likely to emerge when disruptive new technologies begin to be deployed in the drugs industry. For example, major differences are already emerging in US and EU approaches to the regulation of stem cells.

Stem cell technology is highly relevant to future developments in the pharmaceutical industry. For example, if stem cells are effective in providing cures for major diseases such as diabetes, Parkinson's disease, and heart disease, they will undermine many "blockbuster" drug markets. The issue has not yet become controversial, because no company has yet devised a formula for making money out of stem cells. However, as soon as this becomes a possibility, the issue of regulation is likely to attract much more regulatory and public interest.

There are already differences between the United States and the European Union in the emerging regulatory systems for stem cells. In the United States, for example, with strong echoes back to the GM crops debate, stem cells are being treated as a "product." The FDA has claimed jurisdiction over them, with the implication that they will be treated as *drugs* for regulatory purposes (Bonnicksen, 2002). In Europe, on the other hand, stem cells are likely to be treated as *devices*, with the implication that they will be regulated as analogous to surgical procedures (Faulkner et al., 2003).

13.2.2.4 Transboundary risk regulation

Drugs, agrochemicals, and GM crops are produced by multinational companies and traded globally. Tait and Bruce (2004) have extended the product- versus process-based distinction discussed above in the context of transboundary risk regulation. The formal regulatory processes for pesticides and drugs in international trade remain *product-based*, as outlined above. However, there was a need for a second category to describe a new form of non-statutory control exercised by consumers, organized internationally via the internet. The consumer boycott, which we described as a "new instrument of global governance," has, at least in the case of GM crops, exerted more control than governments over international trade, and, because it takes place without formal regulatory systems, this *production-system-based* category raises different issues and requires different approaches to its management.

In this case, the negative risk attribute to which consumers are responding is only symbolically attached to the traded product (e.g. food products that have been produced by intensive farming methods or GM crops that are widely regarded as safe but are stigmatized by environmental and consumer organizations). The production-system-based category is conceptually similar to the process-based category described in section 2.2, but here there is no formal regulatory basis, not even a disputed one such as the EU regulatory system for GM crops, to reject such products in international trade. However, if the origin of the products is identified by labeling, environmental and other pressure groups can campaign against it and individual consumers can exercise their right "not to buy." Even if the product itself is not hazardous, public concerns do cross national boundaries; people want to have an influence on what happens in other countries or on the strategies of multinational companies, often from an altruistic concern for global or local environmental sustainability or for the health or well-being of workers in other countries. These consumer pressures are behind the EU insistence, again in opposition to the United States, that labeling and traceability requirements should be part of EU legislation for GM crops.

13.3 The relationship between regulation and innovation

The extremely demanding regulatory systems to which the agro-biotechnology and pharmaceutical industries are subject have led to the adoption of innovation pathways that are not replicated in other industry sectors. The very high costs and long delays entailed in taking a new product through the regulatory system, regardless of the detailed nature of that system, and related patent protection systems ensure that only large MNCs have the necessary resources to operate through the whole innovation cycle.

This entry barrier for small and medium-sized enterprises (SMEs) has shaped the structure of both industry sectors, leaving the large multinationals in an unassailable position and insulating them from challenges to their supremacy by smaller innovative companies. Small companies rely on MNCs to take their products through to market and thus tailor their innovation strategies to match, rather than to challenge, those of MNCs. Regulation per se has thus been enabling for MNCs, contributing to their global dominance, and highly constraining for small companies. Indeed, where governments promote policies to encourage innovation in SMEs, the usual outcome is the provision of indirect support to the development costs of MNCs (Tait and Williams, 1999). The need for SMEs to make themselves attractive acquisition targets constrains their innovation strategies to products that are compatible with the strategies of MNCs.

Pesticides and GM crops are purchased by farmers, whose products then reach the consumer via an increasingly complex and anonymous distribution and marketing chain. It is difficult for consumers to have a voice in farmers' purchasing decisions, whether for pesticides or for GM crops. However, as outlined in section 2.2.4, the success of production-system-based boycotts, against GM crops in particular, has enabled consumer direct action to leap across several layers in the governance hierarchy and to change the thinking of managers in agro-biotechnology about who their ultimate customers are.

The markets for drugs have been even less open to public direct influence by consumers than those for pesticides, although this has been changing to some extent as a result of internet-based sales of drugs. However, despite growing public concern about some aspects of the behavior of large pharmaceutical companies, no equivalent of production-system-based consumer direct action has yet emerged in the pharmaceuticals sector.

The remoteness of MNCs from any public consumer base, their global reach, their sometimes arrogant attitudes in response to public concerns, and the public perception that they are, in many senses, "beyond governance" have all contributed to the lack of public trust in both agro-biotechnology and pharmaceutical industry sectors.

13.4 The challenge of maturity

The agrochemical industry in the 1980s and the pharmaceutical industry today (Mittra, 2003) have found themselves in an apparently unchallengeable position from the point of view of external threats while being eroded from within.

In each case the most profitable markets, and the products that are easiest to discover and develop, are the first to be exploited. Over time it becomes progressively more difficult to develop new products that are cheap enough and easy enough to exploit, that can comply with increasingly stringent regulatory systems, and that can compete with previous generations of products that are off-patent and relatively cheap. The end result of this process is the need to screen several hundred thousand molecules in order to find one new blockbuster drug or pesticide, with the enormous costs that this implies for R&D budgets.

Regulation, in an *enabling* sense, can provide short-term respite from such pressures. Banning a product found to be defective can open up new markets for patented, and therefore more profitable, drugs or pesticides. However, overall, the trend towards maturity has continued inexorably in the two sectors, although on different timescales, and indeed there are some lessons that the pharmaceutical sector could learn from the earlier maturation trajectory and experience of agro-biotechnology.

The problems of a maturing product range lead companies actively to seek new forms of innovation to take them onto a different, higher-value-added growth curve and to enable them to avoid becoming mere producers of commodity chemicals. The agro-biotechnology sector identified GM crops as filling this role, and the pharmaceuticals sector is looking towards biologicals (small protein-based drugs), pharmaco-genetics, and stem cells.

In the early 1980s, as part of the new governance agenda that was even then emerging, there were several attempts to encourage public engagement with issues related to the new GM crop technology. However, it was not till ten years later, when some products were almost ready to be launched on the market, that the public began to take an interest in GM crops, and by this time environmental and consumer groups had successfully framed the products as components of inherently unsustainable farming systems, leading to higher profits for a demonized industry sector, with no public benefits. There was an opportunity in the 1980s for the agrochemical industry successfully to promote GM crops to the public as an input to more sustainable farming systems that would need less pesticide. However, the industry was not ready at that point to undermine its existing product range to that extent.

These problems of timing for major shifts in agro-biotechnology industry strategies were exacerbated by the maturity-induced increased competition between companies, accompanied by an aggressive round of M&As. Whereas the industry had presented a united front to regulators and the outside world, the pressures of maturity, accompanied by the challenges of coping with a disruptive new innovation pathway, led to the breakup of the old industry hegemony. Just when it most needed it, the sector as a whole was unable to develop a united and strategic approach to the challenge mounted by a coalition of pressure groups making effective use of the new governance agenda that had been emerging in Europe, particularly the predominance of the precautionary principle in Europe's process-based regulatory system.

It is now common to hear scientists and managers in pharmaceuticals and other innovative health-related sectors claim that what happened to agro-biotechnology in Europe could not happen to them, for the reason that new and better health care products will always be seen by the public as desirable, and therefore the public will not actively oppose the industry that produces them. This is a plausible argument, but the pharmaceutical sector could still learn something from the situation in which the agrochemical industry found itself in the 1980s.

For example, new developments in the life sciences could begin to undermine the dominant blockbuster model, which depends for its success on drugs being taken by large numbers of people over long periods of time. If, as is claimed, we will be able to cure – rather than merely treat – major diseases such as diabetes, heart disease, Parkinson's disease, and arthritis, this will remove major blockbuster targets from the current repertoire of pharmaceutical companies. Pharmaco-genetics may also allow a more targeted approach to drug prescribing, which again implies niche, rather than blockbuster, markets. These internal challenges could change the relationships among companies, knowledge providers, policy-makers and markets, leading to radically new models for success and profitability in health care and drug development.

The sector's external operating environment is also becoming increasingly turbulent. Uncertainties about the nature of future governance and regulatory systems has some parallels with the experience of agro-biotechnology. If pharmaco-genetics, as is being predicted, lowers the regulatory hurdle for new drugs targeted to niche markets, this could enable smaller companies to take new drugs through to stage 3 clinical trials and compete directly with multinationals in the market place.

We have also seen several withdrawals of blockbuster drugs recently, either due to the emergence of unsuspected side effects or, more seriously for the industry, after accusations of the concealment of negative results from clinical trials, or of reports of serious side effects. Another major external challenge to the industry comes from public and government demands for cheaper drugs. Adding to pressure from developing countries for cheaper drugs for AIDS, malaria, and tuberculosis, European governments are proposing to make reductions in their drugs budgets, and poorer patients without health insurance in the United States are successfully putting pressure on companies to reduce prices (and hence profits) – and this at a time of increasing overall maturity in the sector. Other relevant changes include major public–private partnership initiatives to develop vaccines against AIDS and malaria, adding to the overall level of turbulence.

Some of these factors are already beginning to undermine public trust in the industry. Pressure groups are emerging with principled objections to high-tech, expensive medicines, favoring instead greater public investment in traditional health care and disease prevention. Such groups are currently counterbalanced by patient groups campaigning for new technology-based treatments, but a series of accidental events that undermines the trust of patient groups in the industry could lead to a wide range of public groups joining a coalition opposed in principle to life-science-based medicine (Tait, 2001).

No single event or outcome is likely to have a catastrophic impact on the current balance of the industry or the strategies of the companies involved. However, a combination of innovation-based challenges to the blockbuster drug model, along with perceived regulatory failures, a failure of public trust in the pharmaceutical industry, and a lack of clarity in policies for the governance and regulation of new technology, could trigger a major change in the fortunes of the sector.

13.5 Governance and innovation in agro-biotechnology and pharmaceuticals

We have argued in this chapter that medical and agricultural innovations are evolving more rapidly than the relevant policy and regulatory systems. Regulatory systems are crucial to the fate of the agro-chemical and pharmaceutical industry sectors that are facing an array of simultaneous challenges:

- maturity in existing product portfolios;
- accommodating disruptive new technologies; and
- increasing public demands for engagement in the development and regulation of innovation through new governance approaches.

Taking agro-biotechnology as an example, the more governance-based European regulatory system has so far failed to provide a viable pathway to market for innovative products from the agro-biotechnology industry. It has also created some unfortunate precedents for the regulation of innovation in other areas, such as nanotechnology. And it has not yet succeeded in providing European consumers with the choice of buying GM foods if they wish to do so. Thus, the shift to a new governance agenda in Europe could not so far be considered as having resulted in better regulation of technology.

The regulatory systems in place for GM crops in the United States correspond more closely to an old-style government approach. There is no doubt that they have successfully fostered innovation in the development of GM crops and given US citizens the option to buy GM foods, although at the same time making it more difficult to avoid buying GM foods – the converse of the choice available in Europe.

However, as GM crop technology becomes more sophisticated and the products being developed become more complex, the current comparative advantage of US over European regulatory systems for effective management may shift. Europe's long and difficult experience in the development of process-based governance systems may yet pay off.

To deliver the most efficient and effective governance systems for agro-biotechnology and for the pharmaceutical industry, the United States and the European Union have much to learn from each other, and the kind of analysis presented here could contribute to that process.

Our analysis would identify constraining, indiscriminate, product-based regulatory policies and instruments as being part of old-style *government*. The *governance* approach, on the other hand, would include policies and instruments that are enabling, discriminating, and process-based. The production-system-based approach, although not part of a formal regulatory system, also plays an informal role in governance processes and, because of that informality, can be difficult for regulators to manage. The precautionary principle could, in theory, be incorporated into either government- or governance-based systems, but its use so far has been mainly to support the incorporation of certain public values into European governance-based regulatory systems.

Important points to emerge from this analysis include the following.
- Some policy and regulatory initiatives can have major, rapid, and positive influences on innovation processes, but they are infrequent and their value and significance may not be recognized.
- It is more usual to find that policies and regulations emerging from one policy area have unexpected negative effects in other areas or are counteracted by constraints that were not previously recognized.
- Good governance is most likely to be achieved by creating a policy and regulatory environment that is enabling in the desired direction, rather than being constraining and restrictive, and also that discriminates among products on the basis of the most relevant criteria.

Whether a policy is enabling or constraining appears to have less of an influence on its ability to encourage "cleaner" innovation than the extent and appropriateness of its discrimination among products. However, having said that, enabling policies are likely to have a more rapid impact and to be less expensive to monitor and enforce.

A recent paper by Perri 6 (2003) has made an important contribution to the integrated governance of technological change. It suggests that there is a need for a substantial research agenda to study the governance of technology, "the shifting kaleidoscope of governance structures and processes," both as an academic study and to inform policy processes. He particularly emphasizes the need to see the whole system, to examine both control and inducement and their interdependencies, and also the interactions between tools and structures.

The challenge for the future is to incorporate the most useful aspects of governance-based approaches into the regulatory systems being developed to accommodate disruptive innovations in the agro-biotechnology and pharmaceutical industry sectors.

References

6, P. (2003), *The Governance of Technology: Concepts, Trends, Theory, Normative Principles and Research Agenda*, paper prepared for the conference on "Human choice and global technological change," 24–25 February, Lisbon.

Agriculture and Environment Biotechnology Commission (2003), *GM Nation? Public Debate*, available at http://www/gmnation.org.uk.

Bache, I. (2003), "Governing through governance: education policy control under New Labour," *Political Studies*, 51 (2), 300–14,

Bemelmans-Videc, M. L., R. C. Rist, and E. Vedung (2003), *Carrots, Sticks and Sermons: Policy Instruments and their Evaluation*, Transaction Publishers, New Brunswick, NJ.

Bonnicksen, A. L. (2002), *Crafting a Cloning Policy: From Dolly to Stem Cells*, Georgetown University Press, Washington, DC.

Bowe, C. (2005), "Vioxx woes help launch a new blockbuster," *Financial Times*, 19 January, 1.

Cabinet Office (1999a), *Modernising Government: White Paper*, Cabinet Office, London.

(1999b), *Professional Policy Making for the Twenty-First Century*, Cabinet Office, Strategic Policy Making Team, London.

(2000), *Wiring it up: Whitehall's Management of Cross-Cutting Policies and Services*, Cabinet Office, Performance Innovation Unit, London.

Chataway, J., J. Tait, and D. Wield (2004), "Understanding company R&D strategies in agro-biotechnology: trajectories and blind spots," *Research Policy*, 33 (6–7), 1041–57.

Commission of the European Communities (2000), *Communication from the Commission on the Precautionary Principle*, COM (2000) 1, European Commission, Brussels.

(2001), *European Governance: A White Paper*, COM (2001) 428 Final, European Commission, Brussels.

Cooke, P., P. Boekholt, and F. Todtling (2000), *The Governance of Innovation in Europe: Regional Perspectives on Global Competitiveness*, Pinter, London.

Davies, H. T. O., S. M. Nutley, and P. C. Smith (eds.) (2000), *What Works? Evidence-Based Policy and Practice in Public Services*, Policy Press, Bristol.

DTI (1998), *Our Competitive Future: Building the Knowledge-Driven Economy*, White Paper Cm 4176, Department of Trade and Industry, London.

Faulkner, A., J. Kent, I. Geesink, and D. Fitzpatrick (2003), "Human tissue-engineered products – drugs or devices? Tackling the regulatory vacuum," *British Medical Journal*, 326, 1159–60.

Giddens, A. (1999), *BBC Reith Lectures*, available at http://news.bbc.co.uk/hi/english/static/events/reith_99/.

Lyall, C., and Tait, J. (2005) *New Modes of Governance: Developing an Integrated Policy Approach to Science, Technology, Risk and the Environment*, Aldershot, Ashgate Publishing.

Mantegazzini, M. C. (1986), *The Environmental Risks from Biotechnology*, Pinter, London.

Mittra, J. (2003), *Innovation Processes in Genomics Industry Sectors*, Working Paper no. 5, Innogen Centre, Edinburgh University, available at http://www.innogen.ac.uk.

Murphy, J., and H. Yanacopulos (forthcoming), "Understanding governance and coalitions: transatlantic networks and the regulation of GMOs," *Geoforum*.

Organisation for Economic Co-operation and Development (1986), *Recombinant DNA Safety Considerations*, Organisation for Economic Co-operation and Development, Paris.

Office of Science and Technology Policy (1990), "Principles for federal oversight of biotechnology: planned introduction into the environment of organisms with modified hereditary traits," *Federal Register*, 55, 31118–21.

Royal Commission on Environmental Pollution (1989), *Thirteenth Report: The Release of Genetically Engineered Organisms to the Environment*, Cmd720, Her Majesty's Stationery Office, London.

Sloat, A. (2002), "Governance: contested perceptions of civic participation," *Scottish Affairs*, 39 (Spring), 103–17.

Stoker, G. (1998), "Governance as theory: five propositions," *International Social Science Journal*, 50, 17–28.

Spinardi, G., and R. Williams (2005), "The governance challenges of breakthrough science and technology," in C. Lyall and J. Tait (eds.), *New Modes of Governance: Developing an Integrated Policy Approach to Science, Technology, Risk and the Environment*, Aldershot, Ashgate Publishing, 45–66.

Tait, J. (2001), "More Faust than Frankenstein: the European debate about risk regulation for genetically modified crops," *Journal of Risk Research*, 4 (2), 175–89.

Tait, J., and A. Bruce (2004), "Global change and transboundary risks," in T. McDaniels and M. Small (eds.), *Risk Analysis and Society: An Interdisciplinary Characterisation of the Field*, Cambridge University Press, Cambridge, 367–419.

Tait, J., and J. Chataway (2000), *Policy Influences on Technology for Agriculture: Chemicals, Biotechnology and Seeds – Zeneca Agrochemicals Monograph*, PITA Report to TSER, project no. SOE1/CT97/1068, available at http://www.technology.open.ac.uk/cts/pita/ and http://www.supra.ed.ac.uk/NewWeb/Reports.htm.

(forthcoming), "Risk and uncertainty in genetically modified crop development: the industry perspective," *Environment and Planning – C* [currently available as Innogen Working Paper no. 1, available at http://www.innogen.ac.uk].

Tait, J., J. Chataway, and D. Wield (2000), *Final Report, PITA Project*, available at http://www.technology.open.ac.uk/cts/projects.htm#biotechnology.

Tait, J., and L. Levidow (1992), "Proactive and reactive approaches to risk regulation: the case of biotechnology," *Futures*, 24 (3), 219–31.

Tait, J., and R. Williams (1999), "Policy approaches to research and development: foresight, framework and competitiveness," *Science and Public Policy*, 26 (2), 101–12.

Yogendra, S. (2004), *FQPA in the United States: A Review of the Dynamics of Pesticide Regulation and Firm Responses*, Working Paper no. 11, Innogen Centre, Edinburgh University available at http://www.innogen.ac.uk.

14 The dynamics of knowledge accumulation, regulation, and appropriability in the pharma-biotech sector: policy issues

LUIGI ORSENIGO, GIOVANNI DOSI,
AND MARIANA MAZZUCATO

THE contributions to this book enrich from a variety of angles our understanding of how the dynamics of knowledge affect the dynamics of firms and industry structures. In this concluding chapter, we discuss some associated policy implications regarding the future of the innovation process in the pharmaceutical industry and the institutional setup supporting it.

14.1 Institutions, industry organization, and innovation: a bird's-eye view

The policy debate in this arena has become extremely intense and often bitter in recent years. The issues at stake concern an area – health care – the importance of which for society is fundamental and rapidly increasing; indeed, they are becoming crucial elements in the very definition of notions such as welfare, justice, and democracy in the new century.

Many fundamental issues in the policy debate on the pharmaceutical industry, however, are certainly not new. Ever since its inception the market for drugs has been (almost) always and everywhere regulated, albeit for different reasons and in different ways. At the same time, the extent and the forms of the regulation have most often sparked discussion and conflict. For example, considerations linked to consumer protection led, throughout most of the twentieth century, to increasingly stringent requirements for the approval of new drugs, and implied larger and more costly clinical trials. The presence of significant information asymmetries in the market for drugs coupled with fundamental considerations of social and economic equity have often been used to justify the introduction of various forms of price regulation. The emergence of the welfare state and the subsequent rise of health care and prescription drug spending have induced, first, a rapid expansion of demand, and then a series of cost containment policies. Developments in legislation and in courts' interpretations of issues concerning intellectual property rights have also had significant impacts on patterns of competition and industrial evolution. Last, but certainly not least, the institutional setups governing the systems of fundamental scientific research have profoundly affected the ability to discover, develop, and commercialize new drugs.

More broadly, it is important to recognize that policies and institutional design have deeply affected innovation and industry evolution, sometimes consciously, sometimes through the unintended effects of interventions taken for reasons by and large independent of considerations related to the performance of the industry. These interventions and the evolving institutional structures have influenced the patterns of accumulation of competencies, the selection mechanisms, and the incentive structures to engage in innovative activities in many different, and sometimes indirect, ways.

Policies have also evolved. Certainly, they often display inertial elements, embedded in and shaped by specific institutional and political environments. However, institutions and policies have been profoundly influenced by technological change – especially at times marked by profound technological discontinuities. And they are also deeply influenced by changes in the broader "political economy" of any one country and the ideological orientation of each historical period. For example, major technological and scientific breakthroughs, such as the "antibiotic revolution" in the 1930s and the emergence of biotechnology in the 1980s, have substantially changed both non-market institutions and industry dynamics. The possibility of discovering and developing new drugs was (and is) not just an occasion for firms to make profit. It has also changed the procedures, forms of organization, and costs of research, as well as the attitudes of people towards health care.

Together, the boundaries between the activities that ought to be regulated explicitly by the public authorities or even undertaken by them directly versus those left in the hand of profit-seeking entities happen to be deeply influenced by the prevailing Zeitgeist on the virtues and shortcomings of market processes, as compared to political decision processes. Indeed, the definition of health care itself – and of people's medical needs – depends both on the status of scientific and technological progress and, together, on fundamental social visions on the very meaning of *citizenship* and *rights*. For example, does every citizen have the "right" to some form of "care"? And, if so, up to what limit? (We return to this question of *rights* in the final section of the chapter.)

In short, technological change has been and remains dynamically coupled with institutional and political change (more on this in Lacetera and Orsenigo, 2002).

The "golden age" of the pharmaceutical industry (from the end of World War II to the 1980s) was supported by a few interrelated factors. The explosion of public support for health-related research provided ample technological opportunities to be exploited. Firms developed highly effective organizational procedures for discovering, developing, and marketing drugs, refining the so-called "random screening" paradigm. The international oligopolistic core of the industry, which is now called "big pharma," came to dominate under such a learning regime. Growing incomes sustained a fast-growing demand, which – especially in "welfare states" characterized by national health care systems (most of them in Europe) – offered a rich, "organized," and publicly regulated market for drugs.

Recently, however, this picture has been drastically transformed. As discussed in several chapters of this book, the industry has experienced a number of interrelated transformations in recent years.

The "molecular biology" revolution has fostered the emergence of a new technological paradigm, potentially involving immense new opportunities for innovation – which, however, have only very partially materialized so far. If anything, the costs of research have been soaring, many analysts claim partly as a consequence of tighter procedures for product approval (this is indeed a matter of controversy, which we shall touch on again below). Meanwhile, the crisis of the welfare state – and, more generally, the explosion of public expenditure for health – has fueled attempts to reduce public outlays in this domain.

The spreading of attitudes and legislation in favor of a tighter intellectual property regime (the Bayh–Dole Act in the United States and international TRIPS [Trade-Related aspects of Intellectual Property Rights] agreements being the sharpest examples), not least in areas traditionally characterized by more lenient appropriability conditions (including, of course, freely accessible publicly funded scientific research), has radically changed the conditions with which knowledge is created, diffused, and accessed.

Newly industrializing countries – such as India and Brazil – have meanwhile emerged as potentially important players in the world pharmaceutical industry, mainly but not exclusively as producers of generics for other developing countries. As a result, the very viability of the business model that dominated the industry in the "golden age" has been called into question. At the same time, the demand for health care and the expectations of the population are continuously rising, and the many humanitarian catastrophes in developing countries highlight the paramount importance of access to drugs by the world's poor.

As already briefly mentioned in the introduction to this book, policy prescriptions crucially depend on a few, very difficult, interpretative issues, including:

- the impact of different patent systems upon (i) the *rates* and directions of innovation, and (ii) their ultimate *distributive* and *welfare effects*;
- the trade-offs and dilemmas between the increasing costs of drugs and health care in general, together with the need to contain public expenditure in this field, versus the demand for (or "right" of) access to drugs by the whole population in rich (and aging) countries, and increasingly so in the poorer countries as well.

The price of drugs is, clearly, one of the crucial issues (albeit not the only one) at stake. To what extent are free (high) prices necessary to sustain innovative activities and the viability of the industry? And how can high prices be reconciled with the need to make drugs accessible to the widest possible share of the population and with the budget constraints of the various states?

It is worth remembering that, perhaps in different forms, these questions have been there throughout most of the history of the industry. The debate around the Kefauver Commission of the US Senate, which around half a century ago investigated monopolistic positions in the American industry, addressed many of these issues, in fact. Nonetheless, the following three issues introduce some elements of genuine novelty into the debate.

(i) The growing and more direct role of scientific knowledge in the process of innovation in the pharmaceutical industry, as a consequence of the "molecular biology revolution." This role is particularly evident in drug discovery, but science is increasingly relevant also in drug development, in the processes of product approval, and in the evaluation of the post-marketing performances of drugs.

(ii) The question marks concerning the very sustainability in the long run of the current structure of the system of innovation and health care provision, which simultaneously faces higher costs, higher expectations, and tighter budget constraints.
(iii) The role of developing countries, both as producers of drugs and as consumers in desperate need, confronted by the challenge of increasingly difficult access to proprietary drugs.

In what follows, we briefly comment upon some of these crucial issues. First, we touch upon the question of price and other forms of regulation. Then, we move to IPR, in particular with regard to patents and the exploitation of basic, publicly funded research in general and developing countries in particular.

14.2 "Market failures" in the pharmaceutical industry

The market for drugs, as mentioned above, has been (almost) always regulated, for a variety of different reasons that may easily be accounted for both in the standard economic framework as well as in more heterodox ones. Many of these reasons can be straightforwardly explained on the grounds of the standard economic tools, in terms of market failures and standard economic efficiency.[1]

The market for drugs is inherently characterized by information asymmetry. Producers inevitably have "more information" on the quality of drugs than consumers do. Moreover, it is the prescribing doctor who makes the decision, but even doctors often do not know in detail the properties of a drug, especially when it is a new one. As a result, there are a number of arguments in favor of regulation.

 (i) It is observed that much of the information available to physicians is provided by the companies themselves. As a consequence, an external assessment of the safety of the drug (and in many places, starting from the early 1960s, of its efficacy) may be necessary to prevent damage to consumers.
(ii) Given the value that users may attribute to the product, especially in extreme cases, demand elasticity tends to be very low – even within the same class of therapeutical products (and, of course, it is zero across them: no one with a kidney problem would accept a

---

[1] See also Comanor (1986), Scherer (2000), and Schweitzer and Comanor (2000) for reviews of these issues.

drug for headache simply because it is cheaper!). Furthermore, most consumers are insured (privately or publicly) against at least a part of the cost of prescription drugs, so they are only marginally interested in drug prices. Similarly, the prescribing physicians are not completely sensitive to prices, both because they will not pay for the prescribed drugs and because the respect of professional norms makes them more attentive to the safety and therapeutic value of medicines. Despite their role of scientific experts, however, physicians' prescribing behaviors do not seem immune to other forces, such as advertising and brand loyalty, and seem to follow routinized patterns. (All this, of course, is premised on the assumption that there are no corrupting linkages between drug producers and prescribing doctors.)

(iii) A related set of reasons for regulation refers to cost containment. In countries with a national health service or where there is a third payer (typically, an insurer), demand elasticity to price tends to be lower than would otherwise have been the case. This may lead to price increases by firms enjoying market power. Moreover, as a consequence, the absence of any countervailing measure is likely to lead to an explosion of public expenditures, because neither the patients nor the physicians ultimately pay for the drugs. Thus, governments may appropriately act as monopsonist and, through various instruments, try to reduce (quasi-monopolistic) profits and the maximization of drug prices.

In fact, on the supply side, the pharmaceutical industry is inherently characterized by non-price competition. Many chapters in this book elaborate on the notion that innovation is a major form of competition in this industry. In turn, producers are attributed (temporary) monopoly power through patent protection. In the absence of such protection, profit-seeking firms – the argument goes – would not invest in research or would underinvest as compared to the "social optimum" (whatever that may mean). Indeed, pharmaceuticals is one of the few industries in which patents are considered very important mechanisms of appropriability for the economic outcomes of innovation. Given the existence of (even temporary) monopoly power, (price) regulation might therefore be justified as a mechanism to counteract monopolistic pricing.

(iv) The pharmaceutical industry is a science-based sector, wherein scientific knowledge plays a central role and is only in part appropriable. Part of the knowledge that is used to produce new drugs is generated by and/or based on publicly funded scientific research, in principle available to everybody through publication. Thus, pharmaceutical companies are at least partly "subsidized" through publicly funded research.

(v) Advertising, wherever it is allowed, might powerfully interact with market power. Most obviously, it might just be misleading. In any case, it tends to generate brand loyalty effects and therefore some positive feedback on profitabilities, which may have little to do with innovation and the intrinsic properties of the various drugs. In turn, both R&D expenditures and advertising involve high fixed, sunk costs, which happen to be powerful mechanisms sustaining oligopolistic/monopolistic positions and, together, provide ample opportunities to exploit them through "excessive" prices (which for our purposes here may be defined as prices in excess of those that would have justified the search investment for the new drug in the first place).[2]

(vi) Even more importantly, a fundamental argument in favor of regulation is based on equity and moral considerations, and makes, to a large extent, the analysis of market processes a social rather than a purely economic issue. Shouldn't everybody have access to drugs, including (new) expensive ones? Regardless of the different attitudes (across time and countries) towards the industry and its regulation, the main goal of state intervention has often been to guarantee the access to safe and (later on) efficacious drugs to the largest possible share of the population (certainly in Europe, but to some extent also in the United States).

The policy-maker thus faces different and possibly contrasting objectives: in brief, the goal of efficacious and safe drugs, and of equity in their availability to the population, has to match the economic incentives to induce investment in research on new medicines by profit-seeking actors in so far as the latter undertake uncertain search, testing, etc.

To sum up, the question is not so much whether to regulate or not but, rather, *what kind of regulation*?

---

[2] Of course, this takes into account the rates of investment in all would-be drugs, most of which turn out to be failures.

14.3 Price controls and product approval regulations

During the "golden age" the pharmaceutical industry was subject to rather tight forms of regulation in most countries concerning prices and the procedures for product approval. Different forms of price controls were adopted in most industrialized countries, the United States and Germany being two major noteworthy exceptions. Conversely, it was the United States that introduced the first severe and strict regulations for product approval with the 1962 amendments.[3] Other countries (primarily the United Kingdom) followed, but the American procedures have remained among the toughest for a long time.

Since the late 1970s, however, two contrasting tendencies can be observed. On the one hand, price regulation has increasingly been considered an inefficient mechanism to protect consumers: the argument here is that not only do they obviously generate hostility in the industry, they also supposedly stifle innovation and introduce distortions in the market. On the other hand, the increasing need to contain public expenditures on drugs has fostered the introduction of drastic cost-cutting measures. The result has been a general move (quite heterogeneous across countries) towards the adoption of more sophisticated and less invasive measures of price control, such as policies aiming at intervening on the demand side of the market to make patients and health providers more cost-conscious and more price-sensitive (e.g. various forms of co-payment, and other interventions attempting to change the behavior of providers through financial incentives and penalties), and the development of the market for generics and systems of cost sharing such as "reference pricing." However, in some countries, notably the United States (somewhat ironically), the arguments in favor of stronger regulation of the price of drugs are increasingly being voiced at the center of the political debate. (Of course, stronger regulation remains both harmful and unacceptable whenever undertaken elsewhere – but this is another story.)

---

[3] The Kefauver–Harris Amendment Act of 1962 introduced a proof-of-efficacy requirement for the approval of new drugs (based on "adequate and well-controlled trials") and established regulatory controls over the clinical (human) testing of new drug candidates. As a result of these amendments, the FDA went from being simply an evaluator of evidence and research at the end of the R&D process to an active participant in the process itself (Grabowski and Vernon, 1983).

The evidence concerning the impact of procedures for product approval and price controls on innovativeness is ambiguous, however. While the 1962 amendments in the United States and the related introduction of tougher requirements for drug approval certainly caused substantial increases in drug development costs, it is much less well established empirically whether they were responsible for lower rates of innovation. Indeed, Thomas (1994) has quite convincingly argued that a less lenient regulatory environment contributed to the take-off of the British pharmaceutical industry as compared to the French one, by conferring an advantage on more innovative firms and penalizing "me too" producers.

The evidence on price controls is far from clear. Several scholars have suggested that in the United States, lacking price regulations, higher profits have led to higher investment in R&D. Symmetrically, along the same line of interpretation, many have suggested that "invasive" command- and control-oriented approaches are likely to generate hostility between regulators and companies, resistance to change, and a low propensity to innovate. Japan, Italy, and France are quoted as examples where the imposition of tough price control mechanisms appears to have introduced strong disincentives to undertake innovative strategies and favored the survival of inefficient companies, undertaking little R&D or none at all and only marginally improving on existing products (often exploiting the absence of product patent protection). However, the British system of price controls, introduced in the 1960s, does not seem to have unduly discouraged innovative activities – possibly the opposite (Thomas, 1994). And even an industry such as the Italian one emerged as a significant innovator in a period of price controls and a lack of patent protection and nearly disappeared under a more laissez-faire regime.

The bottom line, at the very least, is that there seems to be no simple and unambiguous (let alone monotonic) relation between any single aspect of regulation (e.g. free versus controlled prices, the toughness of the product approval procedures, different systems of inducing cost constraints, etc.) and the degree of innovativeness – even abstracting from more demanding measures of collective performance that take into account justice and equity.

Perhaps it might be more useful to think in terms of "systems of institutional governance" rather than isolated policies. Specific combinations of different forms of regulation and competition have, in the past, managed to produce particular competitive environments favoring both the adoption of successful innovative strategies and their fruitful social use.

Moreover, it is worth noting that many of these institutional arrangements were not devised with the explicit aim of supporting innovation, or even industrial prowess. Rather, they resulted from different purposes – such as social policies or science policies – but ended up bearing important consequences for the capacity and willingness to innovate, sometimes after quite prolonged periods of time.

In sum, the evolution of regulatory regimes has interacted throughout the whole history of the industry with the changes in the nature of technological regimes and with the social perceptions of what is considered efficient, just, and fair. It often did so, to repeat, in rather unintended ways, as the outcome of differentiated and sometimes seemingly unrelated small-scale acts of intervention in various domains, more often than not enacted by different agencies for different purposes (e.g. the Department of Health, the Treasury, the Department of Industry and Trade, etc.). In fact, one of the few robust features of the industry is indeed the profound embeddedness of its evolution in the institutions governing the non-profit-motivated generation of knowledge, on the one hand, and those concerned with the public access to health care, on the other.

Given that, what does the evidence suggest on the *technological and social* outcome of different regimes for the private appropriation of technological knowledge?

14.4 The role of patents and recent changes in patenting behavior

It is empirically well established that in pharmaceuticals – differently from several other sectors – patents are a fundamental instrument for protecting innovation from imitators. To recall, patents have a dual role in the innovation process. On the one hand, they are meant to stimulate innovation by guaranteeing the ability of innovating firms to appropriate the rewards/profits by shielding them from imitation. On the other hand, by forcing the patentholder to codify all the relevant information regarding the new (often tacit) knowledge and to make it public, they are meant to foster the eventual diffusion of the knowledge (which could otherwise remain secret) and its actual application in the commercial domain.

The possible welfare gains and losses associated with patents have been discussed extensively in the economic literature.[4] Indeed, the links between patent protection and innovative performance look less direct than is usually assumed. In general, the empirical evidence regarding the relationships between the tightness of the patent regime and the rate of innovation is surprisingly thin. Even abstracting from the intricacies of the theoretical debate, it is worth reminding ourselves that patent laws involve many different aspects and subtleties (e.g. the scope of patents, the costs of litigation and the enforceability of IPR, the rules governing the definition of priority, etc.) that are likely to have fundamental consequences for the actual degree and form of protection for innovators. Furthermore, changes in the degree and in the forms of protection from imitation for innovators are unlikely to have monotonic effects on innovative efforts or, even more so, on the rates of innovation. Putting it more bluntly, there is virtually no robust evidence supporting the idea that higher expected profits translate into higher search efforts and more frequent innovative successes. Of course, if the expected profits are zero, most often search investments by private agents are zero too (but not always: see the open-source software story!). In any case, above some appropriability threshold incentives do not seem to exert any major impact upon the rates of innovation. Rather, the latter seem to be critically affected by the nature of paradigm-specific technological opportunities, the characteristics of the search space, and the capabilities of would-be inventors.

In order to illustrate this point naively, note first that strong innovating companies, active throughout the world, have historically used instruments other than patents to extract profits from their innovations, even in countries where patent protection was low. For example, advertising, direct foreign investment, and licensing have performed as powerful mechanisms for appropriability, especially in an era when generic competition was not as strong as it is today.

Second, the organizational capabilities themselves developed by the larger pharmaceutical firms have acted as a mechanism of appropriability. Consider, for example, the process of random screening (discussed at length in chapter 8) – i.e. the fundamental procedure

---

[4] Two of us discuss at greater length the issue of *appropriability* in Dosi, Orsenigo, and Sylos Labini (2005), where the reader can also find the background references.

governing drug discovery in the era after World War II. As an organizational process, random screening was anything but random. Over time, major incumbents developed highly sophisticated processes for carrying out mass screening programs, which required systematic search strategies as well as the relatively rigorous handling of large amounts of data. Since random screening capabilities were based on organizational processes and tacit skills internal to the firm, they were difficult for potential entrants to imitate, and thus they became a source of first-mover advantages. In addition, under the random screening paradigm, spillovers of knowledge between firms were relatively small, so incumbents having a pre-existing advantage could maintain it over time with respect to firms wishing to enter. Moreover, incumbents relied on the law of large numbers: relatively little could be learned from the competition, but much could be learned from large-scale screening in-house. Each firm needed access to the appropriate information infrastructure for its therapeutic areas (Henderson, Orsenigo, and Pisano, 1999; Pisano, 1996), and that tended to reproduce the advantage of incumbency.

Third, the scope and efficacy of patent protection in pharmaceuticals has varied significantly across countries and over time. Murmann's (2003) comparisons of the role and effects of patent laws in the United Kingdom and in Germany on the emerging synthetic dye industry is quite revealing in this respect. The UK legislation allowed product patents, whereas Germany had no unified patent law until 1877. The Patent Law of that year instituted a rigorous examination by the Patent Office before a patent would be granted, in order to define precisely the scope of the claim. The rigor of the examination process – much tougher compared to that in the United States and, especially, the United Kingdom, at least until 1905 – made a patent legally much more secure once it was granted, by reducing the risk of litigation. In turn, this facilitated the creation of a market for patents, whereas in the United Kingdom patents were often the object of intensive, uncertain, and costly litigation. The German law, however, allowed only *process* patents and required that they be actually applied within the country, whereas this was not the case in the United States and – in practice – in the United Kingdom. As Murmann argues, these features of the patent system were very important in establishing the German dominance in chemicals and in pharmaceuticals. The legal grant of strong *product* patents early on in the history of the British and French industries prevented the entry of new firms and gave few companies monopoly profits without their being forced to develop strong competitive capabilities. Moreover, frequent and costly litigation over patents among British firms further weakened both their ability and the economic interest to search for new products. Conversely, the German system allowed the industry – not simply individual monopolists – to grow and to construct such capabilities, also exploiting the ample possibilities of infringing British patents. As the German industry established itself as the world leader in chemicals, the domestic patent regime began to act as a reinforcing mechanism, providing further incentives to innovate – especially as it concerned processes – and to build systematic R&D efforts. The absence of product patent protection at home, in fact, promoted not only the diffusion of but also, less intuitively, *trade* in knowledge, contributing to the formation of an early (formal and informal) market for technology.

It is worth noting that, while the United States has always provided relatively strong patent protection in pharmaceuticals, many other European countries did *not* offer protection for pharmaceutical *products*; as in Germany, only process technologies could be patented. France introduced product patents in 1960, Germany in 1968, Switzerland in 1977, and Italy and Sweden in 1978 (and Japan in 1976).

More recent experiences of changes in IPR regimes raise further questions about the actual effectiveness of stronger or weaker patent protection on innovativeness and industrial growth.

The United States and Japan represent two important cases where patent legislation has been strengthened. In the United States, over the past twenty years, extremely aggressive institutional changes in the IPR system have been introduced. These reforms, taken together, involved cost and time reductions in patent applications and evaluation; the extension of patent duration for some classes of products (primarily chemical and pharmaceutical technological classes); and encouragement for non-profit research institutions to patent and market technologies developed through public funding. The Bayh–Dole Act of 1980 clearly falls into this latter category. Also, in Japan, a 1988 reform introduced significant changes in patent laws, implying broader patent scope coupled with an extension of the protection period for pharmaceutical products. In both cases, however, the evidence does not seem to support unambiguously the hypothesis that a tighter IPR regime automatically leads to an increase in innovative activities.

In the case of the United States, Mowery et al. (2001) have shown that the emergence of the "industry–university complex" (Kenney, 1986) and of the "entrepreneurial university" well pre-dates Bayh–Dole, while depending critically on the rise of the two main technological revolutions of the second half of the twentieth century, namely microelectronics and, more so, biotechnology. Moreover, Mowery et al. (2001) emphasize that much of the university patenting activity observed after the Bayh–Dole Act stems from long-standing characteristics – in terms of scale and structure – of the US academic system. In a somewhat different vein, Kortum and Lerner (1997) have investigated the reasons for the observed massive increase in the number of patents that has occurred in the United States in the preceding ten years. Their results seem to support the so called *fertile technology hypothesis* – i.e. that the strong increase in the number of patents is not the effect of a stronger IPR regime but, rather, a consequence of a wider set of technological opportunities, and improvements in the management of the innovative processes.

In addition, in the case of Japan, the evidence for the actual effects of reform on innovative efforts is quite mixed. In particular, Sakakibara and Branstetter (2001) show that, after 1988, there has been no substantial increase in R&D efforts. The observed rise of R&D spending actually started at the beginning of 1980 – i.e. much earlier than the year of the reform. If anything, in 1988/9 R&D expenditures showed a relative decline. Also, in the specific case of pharmaceuticals, the authors do not find any significant correlation between the increase in intellectual protection and R&D efforts.

Conversely, consider India, which is one of the few cases where there has been a weakening of IP protection: in the last twenty years, despite strong international political pressures, patent protection has actually been lowered. After these reforms, a significant *growth* in industries such as pharmaceuticals and chemicals is observed. Almost all the empirical studies on the Indian case agree that a weaker IP protection system has encouraged the development of indigenous technological capabilities and promoted catching up (see, among others, Lanjouw, 1998; Kumar, 1998, 2002).

Obviously, strong patent laws do indeed confer an advantage to innovators, but – the evidence seems to suggest – they are certainly not enough to promote innovation in contexts where innovative capabilities are low or missing altogether. On the other hand, weak patent protection might constitute a fundamental mechanism for learning and developing domestic capabilities in laggard countries, when coupled with a complementary emphasis on pre-production research and reasonable incentive systems favoring innovative and imitative activities.

14.5 Intellectual property rights and open science

Besides the foregoing evidence (or lack of it) on the effects of IPR upon the micro-incentives to innovate, serious worries have been raised that the spread of an excessively permissive attitude towards the granting of broad claims on patents might actually slow down the processes of the diffusion and circulation of knowledge and hence the future rate of technological advance, especially as it concerns publicly funded research. More generally, several observers (e.g. Dasgupta and David, 1994; Merges and Nelson, 1990) have argued that this trend can end up seriously undermining the norms and rules of "open science."

There is little question that science has played a crucial role in opening up new possibilities for major technological advances in biomedical research, as in most other technological fields. If anything, the role of science has been more direct and immediate in pharmaceuticals than in most other technologies. Notwithstanding the diversities across countries in the institutional systems governing the interactions between scientific and technological research (see Gambardella, 1995; Henderson, Orsenigo, and Pisano, 1999; Lacetera and Orsenigo, 2002; and McKelvey, Orsenigo, and Pammolli, 2005, for a discussion centered on biomedical research), in almost all cases (the former Soviet Union and China being notable exceptions) publicly funded science, largely undertaken in universities and in national laboratories like those of the NIH in the United States, appears to be complementary to that undertaken in private corporations. And the interactions between them have, typically, resulted in a fuzzy area at the boundaries between public and private research.

Open science has been a fundamental component of such a system, and it is responsible for the productive, yet serendipitous, two-way feedback between technological innovations and scientific knowledge. The open science (OS) paradigm (see Nelson, 2004, and David, 2004) is based on an open, accountable scientific system involving the free dissemination of results (via publications open to the public), peer review, and rewards tied to recognized contributions to the communal scientific effort. The emphasis on serendipity – that is, the radical unpredictability of the ultimate application of any advance in scientific knowledge – also highlights, in terms of eventual technological fallout, the importance of government support for fields where practical pay-offs are less certain and direct (e.g. theoretical physics). More generally, when scientists are not constrained (by the nature of funding, and by patent dynamics) to produce knowledge that has direct and immediate practical pay-offs, the chances of this serendipity are greater (Nelson, 2004). For this reason, many students of the history of science have concluded that it is fundamental that neither the market nor the government influence too much the areas in which scientists pursue their quests for knowledge (Polanyi, 1967).

In the words of Nelson (2004):

[I]n all the fields of technology that have been studied in any detail, including those where the background science is very strong, technological advance remains an evolutionary process. Strong science makes that process more powerful, but does not reduce the great advantages of having multiple paths explored by a number of different actors. From this perspective, the fact that most of scientific knowledge is open, and available through open channels (e.g. publications), is extremely important. This enables there to be at any time a significant number of individuals and firms who possess and can use the scientific knowledge they need in order to compete intelligently in this evolutionary process. The "commitarianism" of scientific knowledge is an important factor contributing to its productivity in downstream efforts to advance technology.

In turn, as David (2004) argues, such an OS system is relatively *recent* and relatively *fragile*. It dates to the break, during the sixteenth and seventeenth centuries, from a system of knowledge pursuit domin-ated by secrecy and the quest for "Nature's secrets" (e.g. the medieval and Renaissance traditions of alchemy; the medieval guilds preserving the secrets of certain trades; and mercantile secrets on trade routes, etc.). The new set of norms, incentives, and organizational structures (such as the use of peer review) reinforced scientific researchers' com-mitments to the rapid disclosure of new knowledge, and to a painstak-ing process that developed into the research system that we experience now in the early twenty-first century.

It is, however, a delicate system, which remains "vulnerable to destabilizing and potentially damaging experiments undertaken too casually in the pursuit of faster national economic growth or greater military security" (David, 2004), or – we would add – excessive greediness in the appropriation of the returns of the knowledge quest itself (in that, going back to the older "feudal" ethos).

Changes in patent laws and practices might constitute one glaring example of those "experiments" threatening OS. The already mentioned Bayh–Dole Act in the United States is possibly the best-known example of partly unintended consequences from the reckless manipulations of such a fragile system. To recall, the Act allowed universities and small businesses to patent discoveries emanating from NIH-sponsored research, and then grant exclusive licenses to drug companies. Along with it, a series of court cases in the mid-1990s overturned previous practices, granting patents on upstream research and significantly extending patents' scope, even to cases in which the practical application of the objects had not been demonstrated (e.g. the BRCA1 gene discussed in chapter 11).

Many analyses (including chapters 10 and 11 in this book, and Dosi, Llerena, and Sylos-Labini, 2005) warn about the dangers of these trends. One of the problems here is that, since scientific research is usually not the final product, by strongly enforcing patents on research outputs one is potentially preventing the exploration of new outputs and products based on that research (Nelson, 2004). There are, indeed, strong reasons to conjecture that the strong enforcement (and misuse) of patent rights can stunt future innovations.

First, bringing science into the "market" is likely to distort incentives away from basic research and into specific, practical areas that promise commercial rewards.

Second, science "proceeds most effectively and cumulatively when those who do science are part of community where open publication and access to research results is the norm, and rewards are tied to recognized contributions to the communal scientific effort" (Nelson, 2004). But widening the scope of appropriability runs completely counter to this principle.

Historically, the reason for *not* granting patents on upstream research has been precisely that this could prevent the circulation of basic knowledge within the community of inventors. Similarly, granting patents on objects where the practical or commercial utility has not been proved might induce discriminatory practices that would (a) prevent the public from benefiting from the inventions and (b) prevent future innovation (Arrow, 1962; Nelson 1959). There is little to support the idea that these reasons have ceased to apply nowadays.

Other sources of worry relate to the "anticommons problem" (Heller and Eisenberg, 1998), discussed in chapter 10, concerning the possibility that the extension of patents into research tools will limit innovation due to the *numerous property right claims to separate building blocks for some product or line of research*.

Thus, critics of the current policy trends suggest that, at the very least, one ought to enforce legally those parts of the Bayh–Dole Act that require that knowledge that emerges from publicly funded research remain in the hands of the public – one of the main points being that university research must stay in the hands of the public, regardless of whether it is patented. In fact, the Bayh–Dole Act stated that its purpose is to "ensure that inventions made by nonprofit organizations ... are used in a manner to promote free competition and enterprise without unduly encumbering future research and discovery" (Nelson, 2004). The problem is that those provisions aimed at preventing the "encumbering" are not enforced. Hence, to make sure that university research contributes to the "scientific commons," many suggest that the law must urgently be *enforced* fully and possibly reformed, to prevent, for example, exclusive and narrow licensing by universities. Unfortunately, as Nelson admits, the universities, with their new, profit-seeking goals, have become one of the main problems.

In addition to the reform of the Bayh–Dole Act, Nelson (2004) proposes to

(i) limit the scope of patents to those that imply "substantial transformation" of the natural – as opposed to proprietary – claims stemming just from having discovered "how nature works" (on these grounds, Newton could very probably have claimed a rent on every reference to gravitation!); and

(ii) adopt much stricter and more precise interpretations of the meaning of "utility" and "usefulness" with respect to patents (see the discussion in chapter 11).

Conversely, supporters of this "new" IPR regime argue that patents on publicly funded research serve the purpose of creating markets for knowledge. The establishment of property rights on the outcomes of research facilitates the economic exploitation of such knowledge (in the absence of patents, firms would not invest in R&D based on the new discovery because everybody could have access to it) and allows an "ordered" path for the exploitation of such knowledge, avoiding waste-ful duplication of effort. The boom in biotech companies (often founded by university scientists) is typically cited as an example of the effects of the "new" IPR regime on the commercial exploitation of basic scientific research.

Be that as it may, notice all the same that this argument is profoundly different from the standard argument, recalled above, that patents represent an incentive to innovation. In fact, this incentive is not needed in the case of publicly funded scientific research: the invention has already been paid for (by the public) and has already been realized. Moreover, the argument in favor of the imposition of property rights on otherwise open science rests on a series of specific assumptions about the mechanisms underlying the generation and economic exploi-tation of knowledge that – as argued by Mazzoleni and Nelson (1998) – make it very hard to accept them in general.

Clearly, these issues and their industrial policy implications are also at the core of the policy controversies regarding the institutional gov-ernance of the pharma-biotech sector. For example, what arrange-ments ought to govern the acquisition of knowledge by profit-seeking firms generated with public resources? What disclosure arrangements should be made? To what degree should one adjust the boundaries between what is publicly paid and universally available, on the one hand, and what is privately financed and privately appropriable, on the other?

14.6 Trends in innovative opportunities and behaviors

The concerns over the future of the pharmaceutical industry stemming from the debates on regulation and IPR are compounded by the obser-vation that innovation in the pharmaceutical industry actually seems to be slowing down, despite the promise of the "biotech revolution" (Angell, 2004; Nightingale and Martin, 2004; *Economist*, 2005).

Nightingale and Martin (2004) suggest that the biotechnology "revolution" has not, in fact, increased the observed productivity of R&D, because of the inability of drug firms to keep pace with the increased intrinsic complexity of the biochemical problems that inno-vative search is addressing. Over the period 1978–2003, research "productivity," measured by the number of patents per dollar of R&D

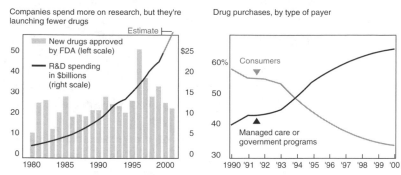

Figure 14.1. R&D expenditure, FDA drug approvals, and drug purchases, 1980–2000
Source: Harris (2002).

expenditure, actually fell: R&D expenditure increased tenfold, while patenting output increased only sevenfold. This is further corroborated by the number of new chemical entities (a much more demanding measure of innovativeness than patents)[5] approved by the FDA in the United States over the period 1983–2003: some increase is recorded until the mid 1990s, followed by a sharp decline subsequently. So, in 2002, US R&D expenditures in pharmaceuticals were thirty times greater than in the early 1980s, while roughly the same number of drugs were approved annually (see figure 14.1). Considering that in recent years regulations have become more relaxed and approval times have been shortened (due to the Prescription Drug User Free Act of 1992 and the FDA Modernization Act of 1997), the fall in R&D productivity is even more surprising.

Are these patterns due to the progressive drying out of innovation *opportunities*? Arguing against this possibility, however, is the difficulty of reconciling it with the novel horizons of discovery commonly associated with the "bioengineering revolution."

Conversely, could these patterns be the outcome of changing *directions of search* by many pharmaceutical firms, increasingly favoring

---

[5] Patents are of orders of magnitude greater than NCEs because they may well concern new ways of delivering existing drugs or new combinations thereof, and, even more importantly, they include *potential* NCEs that do not achieve FDA approval because of failed efficacy tests, harmful side effects, etc.

incremental improvements on existing drug families while penalizing
more uncertain long-term search? Were this the case, a challenging
paradox could appear, with (i) private firms increasingly relying on
non-profit institutions – *in primis*, public labs and universities – as
long-term suppliers of novel basic science, while, at the same time,
(ii) the same supposedly non-profit institutions are increasingly meeting incentives to undertake research projects with plausible down-the-
line profitability.

Of course, there are many more sophisticated interpretations that
fall between these two extremes; flagging them helps set the terms of
the debate. Are we basically talking about decreasing returns to search?
Is the lack of radical innovation typical of the mature phase of the
industry life cycle? Is it a case of private strategic myopia? Or short-
term free-riding upon past knowledge that is still untapped? Or is the
growth of search/testing costs "intrinsic" to the technology-driven
approach?

A complementary issue in the contemporary discussion concerns the
cost of regulation (in part mimicking a similar debate in the 1960s and
1970s) and its impact upon the *observed* rates for the successful introduction of new products to the market (most obviously, a new drug
approved by the FDA under permissive side effect rules would not be
introduced under stricter ones, thus lowering the *apparent* rate of
innovation).

In the multiple controversies over all these issues, Angell (2004)
offers an interpretative benchmark on one side of the debate. Many
economists provide the opposite view. In brief, Angell (not an economist, but a top-level practitioner in the field and a recent editor of
the *New England Journal of Medicine*) makes two main suggestions.

(i) The research reliance of the pharmaceutical industry is vastly over-
    stated and the (corrupting) importance of its efforts on sheer market
    penetration are symmetrically overlooked. For example, she points
    out that in 1990 marketing expenditures in the US drug industry
    were equivalent to 36 percent of sales, compared to an 11 percent
    R&D allocation (subject to further caveats as to whether the "dis-
    semination of knowledge" on a particular proprietary drug offer-
    ing benefits to doctors, such as conferences located in particularly
    enticing environments, should be put under the "R&D" heading,
    as is usually the case in this industry). In this respect, note that the
    ratio of R&D to marketing expenditures is, in general, of no great

significance. It becomes more worrying whenever the outcomes of such marketing efforts end up being paid, directly or indirectly, by taxpayers.

(ii) While there is little doubt that there is no relation (nor should there be) between the production costs of any particular drug and its price, Angell – along with other critics – claims that a good deal of the upfront search and innovation costs are actually borne by the public sector (e.g. in the United States by NIH-funded research).

According to Angell, big pharma concentrates on "me too" drugs (drugs almost identical to existing drugs – sometimes differing only in terms of dosage – extending the monopolistic profits of the old drug under a different name).[6] Conversely, truly innovative drugs (i.e. new molecular entities with priority ranking under FDA procedures) almost always trace their origin back to publicly funded laboratories (either NIH labs or publicly funded universities). More than one-third of the drugs marketed by big pharma are either licensed from universities or small biotech companies. And this third comprises the most innovative element of all new marketed drugs.

Again, the issue is ultimately empirical. It would certainly be helpful if drug producers convinced of their continuing innovativeness allowed independent researchers to browse through their R&D investment portfolios and their product selection strategies.

Angell's point can easily be rephrased in terms of the threat to open science posed by the private funding of basic research, as discussed above. Since Bayh–Dole, it is the "market," more than open scientific priorities, that determines what type of research is pursued, and funded, in the pharmaceutical/medical/biological fields. Given the subtle synergies, complementarities, and overlappings of interests between public and private research (see Nelson, 2004), the novel, emerging institutional arrangement is likely to jeopardize not only the openness of science (discussed above) but even the technological productivity of science itself.

These criticisms are fiercely rebutted by the industry, and by many economic analysts. For example, it is pointed out that R&D intensity

---

[6] The FDA allows the approval of new drugs if they are better than a placebo, not requiring the applicant to test the drug on an older, incumbent drug. In fact, many new drugs differ solely in terms of dosage, yet the millions of dollars spent on marketing make them "seem" new and better (Angell, 2004).

is, in any event, much higher than in any other R&D-intensive industry (which, under the caveats briefly listed above, is a robust fact). Given that, a possible argument in support of the continuing "progressiveness" of the industry against the unfavorable evidence regarding innovative output is in terms of the deep organizational transformation that the pharmaceutical industry is now undergoing. Following the biotechnology revolution, the role of scientific, academic, research has increased dramatically. The process by which the new – biology-based – knowledge is absorbed and adopted has been slow and painful for large established corporations (Henderson, Orsenigo, and Pisano, 1999). By the mid-1990s, however, some of them had successfully adapted, and they have again become leading innovators (Galambos and Sturchio, 1998). All in all, the structure of the industry has changed, with a stronger division of labor between the industry, on one side, and academia and the biotech companies, on the other. This phenomenon is partly due to the fact that, given the tumultuous rate of advance in biomedical research, no single organizational entity can survey and control – let alone produce – technological opportunities from relatively basic scientific advances all the way to final approved products. Moreover, as noted previously, the new IPR regime has introduced strong incentives for the creation of markets for technology and the division of labor between large companies (specializing in the development and marketing of drugs) and biotech companies (specializing in drug discovery) (Arora, Fosfuri, and Gambardella, 2001).[7] And the slowdown in innovation could result, as suggested by Nightingale and Martin (2004), from the difficulties and the time lags involved in mastering the new scientific and technological knowledge base.

This interpretation is intuitively in tune with the more general observation that new technological paradigms take time to establish themselves, and their diffusion into the economy requires concomitant changes in the whole organizational and institutional structure of the economy (Freeman, 1995; David, 1990). In the case of pharmaceuticals,

---

[7] It has to be noted, however, that, while the tendency towards the division of labour is certainly undeniable and very strong, specialization is far from complete, especially as far as large corporations are concerned. For example, interactions between big companies and biotech firms do not occur solely and simply via market-mediated transactions in which the latter sell a discovery to the former. Rather, a large fraction of these interactions take place at the preclinical stage. See, for example, Orsenigo, Pammolli, and Riccaboni (2001).

it may well be that new products are still in the phase of infancy with respect to their full potential uses throughout the economy (agriculture, medicine, life science research, etc.) – as happened with electricity, cars, and the PC, when it took almost thirty years for the new product to be adopted by mainstream businesses and consumers (Wong, 2005).

Clearly, the foregoing alternative scenarios depend critically on whether, and how, the biotechnology revolution will deliver its promises. But in all cases the question remains open as to whether the traditional business model as incarnated in big pharma is still viable, and/or what the functions are that it might perform. Can big pharma companies continue to be crucial agents in the innovative process, along – and interacting – with academia and biotech companies? Will they perform the function of integrators of the different fragments of knowledge and capabilities that are required to produce a new drug? Or, as the division of innovative labor deepens, will they progressively lose their innovative capacities, and even their "absorptive capacities" – i.e. the ability to understand, evaluate, and absorb new, externally created knowledge? Will the large established pharmaceutical companies become (or, as Angell suggests, remain) essentially marketing-based organizations, the function of which is "simply" to conduct clinical trials, get approval for the products, and sell them?

Yet another interpretation is the view that, in fact, biotechnology is not a "revolution" by any means, but, rather, that (as suggested by Nightingale and Martin, 2004) the stagnation, or even fall, in innovative output is the outcome of an incumbent "maturity" of the industry characterized by a *fall* in innovative *opportunities* – a little like the mature phase in the life cycle of such industries as steel or automobiles (Klepper and Simons, 1997).

As yet there are no clear-cut answers to these questions. Whatever the answers are, though, they bear fundamental implications in terms of policies.

14.7 Some provocative policy issues by way of a conclusion

First, note that a good deal of the debate on patents and the regulation of drug prices boils down, from a normative point of view, to the relationships (a) between the (actual and expected) profitabilities and rates of investments in innovative search and (b) between the latter and actual rates of discovery.

Concerning point (a), to our knowledge there is no clear evidence either way. The (cross-sectional) evidence on different firms simply confirms their different propensities to undertake R&D – that is, different innovation/imitation strategies amongst firms. Conversely, over time, the observational windows are too short to infer anything whatsoever about the strategic reactions of various firms to changes in appropriability regimes and profitabilities (while, of course, self-serving claims should not be taken too seriously).

Concerning point (b), let us just remark that one cannot claim *at one and the same time* that the industry invests a great deal in truly innovative search, that opportunities for innovation have increased due to the biotech revolution, and that the rates of successful innovation have actually remained stagnant or fallen (unless one claims, as is often done by big pharma, that *increasing* testing requirements and *increasingly* stringent selection criteria based on safety grounds are the major cause of the observed patterns – which is, frankly, a far-fetched claim, given that in recent years regulations have become more relaxed and approval times have been shortened, as a result, for example, of the Prescription Drug User Free Act of 1992 and the FDA Modernization Act of 1997).

Clearly, the outcome of such controversies entails big economic stakes. For example, if much of the search and preclinical test discovery occurs "upstream," within non-profit, publicly funded institutions, the argument in favor of OS institutional arrangements is tremendously strengthened. Conversely, if large corporations become specialized in product development, approval, and marketing, one reasonable scenario would be for non-profit, mostly public, agencies to move downstream one step further into clinical trials. What would the economic arguments be for and against having pharmaceutical companies mainly *producing and distributing* drugs, as they already largely do in the case of vaccines, on (quasi-)marginal cost rules? Why not have the whole range of search/development/screening/testing activities in the hands of non-profit organizations or ad hoc subcontractors thereof, given that the public, one way or another, pays for it in any case?

Second, let us end with an even broader and more provocative suggestion, concerning the very notion of "universal rights" for decent health care. In this respect, the notion of "market failures" misses, perhaps, a fundamental dimension of the problem. Should public support for scientific research be justified (or criticized) only in terms of a market outcome? And is therapeutical knowledge only a "public good" – i.e. non-rival in nature and freely accessible at zero cost? In all probability, most would consider health to be a value in itself, for individuals and for society as a whole, at least at a "minimum and decent" level. If so, what should that level be, in so far as it is constantly redefined by the interaction between technological opportunities, expectations, and perceptions of what is right and wrong, as well as its costs? Consider an extreme but revealing example. Should one define the access to drugs and treatment for HIV/AIDS by people in Africa as a public good? This sounds genuinely awkward. Many of us would rather consider it a basic human right.

In economics, the concept of human rights looks much like an oxymoron: something that is (should be) priceless, but costly. The "economics of human rights" is a vast and unexplored field of analysis. However, in all likelihood, it is going to be quite different from the standard analysis of public goods. A good starting point might be the observation that "goods" such as education, defense, environmental preservation, etc. should be funded and supported by the state because of their sheer importance to the social fabric and government's responsibility towards its citizens – i.e. societal "values" as to what is right or wrong, justice, fairness, etc. – rather than the state's role simply being one of fixing "market failures" (Nelson, 2004).

There are sound reasons to believe that science and the preservation of an "open" and accountable scientific system, based on the widespread dissemination of results, also fall into the category of universal entitlements *in their own right*, precisely because open science is grounded in values that go beyond their immediate practical and economic function, and is part of the "vital infrastructure" of society and hence the "responsibility" of society and the state.

Seen from this angle, one certainly continues to face all the economic and organizational issues insightfully addressed in many chapters of this volume concerning topics such as incentive compatibilities, organizational learning, and the strategic management of innovation. However, from a normative point of view, the division of labor between an OS system of scientific and technological discovery, on the one hand, and private profit-seeking actors, on the other, ought to be assessed under the criteria not just of economic viability but also of social "rights of access."

Moreover, it is certainly misleading to think that "production" and "distribution" can be delinked. Pushing it to the caricatural extreme, it would be like saying that one could innocently privatize the exercise of criminal justice, with public authorities able to influence the access to fair (and expensive) producers simply by paying for private sheriffs. Rather, we are currently witnessing a period of profound institutional transformation wherein the capabilities to generate new therapeutical advances are coevolving with the distribution of their costs, the ensuing rents – all with profound effects on the very structure of the social fabric.

If this book has contributed, even if only marginally, to the understanding of such dynamics, we would consider it a significant success.

References

Angell, M. (2004), *The Truth About the Drug Companies: How They Deceive Us and What to Do About It*, Random House, New York.

Arora, A., A. Fosfuri, and A. Gambardella (2001), *Markets for Technology*, MIT Press, Cambridge, MA.

Arrow, K. (1962), "Economic welfare and the allocation of resources for inventions," in R. R. Nelson (ed.), *The Rate and Direction of Innovative Activity: Economic and Social Factors*, Princeton University Press, Princeton, NJ, 609–26.

Comanor, W. S. (1986), "The political economy of the pharmaceutical industry," *Journal of Economic Literature*, 24, 1178–217.

Dasgupta, P., and P. A. David (1994), "Toward a new economics of science," *Research Policy*, 23, 487–521.

David, P. A. (1990), "The dynamo and the computer: an historical perspective on the modern productivity paradox," *American Economic Review*, 80 (2), 355–61.

(2004), "Understanding the emergence of open science institutions: functionalist economics in historical context," *Industrial and Corporate Change*, 13 (3), 571–89.

Dosi, G., P. Llerena, and M. Sylos Labini (2005), *Science–Technology–Industry Links and the "European Paradox": Some Notes on the Dynamics of Scientific and Technological Research in Europe*, Working Paper no. 2005/02, Laboratory of Economics and Management, Sant'Anna School for Advanced Studies, Pisa.

Economist (2005), special report on "The drugs industry," 19–25 March, 89.

Dosi, G., L. Orsenigo, and M. Sylos Labini (2005), "Technology and the economy," in N. J. Smelser and R. Swedberg (eds.), *The Handbook of*

Economic Sociology, 2nd edn., Princeton University Press and Russell Sage Foundation, Princeton, NJ, 678–702.

Freeman, C. (1995), "The national systems of innovation in historical perspective," *Cambridge Journal of Economics*, 19, 5–24.

Galambos, L., and J. L. Sturchio (1998), "Pharmaceutical firms and the transition to biotechnology: a study in strategic innovation," *Business History Review*, 72, 250–78.

Gambardella, A. (1995), *Science and Innovation: The US Pharmaceutical Industry in the 1980s*, Cambridge University Press, Cambridge.

Grabowski, H., and J. Vernon (1983), *The Regulation of Pharmaceuticals*, American Enterprise Institute for Public Policy Research, Washington, DC, and London.

Heller, M. A., and R. S. Eisenberg (1998), "Can patents deter innovation? The anticommons in biomedical research," *Science*, 280, 698–701.

Harris, G. (2002), "Why drug makers are failing in quest for new blockbusters," *Wall Street Journal*, 18 March.

Henderson, R. M., L. Orsenigo, and G. Pisano (1999), "The pharmaceutical industry and the revolution in molecular biology: exploring the interactions between scientific, institutional, and organizational change," in D. C. Mowery, and R. R. Nelson (eds.), *Sources of Industrial Leadership: Studies of Seven Industries*, Cambridge University Press, Cambridge, 267–311.

Kenney, M. (1986), *Biotechnology: The Industry–University Complex*, Cornell University Press, Ithaca, NY.

Klepper, S., and K. Simons (1997), "Technological extinctions of industrial firms: an inquiry into their nature and causes," *Industrial and Corporate Change*, 6 (2), 379–460.

Kortum, S., and J. Lerner (1997), *Stronger Protection or Technological Revolution: What is Behind the Recent Surge in Patenting?*, Working Paper no. 6204, National Bureau of Economic Research, Cambridge, MA.

Kumar, N. (1998), "Technology generation and transfers in the world economy: recent trends and prospects for developing countries," in N. Kumar (ed.), *Globalization, Foreign Direct Investment and Technology Transfer: Impacts on and the Prospects for Developing Countries*, Routledge, London and New York, 11–42.

(2002), *Intellectual Property Rights, Technology and Economic Development: Experiences of Asian Countries*, paper prepared for the Commission of Intellectual Property Rights, London.

Lacetera, N., and L. Orsenigo (2002), *Political and Technological Regimes in the Evolution of the Pharmaceutical Industry in the USA and in Europe*, 2nd draft of paper prepared for the "Conference on

Evolutionary Economics," Johns Hopkins University, Baltimore, 30–31 March 2001.

Lanjouw, J. O. (1998), *The Introduction of Pharmaceutical Product Patents in India: Heartless Exploitation of the Poor and Suffering?*, Working Paper no. 6366, National Bureau of Economic Research, Cambridge, MA.

Mazzoleni, R., and R. R. Nelson (1998), "The benefits and costs of strong patent protection: a contribution to the current debate," *Research Policy*, 27, 273–84.

McKelvey, M., L. Orsenigo, and F. Pammolli (2005), "Pharmaceuticals analysed through the lens of a sectoral innovation system," in F. Malerba (ed.), *Sectoral Systems of Innovation*, Cambridge University Press, Cambridge, 11–42.

Merges, R., and R. R. Nelson (1990), "The complex economics of patent scope," *Columbia Law Review*, 839, 890–4.

Mowery, D. C., R. R. Nelson, B. Sampat, and A. Ziedonis (2001), "The growth of patenting and licensing by US universities: an assessment of the effects of the Bayh–Dole Act of 1980," *Research Policy*, 30, 99–119.

Murmann, J. P. (2003), *Knowledge and Competitive Advantage*, Cambridge University Press, Cambridge.

Nelson, R. R. (1959), "The simple economics of basic scientific research," *Journal of Political Economy*, 67, 297–306.

(2004), "The market economy and the scientific commons," *Research Policy*, 33, 455–72.

Nightingale, P., and P. Martin (2004), "The myth of the biotech revolution," *Trends in Biotechnology*, 22 (11), 564–9.

Orsenigo, L., F. Pammolli, and M. Riccaboni (2001), "Technological change and the dynamics of networks of collaborative relations: the case of the bio-pharmaceutical industry," *Research Policy*, 30, 485–508.

Pisano, G. (1996), *The Development Factory*, Harvard University Press, Cambridge, MA.

Polanyi, M. (1967), *The Tacit Dimension*, Anchor Books, New York.

Sakakibara, M., and L. Branstetter (2001), "Do stronger patents induce more innovation? Evidence from the 1988 Japanese patent law reforms," *RAND Journal of Economics*, 32, 77–100.

Scherer, F. M. (2000), "The pharmaceutical industry," in A. J. Culyer, and J. P. Newhouse (eds.), *Handbook of Health Economics*, Vol. I, Elsevier, Amsterdam, 1297–336.

Schweitzer, S. O., and W. S. Comanor (2000), "Pharmaceutical prices and expenditures," in R. M. Andersen, T. H. Rice, and G. F. Kominski (eds.), *Changing the U.S. Health Care System*, 2nd ed., Jossey Bass Publishers, San Francisco, 100–24.

Thomas, L. G. (1994), "Implicit industrial policy: the triumph of Britain and the failure of France in global pharmaceuticals," *Industrial and Corporate Change*, 3 (2), 451–89.

Wong, J. F. (2005), "Is biotech in the midst of a fifty-year cycle?," *Genetic Engineering News*, 25 (5), 60.

Index